Optical Properties of Functional Polymers and Nano Engineering Applications

NANOTECHNOLOGY AND APPLICATION SERIES

Series Editor
Muhammad H. Rashid

Optical Properties of Functional Polymers and Nano Engineering Applications,
edited by Vaibhav Jain and Akshay Kokil

Optical Properties of Functional Polymers and Nano Engineering Applications

Vaibhav Jain

OPTICAL SCIENCES, NAVAL RESEARCH LAB
WASHINGTON DC, USA

Akshay Kokil

UNIVERSITY OF MASSACHUSETTS,
LOWELL, MA, USA

CRC Press
Taylor & Francis Group
Boca Raton London New York

CRC Press is an imprint of the
Taylor & Francis Group, an **informa** business

Cover art by Akshay Kokil and John Gandour

CRC Press
Taylor & Francis Group
6000 Broken Sound Parkway NW, Suite 300
Boca Raton, FL 33487-2742

First issued in paperback 2017

© 2015 by Taylor & Francis Group, LLC
CRC Press is an imprint of Taylor & Francis Group, an Informa business

No claim to original U.S. Government works

ISBN-13: 978-1-4665-5690-4 (hbk)
ISBN-13: 978-1-138-74964-1 (pbk)

**Visit the Taylor & Francis Web site at
http://www.taylorandfrancis.com**

**and the CRC Press Web site at
http://www.crcpress.com**

Contents

Preface

Functional polymers offer a unique set of properties according to their chemical structure, even a slight alteration to which brings a significant difference in the material properties of the polymers and hence the applications. The study of their optical properties has emerged as a fast-moving field for nanoengineering applications in scaling up and enhancing various properties of polymers. Although there is literature available to readers that covers the basic optical properties of functional polymers, there has been very little that actually presents a complete and comprehensive overview of the fast-developing field of nanoengineering applications. This book presents a good understanding to readers and gives a basic introduction to the optical properties of polymers, along with providing a systematic overview of the latest developments in their applications in nanoengineering.

The authors have done a tremendous job elaborating on the high-level research along with explaining the basics very well. The introductory chapter (Chapter 1) will broadly discuss the importance of nanoengineering in improving the fundamental optical properties of the functional polymers and will be a preamble for subsequent chapters. Chapter 2 discusses the background and advantages of the liquid gradient refractive index (L-GRIN) lens over the conventional solid lens in various low-cost, reliable medical imaging devices for endoscopy and confocal microscopy. Chapter 3 details photorefractive polymers for 3D display application, providing a good introduction to the electrochemistry of photorefractive polymers, the molecular structure of commonly used polymers, and device manufacturing techniques along with a detailed description of the application of the photorefractive polymers to 3D displays. Chapter 4 focuses on optical gene detection using the optical properties of conjugated polymers. The chapter provides a brief introduction to the physics of fluorescence in photoluminescent polymers, energy and electron transfer mechanisms, along with a detailed explanation of various kinds of DNA sensors.

Chapter 5 contains an introduction to conventional polymer ion sensors based on the optical sensors of conjugated polymers prepared by click chemistry reaction, followed by a detailed explanation of colorimetric visual detection of ion recognition behaviors by using nonplanar donor–acceptor chromophores. Chapter 6 presents details about optical sensors that are based on fluorescent polymers and are used for the detection of explosives and metal ion analytes. Chapter 7 discusses holographic polymer-dispersed liquid crystal technology, its optical setups, and its applications in areas such as organic lasers. Chapter 8 introduces electrochromic phenomena with various organic and organic–inorganic hybrid materials and devices made from them. It presents a detailed background on the research of electrochromic devices, along with new concepts, prototypes, and commercial products available in the market, and future prospects of this technology. Chapter 9, on polymer nanostructures through packing of spheres, demonstrates new techniques for creating nanoscale morphologies through self-assembly, which were not previously possible using known techniques. This approach to assembling polymer nanoparticles into various structures provides distinctive nanoscale morphologies, which in turn affects the optical properties of the functional polymers.

It has been a pleasure working and interacting with the authors and we thank all of them for their efforts. The editors also thank Taylor & Francis for agreeing to publish this book. Special thanks go to Nora Konopka, Joselyn Banks-Kyle, and all of the other supporting staff of Taylor & Francis, Boca Raton, Florida, in producing this book. As editors of this book, we are grateful to the authors, our collaborators, advisers, and colleagues who have provided their valued input and discussions in making this book a success.

Vaibhav Jain
Optical Sciences, Naval Research Lab
Akshay Kokil
University of Massachusetts, Lowell

Editors

Dr. Vaibhav Jain currently works in the private industry and is a subject matter expert in materials science and engineering. Before that, he was a materials engineer at the Optical Sciences division at the Naval Research Lab from 2011 to 2013. From 2009 to 2011, he worked in the same division as a National Research Council postdoctoral researcher. Dr. Jain's most recent research focuses on plasmonics, scanning probe microscopy, solar cells, chemical and biological sensors, optical materials, and surface science. He has worked in single-wire spectroscopy to develop electrical contact fabrication on vertically aligned zinc oxide nanowires (NWs) and to investigate them using current-sensing atomic force microscopy (CS-AFM) for photodetectors and solar cell applications. This study also helped in understanding the basic electrical properties of semiconductor NWs with different metals to produce an economical way of developing improved metal–semiconductor contacts (MSCs) for the organic electronics industry.

Dr. Jain received his PhD at Virginia Tech in macromolecular science and engineering in 2009. His graduate research work spans a range of topics in polymer science, applied physics, and optoelectronics, including the fabrication of fast-switching electrochromic devices for flat-panel displays using thin film deposition by layer-by-layer (LbL) and self-assembly techniques, deposition of the nanoparticle stationary phase in microelectromechanical system (MEMS)–based columns for high-speed gas chromatography, and fabricating bending and linear actuators for artificial muscles by LbL assembly. Dr. Jain's expertise also includes synthesizing and characterizing different kinds of nanoparticles (metal nanoparticles, such as Au, Ag, Cu; and inorganics, such as ruthenium purple or Prussian blue), NWs, quantum dots, and depositing thin films from these for a wide variety of applications. He has coauthored 35 publications, one review article, and more than 50 conference presentations. He did his undergraduate education in polymer science and chemical technology at Delhi College of Engineering, New Delhi, India.

Dr. Akshay Kokil is a materials scientist with diverse academic and industrial research experience in design, synthesis, characterization, and processing of novel materials for applications in new technologies. He is actively involved in research on solar cells, organic electronics, sensors, and carbon nanomaterials. He is also experienced in research on environmentally benign synthesis and processing of advanced materials, including electrically conducting, flame-retardant, and self-assembled polymers. He has authored more than 30 publications and 11 conference preprints, and he has presented his research at multiple international research conferences.

Several of his papers have been featured on journal covers and highlighted in research sections of magazines and websites. He is also the author of the book *Conjugated Polymer Networks: Synthesis and Properties* published by VDM Verlag.

Dr. Kokil received a PhD degree in macromolecular science and engineering from Case Western Reserve University, Cleveland, Ohio. He did his undergraduate education in polymer engineering at University of Pune, Pune, India.

Contributors

Pierre-Alexandre Blanche
College of Optical Sciences
University of Arizona
Tucson, Arizona

Yuchao Chen
Department of Engineering Science
and Mechanics
Pennsylvania State University
University Park, Pennsylvania

Timothy S. Gehan
Department of Chemistry
University of Massachusetts
Amherst
Amherst, Massachusetts

Anupama Rao Gulur Srinivas
Polymer Electronics Research
Center
School of Chemical Sciences
University of Auckland
Auckland, New Zealand

and

MacDiarmid Institute for
Advanced Materials and
Nanotechnology
Wellington, New Zealand

Feng Guo
Department of Engineering Science
and Mechanics
Pennsylvania State University
University Park, Pennsylvania

Tony Jun Huang
Department of Engineering Science
and Mechanics
Pennsylvania State University
University Park, Pennsylvania

Wenbin Huang
State Key Laboratory of Applied
Optics
Changchun Institute of
Optics, Fine Mechanics and
Physics, Chinese Academy
of Sciences
Changchun, China

and

University of Chinese Academy of
Sciences
Beijing, China

Tsuyoshi Hyakutake
Advanced Materials Research
Team
Public Works Research Institute
Ibaraki, Japan

Vaibhav Jain
Optical Sciences
Naval Research Lab
Washington, DC

Justin Kiehne
Department of Engineering Science
and Mechanics
Pennsylvania State University
University Park, Pennsylvania

Akshay Kokil
Center for Advanced Materials
University of Massachusetts
   Lowell
Lowell, Massachusetts

Jayant Kumar
Center for Advanced Materials
University of Massachusetts Lowell
Lowell, Massachusetts

Peng Li
Department of Engineering Science
   and Mechanics
Pennsylvania State University
University Park, Pennsylvania

Ji Ma
Liquid Crystal Institute
Kent State University
Kent, Ohio

and

State Key Laboratory of Applied
   Optics
Changchun Institute of Optics, Fine
   Mechanics and Physics, Chinese
   Academy of Sciences
Changchun, China

Xiaole Mao
Department of Engineering Science
   and Mechanics
Pennsylvania State University
University Park, Pennsylvania

Tsuyoshi Michinobu
Department of Organic and
   Polymeric Materials
Tokyo Institute of Technology
Tokyo, Japan

Ahmad Ahsan Nawaz
Department of Engineering Science
   and Mechanics
Pennsylvania State University
University Park, Pennsylvania

Lawrence A. Renna
Department of Chemistry
University of Massachusetts
   Amherst
Amherst, Massachusetts

Prakash R. Somani
Applied Science Innovations
   Private Limited
Maharashtra, India

and

Department of Physics
Banasthali University
Rajasthan, India

and

Center for Applied Research and
   Nanotechnology
Siddaganga Institute of Technology
   (SIT)
Karnataka, India

Jadranka Travas-Sejdic
Polymer Electronics Research
    Center
School of Chemical Sciences
University of Auckland
Auckland, New Zealand

and

MacDiarmid Institute for
    Advanced Materials and
    Nanotechnology
Wellington, New Zealand

D. Venkataraman
Department of Chemistry
University of Massachusetts
    Amherst
Amherst, Massachusetts

Li Xuan
State Key Laboratory of Applied
    Optics
Changchun Institute of Optics, Fine
    Mechanics and Physics, Chinese
    Academy of Sciences
Changchun, China

Hiroshi Yokoyama
Liquid Crystal Institute
Kent State University
Kent, Ohio

Yanhui Zhao
Department of Engineering Science
    and Mechanics
Pennsylvania State University
University Park, Pennsylvania

1

Optical and Optoelectronic Properties of Polymers and Their Nanoengineering Applications

Akshay Kokil and Vaibhav Jain

Polymers consist of multiple small repeating units that are linked together. The small repeating units are called monomers. A number of synthetic strategies have been developed for connecting the monomeric units to obtain the polymers. As expected, the utilized parent monomers govern various properties of the obtained polymers. However, polymers also display a matrix of distinct properties that has resulted in them being widely investigated and utilized in a number of applications. These properties can be tailored by readily altering their chemical structure. The chemical structure of the polymers also has a considerable effect on the organization of the individual polymeric units in bulk, which in turn governs some of the obtained materials' properties.

Along with a number of favorable attributes, the polymers display interesting optical characteristics. Many of the polymers only absorb light in the ultraviolet region, which renders these materials transparent in the remaining visible spectrum. Such transparent polymers have been blended with a number of dyes and pigments to impart distinct photophysical properties to the blend. This has resulted in their utilization in various applications ranging from packaging materials to contact lenses. Alteration of the chemical structure to introduce conjugation in the backbone extends the absorption of light in the visible region and in some cases even the near infrared. On absorption of photons, some of these polymers also emit light as fluorescence or phosphorescence. Owing to their unique chemical structure and photophysical properties, such polymers have been utilized in recent optoelectronic devices and sensors. The underlying principles governing the optical characteristics of polymers are discussed in detail in each chapter of this book.

Confining polymers to nanometer-sized architectures also results in unique properties, which either makes their use in certain applications possible or improves the performance of the applications they are used in. The utilization of architectures on the nanometer scale consisting of polymers with interesting photophysical properties is one of the focuses of this book.

Owing to the difference in their chemical structure, every polymer displays a different refractive index. The organization of the polymer chains in the solid also has a high impact on the refractive index. For example, the refractive index of the crystalline region is different from that of the amorphous region in most of the polymers. If the crystallites are larger than the wavelength of light, the difference in refractive index leads to scattering of light, consequently rendering the polymer either opaque or translucent. The utilization of change of refractive index has been adopted in a variety of applications. Multilayer polymer films, due to the distinct refractive indices of the polymers used in individual layers, display controlled refraction or reflection of the incident light. Chapter 2 discusses in detail the novel application of such gradient refractive index assemblies. The refractive index of the material can also temporarily be altered on application of an external stimulus, such as an intense light beam from a laser source. Such materials have been utilized for applications in holographic displays, which will be discussed in Chapter 3. The use of polymer-dispersed liquid crystals for fabrication of holographic displays is also discussed in Chapter 7.

Polymers that display an alternating sequence of single and multiple bonds in the backbone are called conjugated polymers. Similar to many of the organic dyes, due to the conjugated structure, these polymers display strong absorption of light in the visible region of the electromagnetic spectrum. Many of these polymers are semiconducting and also display emission of light either through fluorescence or phosphorescence. The emission from these polymers can also be altered through various energy/electron transfer processes. The fluorescence in many of the conjugated polymers is dependent on the chemical environment; hence, it has been used as the detectable response in various optical sensors. Multiple analytes such as genes, explosive vapors, metal cations, and corrosive anions have been detected using the change in fluorescence of the conjugated polymers. Such sensors based on conjugated polymers are discussed in Chapters 4, 5, and 6. Because these polymers are semiconducting, application of an external field also causes a change in their photophysical properties such as absorption and fluorescence. Such changes in the optical properties have been utilized in optoelectronic devices, which are also discussed in Chapters 2, 3, 7, and 8.

The fabrication of architectures on the nanometer scale can result in enhancement of performance of many of the applications discussed in this book. Hence, fabrication and assembly of the nanoarchitectures and their utilization in the discussed applications are also presented in Chapter 9.

2

Liquid Gradient Refractive Index Lenses

Yanhui Zhao, Ahmad Ahsan Nawaz, Peng Li, Justin Kiehne, Yuchao Chen, Feng Guo, Xiaole Mao, and Tony Jun Huang

CONTENTS

2.1 Introduction

Optical materials, devices, and systems are vital in many applications in modern technology, from entertainment to consumer electronics to scientific research to medical diagnostics [1,2]. Over the past few decades, much academic and industry attention has been directed at miniaturizing optical systems. Making the optical systems small aims to achieve portability (easily move a device from one location to another), maneuverability (easily change the orientation or configuration of a device), and low cost [3]. Miniaturized optical systems have already found applications in personal electronics (e.g., cameras, cell phones, and tablets), in vivo bioimaging devices such as endoscopes, surveillance/security systems, miniaturized microscopes, and point-of-care diagnostics [4–6].

Most optical systems require precise fabrication, alignment, and actuation of various optical components. This is a challenging enough task at the traditional scale, and has proven to be only more difficult within the constraints enforced by the extremely limited space of miniaturized systems. Thus, in recent years, researchers have been seeking new approaches to the miniaturization of optical systems.

In their efforts to miniaturize optical systems, academics turned to the discipline of microfluidics [7–9]. Years of research effort have yielded a large catalogue of clever microfluidic devices and elements that can manipulate light at the microscale—a discipline now termed optofluidics [10,11]. Some of these elements manipulate light traveling in the device plane [12–14], some manipulate light traveling perpendicular to the device plane [15–17], and others serve to redirect light from a perpendicular path of travel into the device plane [18,19].

This chapter will cover one promising optofluidic component, the polymer (polydimethylsiloxane in specific)-based liquid gradient refractive index (L-GRIN) lens [20,21]. The L-GRIN lens can be readily fabricated via the standard soft-lithography technique and is highly compatible with other microfluidic components. It shows a high tunability and strong focusing performance. The operational flow rate of the L-GRIN lens is on the order of a few microliters per meter, a range two orders of magnitude lower than those in the previously reported microfluidic lenses [22,23]. Such a significant reduction in liquid consumption leads to sustainable operation of the lens and much less stringent requirements in the future on-chip pumping systems for lens control. The L-GRIN lens, which delivers flexibility, performance, and compatibility, will greatly benefit a wide variety of optics-based lab-on-a-chip applications.

2.2 Overview of PDMS-Based Microfluidic Lenses

Generally, microfluidic-based devices are constructed on either silicon or glass substrates, resulting in difficulties in integration of on-chip optical components with the microfluidic systems. Silicon is not a visible-light friendly material, absorbing most light on incidence. Glass, on the other hand, is a transparent material that allows high transmittance for a wide optical frequency range. However, engineering glass microstructures with optical smoothness is extremely challenging. In the past several years, the fabrication or replication of micro- and nanostructures using elastomeric stamps, molds, and conformable photomasks have been enabled with the emergence of soft-lithography techniques [24–28]. Soft lithography has provided a new technical route for the development of polymer-based on-chip optical systems and components. One of the most commonly used polymeric materials for soft lithography is polydimethylsiloxane (PDMS) [29–34]. PDMS has unusual flow properties as a gel material. It can be cured on heating and is optically clear, inert, nontoxic, and nonflammable. With such properties, PDMS is now the most commonly used material for microfluidic channel fabrication. The simple fabrication process and its unique properties provide enormous potential for developing PDMS-based on-chip optical devices, especially PDMS-based tunable microfluidic lenses for on-chip imaging and on-chip light manipulations.

Microfluidics, with the use of polymer (such as PDMS) and soft-lithography techniques, has rendered new techniques for production of curved-surface microfluidic lenses. Light, when incident onto these curved surfaces, would bend following conventional light refraction laws. Alternatively, optical lenses can be created in microfluidics by exploiting the fact that light bends differently in liquids of different refractive index. This allows liquids to replace some or all of the solid materials that would otherwise be used [35–41]. As such, a large category of polymer-based microfluidic lenses has been developed using the combined strength and flexibility offered by both microfluidics and transparent elastic polymer containers [42–45]. Microfluidic lenses have unique properties that are not seen in solid lenses, such as real-time configuration of lens type and focal length. A liquid–liquid lens consists of two fluids with different refractive indices and makes use of their interface as the lens surface to redirect light by altering the curvature at the interface of the two fluids. In this way, the shapes (curvatures of the lenses) and material compositions (refractive indices) of liquid lenses can be conveniently manipulated, thus allowing us to adjust a liquid lens to suit a variety of applications. Furthermore, the combination of microlenses with microfluidic channels provides additional adaptability to the microfluidic lenses: the polymeric microchannels not only work as a container for the liquids, but can also act as part of the microfluidic lenses due to their transparency and deformability.

One well-developed technique to fabricate a polymer-based microfluidic lens is to trap liquids within an elastic-membrane-sealed cavity. The sealed membrane is usually thin and possesses the freedom to bend outward or inward with the application of positive or negative pressure, allowing us to produce and maintain temporary convex (focusing) or concave (divergent) structures. Wide selections of liquids with various refractive indices bring more potential functionality and diversity to the lens. One tunable compound microfluidic lens, reported by Fei et al., used a multilayered PDMS structure to simulate a self-aligned lens group consisting of one biconvex and two plano-concave lens chambers, independently controlled by pneumatic valves programmed to be switched on and off digitally [46]. The valves can be precisely controlled to push exact volumes of fluids of varying refractive indices into multiple lens chambers, thereby adjusting the imaging power/focal length from centimeters to submillimeters on demand. Such lens configurations also make it possible to change lens types (from convex to concave) with adjustments of the liquids used. Though most other tunable microlenses can only change their focal lengths, liquid adjustment would allow the creation of different lens curvatures. Another microlens array design uses elastic-membrane-sealed fluidic chambers with an underlying microfluidic network (Figure 2.1a). The focal lengths of all the microlenses can be simultaneously controlled by pneumatically regulating the pressure

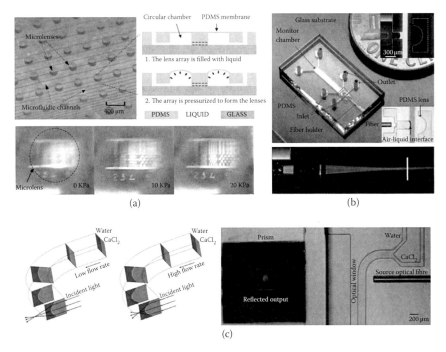

FIGURE 2.1

(See color insert.) (a) Microscopic image of the microlens array with underlying microfluidic network for pressure regulation (left) and schematic of the microlens actuation via pressure control. The focusing results can be varied due to change of the pressure. (b) A microfluidic lens that is tunable through active pressure control on an air–liquid interface. The experimental results of the device demonstrate good focusing and allow focal length adjustment through pressure control. (c) Mechanism of in-plane tunable fluid–fluid lens for focusing light in the z-direction with respect to device plane (where the plane of the device is the x–y plane). Change of the lens curvature and focal distance can be achieved by changing the flow rate. (Images are recompiled with permission from *Optics Express*, The Optical Society of America; *Microfluidics and Nanofluidics*, Springer; and *Lab on a Chip*, Royal Society of Chemistry.)

inside the microfluidic network via a pneumatic pump with a pressure regulator. By pumping high-refractive-index oil into the fluidic chamber, the focal length can be tuned from several hundred microns to several millimeters.

PDMS channels of differing shapes and designs can be used to confine fluids in such a way that they produce new tunable microfluidic lens configurations. One such example is a dynamic microfluidic lens with a liquid core–liquid cladding configuration reported by Tang et al. [22]. The flow pattern of liquid core and liquid cladding will change under different flow rates in a rectangular chamber fluidic channel. This interface change between the two fluids with different refractive indices provides a lens-like modulation of incident light from an optical fiber, with the focusing effect and focal length dynamically controlled by changing flow rates. A similar design concept has been applied in another lens configuration using an air–liquid interface inside a PDMS channel, as shown in Figure 2.1b [47]. The PDMS microchannel is specially designed such that the interface is in-line with the optical fiber and modulates the laser beam coming out from the fiber tips. The air–liquid interface can be finely adjusted by simply changing the flow rate of the liquids inside the chamber. The air–liquid interface is not as stiff as the channel wall and will not withstand the pressure from the high flow rate applied on the interface. As a result, the interface will be pushed toward the air side, causing it to deform and thereby modulating the light beams impinged on the interface. Together with another PDMS curvature design as beam collector, the air–liquid interface and PDMS channel design form a micro-fluidic lens system with tunable lens performance through adjustable flow rate.

A final example of tunable microfluidic lenses enabled by PDMS channel designs, illustrated in Figure 2.1c, is a hydrodynamically tunable cylindrical microlens [48,49]. This lens is unique in that the light it focuses travels in the device plane, but the plane of focusing is orthogonal to the device plane. The lens is formed by two laminar flow streams of different refractive indices moving within a microfluidic channel that is brought into a 90° bend. When the two laminar fluids flow around the 90° bend, centrifugal force is induced. Along the curvature, the fluid elements near the inner channel wall experience greater centrifugal force than the fluid near the outer channel wall. This leads to secondary flow in the form of two counter-rotating vortices with the middle plane at the channel middle cross-sectional plane. As a result, the fluid near the channel inner wall (of the curvature) bulges outward toward the outer channel wall. Hence, the originally flat fluidic interface bows outward, with the fluid of higher refractive index bowing into the fluid of lower refractive index, creating a cylindrical lens. The amount of interface bowing, and, consequently, the focal point of the lens, can be adjusted by changing the flow rates of the two fluids or the curvature angle of the microfluidic channel. Higher flow rates generate a lens with greater curvature and shorter focal length and vice versa.

All the previously mentioned tunable microfluidic lenses are capable of light modulation in one dimension (i.e., focal length tuning along the optical axis). However, in many circumstances, such as confocal imaging and other scanning-based imaging and detection methods, light modulation in one dimension is insufficient. The L-GRIN lenses discussed in Section 2.3, which can achieve light modulation in two dimensions, are excellent candidates to address this challenge.

2.3 In-Plane Liquid Gradient Refractive Index Lens

Nearly all of the tunable microfluidic lenses that have been developed so far are classic refractive lenses that share a similar working mechanism: a curved refractive lens surface formed at liquid–air, liquid–liquid, or liquid–solid interfaces that refracts incident light, focusing it at a point that varies with the radius of curvature of the interface. The tunability of these devices is usually achieved through curvature changes at the interface by means of liquid interactions with air, liquid, or solid. Numerous ways exist to modify those interfaces, the most widespread of which are electrowetting [50], electroosmosis [51,52], and pneumatic pressure controls [35,37]. It is relatively easy to adaptively reshape the microfluidic lens in the one-dimensional region. However, the aforementioned approaches cannot generate lenses with tunability in two dimensions. The limitations on classic curvature-based refractive microfluidic lenses challenge us to search for alternative solutions and other lens configurations. In this section, we will focus our discussion on one particularly promising alternative: the L-GRIN lens.

2.3.1 Design Concept and Working Mechanism of the Liquid Gradient Refractive Index Lens

The novel lens configuration used by the L-GRIN lens got its name from its focusing mechanism, which is similar to that of the solid-based gradient refractive index (GRIN) lens that is commonly used in the optical fiber industry for light collimation and coupling. A typical GRIN lens structure consists of a flat lens surface and a nonuniform refractive index distribution along the transverse direction, perpendicular to the optical axis as shown in Figure 2.2a [20]. In this case, there will be no curved surface to refract incident light abruptly. Instead, the light traveling inside the GRIN lens, along the optical axis, will experience small refractions at each infinitely small layer along the refractive index distribution and get bent toward the

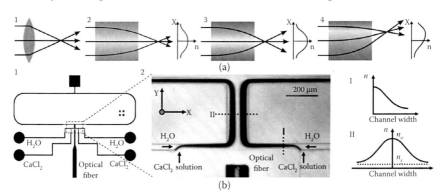

FIGURE 2.2
(a) A schematic diagram showing the difference between the classic refractive lens (a1) and gradient refractive index (GRIN) lens (a2). Change of the refractive index contrast in a GRIN lens can result in change of focal distance (a2–a3), and shift of optical axis can result in change of output light direction (a4). (b) Schematic of the liquid gradient refractive index (L-GRIN) lens design (b1), microscopic image of the L-GRIN lens in operation (b2), and the expected refractive index distribution at two locations (I and II) inside the lens (b2). High optical contrast areas (dark streaks) were observed near the fluidic boundaries, suggesting significant variation of refractive index due to $CaCl_2$ diffusion. (Images are recompiled with permission from *Lab on a Chip*, Royal Society of Chemistry.)

optical axis gradually. This forms a focusing point on or away from the main focal axis, depending on the refractive index profiles. In designing such a GRIN lens, the most important step is to generate the refractive index profile within the lens structure. Establishment of such refractive index gradient profile in solid materials requires complicated, costly, and time-consuming fabrication processes, including microcontrolled dipcoating, field-assisted ion exchange, and vapor deposition. More importantly, the focusing cannot be tuned due to fixed refractive index gradient.

By introducing microfluidics into GRIN lens design, Mao et al. have been able to utilize some unique properties of fluids at the microscale to form a liquid-based GRIN lens. Fluids on the microscale act differently from their counterparts on the macroscale. Two fluids flowing together in a specially designed microchannel tend to maintain *immiscible* liquid properties due to the high Reynolds number. When they slow down, they tend to mix together through diffusions. Diffusion speed and range can be precisely controlled using methods like heat, concentration gradient, and liquid flow rates. These microscale behaviors, along with liquids of varying refractive indices, are exploited to create a liquid-based GRIN lens. By manipulating the progress of diffusion, the refractive index profile and the focus pattern can be conveniently adjusted. More importantly, the gradient refractive profile can be finely adjusted asymmetrically through controlled diffusion of the fluids. This provides two degrees of freedom toward on-chip light manipulation, scanning, and imaging for many potential applications.

A parabolic profile and a hyperbolic secant (HS) profile are two of the several refractive index profiles that can be used in constructing a focusing L-GRIN lens. In the following L-GRIN lens design, an HS profile is achieved when solely relying on diffusion of a side-by-side laminar flow configuration because it mathematically resembles the analytical solution to a linear diffusion problem. The refractive index profile of an HS distribution can be expressed as:

$$n^2(x) = n_s^2 + (n_0^2 - n_s^2)\sec h^2(ax) \tag{2.1}$$

where $n(x)$ is the position-sensitive refractive index of the GRIN material at transverse position x, n_0 is the refractive index maximum contributed by the material at the center axis, n_s is the refractive index minimum that is counted as the refractive index of the background, and a is the gradient parameter. The trajectory of the light travelling inside the GRIN lens can be analytically solved. The selection of different solutions based on difference in refractive index is important. The L-GRIN lens requires that the fluids chosen are miscible through diffusion under low flow rate. If two fluids are immiscible, there is no effective diffusion at the interface and this means there is no GRIN profile to focus light beams.

An example of a liquid-based GRIN lens based on those design concepts is shown in Figure 2.2b. The two solutions used for generating the GRIN are

CaCl$_2$ and water. The refractive index of CaCl$_2$ solution can be adjusted by changing the concentration of solute (e.g., 3.5 M, $n \sim 1.41$). Conversely, we considered the refractive index of water to be constant ($n \sim 1.33$). The diffusion of CaCl$_2$ solution into water will produce a gradient concentration profile of CaCl$_2$ that linearly translates onto the refractive index profile. The microfluidic channel that constitutes the lens design consists of four inlets and two outlets. To maintain the equal back pressure at both sides, the two outlets are directed to a common outlet further downstream. On each side, the fluids are merged to form coinjected laminar flows and establish a CaCl$_2$ concentration distribution that resembles half of an HS profile, which can be correlated to Equation 2.1. The convergence of coinjected streams from both sides results in a complete HS profile refractive index pattern inside the main diffusing channel. A steady HS profile can be generated and maintained with a CaCl$_2$ flow rate of 3 µL/min and a water flow rate of 1.8 µL/min on each side of the two channels. By altering flow rates of the channels at both sides, the refractive index profile can be distorted while maintaining an asymmetric gradient profile. This will bend the light, leading to a focusing point that is out of the optical axis. This method allows us to achieve a scanning approach based solely on controlling the flow rates of the two fluids.

2.3.2 Device Fabrication and Experiment Setup

The device fabrication follows the standard procedure of soft lithography involving mask design, lithography, etching, stamping, and molding. A design mask of the L-GRIN lens is shown in Figure 2.2b1, which is generated using computer-aided design (CAD) software L-Edit or using AutoCAD. After the mask is designed, a standard lithography procedure is utilized to replicate the mask design into a photoresist pattern on a silicon wafer. The silicon master mold is created using deep reactive ion etching. This master mold has the pattern created in previous step via CAD and can be used hundreds of times. The silicon mold is cleaned and coated with 1H, 1H, 2H, 2H-perfluorooctyltrichlorosilane (Sigma-Aldrich, St. Louis, Missouri) to reduce the surface energy. This treatment protects the PDMS from deformation during the following detachment process from the silicon mold by significantly reducing the surface roughness and creating hydrophobicity. Next, Sylgard 184 Silicone Elastomer base and curing agent (Dow Corning, Midland, Michigan) mixed in a 10:1 weight ratio is poured onto the pretreated silicon wafer. The mixture is then cured at 70°C for 20 minutes. Once cured, the PDMS has the negative impression of the microstructures that were present on the silicon master mold. The PDMS is then peeled from the silicon mold and goes through mechanical punching for channel inlets and outlets. Before the PDMS channel is attached to a clean glass slide surface to form a sealed microfluidic channel chamber, oxygen plasma treatment on both the PDMS channel and glass substrate is performed to enhance the bonding between the two materials and prevent possible fluidic leaking, which is the cause of failure for many microfluidic experiments.

All of the inlets and outlets are designed to be 50 μm in width. The distance between optical fiber aperture and main flow channel is 100 μm and the main diffusion channel is 160 μm in width, putting it at the same scale range as the diameter of the optical fiber ($D_f = 155$ μm). Different length designs for the main channels are tested. The first of the two designs is a *long version* with channel length of 400 μm. The second is a *short version* with channel length of 250 μm. The channel height is set at ~157 μm (few micrometers more than optical fiber diameter) and controlled through a deep reactive ion etching process so that the optical fiber can be easily inserted into the fiber sleeve. The light beam is introduced into the L-GRIN lens setup using an optical fiber. Two light sources, a semiconductor laser diode and a halogen white-light source (Ocean Optics), are employed to test the performance of the L-GRIN lens. The optical fiber is a multimode fiber that supports multiple transmission modes and has a higher efficiency when compared to a single mode optical fiber. The diameter of the optical fiber used in the L-GRIN lens experiment is 155 μm with a center core diameter of 50 μm, providing a nominal numerical aperture (NA) around 0.22. The inserted optical fiber is aligned precisely with the center of the main channel through engineered channel confinement (Figure 2.2b1). The injection of the fluids is carried out using precision syringe pumps (KD Scientific 210; KD Scientific, Holliston, Massachusetts). The performance of the L-GRIN lens is characterized using an inverted optical microscope system (Nikon TE 2000U) connected to image acquisition devices of one 16-bit monochromatic charge-coupled device (CCD) camera (CoolSNAPHQ2) and one color digital camera (Nikon).

2.3.3 Operation Modes of the In-Plane Liquid Gradient Refractive Index Lens: Translation Mode and Swing Mode

As previously mentioned, L-GRIN lenses are designed to achieve light modulations with two degrees of freedom. The two degrees of freedom can be further divided into two operation modes: the *translation mode* and the *swing mode* (Figure 2.3). All of the light manipulation through the L-GRIN lens will fall into either one or both of the two operation modes. In both modes, the flow rate of $CaCl_2$ solution remains unchanged. However, in the translation mode, the GRIN profile can be adjusted by changing the flow rate of water simultaneously and symmetrically from both sides. This adjustment will lead to the dynamic realization of different focal lengths, ranging from no focusing to a long focal length, and finally to a short focal length. This is all accomplished with a single setup by simply varying the flow rates of the two fluids. In the case of swing mode, the direction of the output beam can be adjusted continuously through asymmetrical adjustment of water solution from each side. The uneven $CaCl_2$ distribution affected by water solution with different flow rates will alter the optical axis of the L-GRIN lens, making it deviate from the center optical axis and achieving focusing off-axis at the transverse direction.

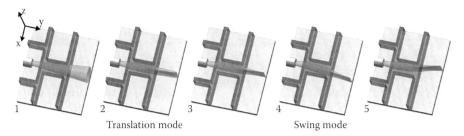

FIGURE 2.3
Schematic drawing showing the two operation modes of the liquid gradient refractive index lens: the translation mode with variable focal length, including no-focusing (1), a large focal distance (2), and a small focal distance (3), and the swing mode with variable output light direction (3–5). (Images are reproduced with permission from *Lab on a Chip,* Royal Society of Chemistry.)

2.3.3.1 Translation Mode

In the case of translation mode, the *long version* main channel is used to study the translation of light beams in order to cover a large focal length range. Computational fluid dynamics (CFD) and optical ray-tracing simulations are first conducted to optimize the refractive index distribution and light propagation in the L-GRIN lens at different flow conditions. The simulation helps to estimate the lens performance at different flow rate combinations. Figure 2.4a shows a simulated $CaCl_2$ concentration distribution at a representative flow condition ($CaCl_2$ flow rate = 3 µL/min, H_2O flow rate = 1.8 µL/min). The simulations strongly agree with the experimental results. Further simulation indicates that the concentration profiles are stable along the length of the main channel of the L-GRIN lens with uniform refractive index distributions at the entrance and exit of the main channel before and after diffusion occurs.

The horizontal line in Figure 2.4b is caused by the slight refractive index difference between $CaCl_2$ and PDMS ($n \sim 1.41$). Different flow conditions will lead to different refractive index profiles, as discussed previously. Here, ray-tracing simulation combined with CFD simulations are again used to estimate the L-GRIN lens performances. The flow rate for $CaCl_2$ is set to 3 µL/min and the water flow rates vary from 0.6 µL/min, 1.2 µL/min, 1.8 µL/min, and 2.4 µL/min to 3 µL/min. The CFD simulations indicate an excellent fit between the refractive index distribution and HS curves that are obtained through analytical solutions from Equation 2.1 (Figure 2.4c). Additional ray-tracing simulations based on the refractive index profile from the CFD simulation clearly show that the translation modes change along with the flow rate change of water. Higher water flow rates lead to larger refractive index contrast; this is caused by less diffusion occurring at the interface of the two fluids as they travel inside the main channel. The large refractive index contrast will make the light bend toward the optical axis more significantly within a short distance, thus reducing the focal length of the L-GRIN lens. Confirmation of this working principle is shown in the ray-tracing simulation results in Figure 2.4d.

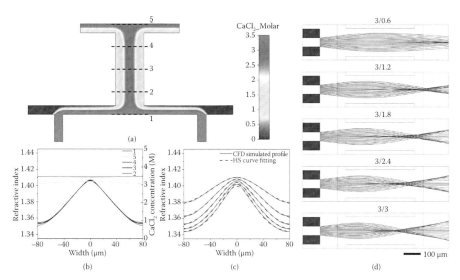

FIGURE 2.4
Computational fluid dynamics (CFD) and ray tracing simulation of the translation mode. (a) Simulated $CaCl_2$ concentration distribution in a long-version liquid gradient refractive index (L-GRIN) lens ($CaCl_2$ flow rate = 3 µL/min and H_2O flow rate = 1.8 µL/min). The color bar represents the molar concentration of $CaCl_2$. (b) Cross-sectional $CaCl_2$ concentration/ refractive index profiles at different locations (top to bottom: cross sections 1, 5, 4, 3, and 2, respectively, as defined in Figure 2.2a). (c) Refractive index profile at the middle of the L-GRIN lens (cross-section 3 defined in Figure 2.2a) for different flow conditions (top to bottom, the $CaCl_2$ flow rates were fixed at 3 µL/min, and the H_2O flow rates were 0.6, 1.2, 1.8, 2.4, and 3 µL/ min, respectively). Dotted lines are hyperbolic secant curve fitting. (d) Ray-tracing simulation in a long-version L-GRIN lens using the parameters obtained from the CFD simulation. The flow conditions are indicated in the graph (e.g., 3/3 represents $CaCl_2$ flow rates = 3 µL/min, and H_2O flow rates = 3 µL/min, respectively). (Images are reproduced with permission from *Lab on a Chip*, Royal Society of Chemistry.)

Experimental verification of the L-GRIN lens in the translation mode requires ray-tracing trajectories to be visible to image detectors like the naked eye or CCDs. Given the transparent nature of both $CaCl_2$ and water solutions to the excitation laser at wavelength 532 nm, fluorescent dye Rhodamine B is added to both solutions to show a clear contrast between the light path and the background. The light path within the excitation region will display a reddish color due to the fluorescent emission of the rhodamine when the solution has been dyed. The effect of introducing rhodamine dye to the concentration profile of the L-GRIN lens can be regarded as negligible due to its low concentration. The excitation laser will experience much more scattering when travelling inside PDMS walls due to the higher material density of PDMS as compared with solutions, and impurities such as small particles and air bubbles that are trapped inside the PDMS wall during the fabrication process. A greenish color due to such effect can be clearly observed inside PDMS channel walls with high intensities observed in Figure 2.5.

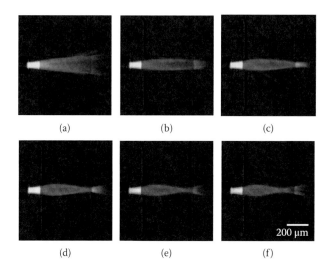

FIGURE 2.5
(See color insert.) Ray-tracing experiments for the translation mode to characterize the variable focal length at different flow conditions. (a) Ray tracing for stagnant flow (homogenous refractive index). (b–f) Ray tracing for dynamic flows. The $CaCl_2$ flow rates were fixed at 3 µL/min and the H_2O flow rates were (b) 0.6, (c) 1.2, (d) 1.8, (e) 2.4, and (f) 3 µL/min. (Images are reproduced with permission from *Lab on a Chip*, Royal Society of Chemistry.)

The experimental control consisted of a homogenous mixture of both fluids and dyes. In this case, the light path through the L-GRIN lens will follow a cone-shaped divergence output without significant lens effect due to the homogenous concentration distribution. Further experimental setup followed the optimized flow rate parameters from CFD and ray-tracing simulation, with experimental results shown in Figure 2.5. The change of focusing position is obvious from the experimental results and the transition from one translation mode to another is smooth. The tuning range of focal length is estimated around 700 µm (from 500 to 1200 µm), which is accomplished in the current design by changing flow rates. The dynamic tuning of the focal lengths also implied the availabilities of different NA, from almost 0 (Figure 2.5a) to ~0.3 (Figure 2.5f). A high NA is important in many applications, such as high-resolution on-chip imaging, optical trapping, and manipulation. It is quite challenging to achieve a high NA in many on-chip microfluidic lens designs. With further optimization, an L-GRIN lens design with an NA larger than 0.8 is quite possible by increasing the refractive index contrast, for example, by replacing $CaCl_2$ with other fluids with higher refractive index or simply increasing the concentration of $CaCl_2$.

In addition to focal length tuning, additional characterizations on the focused beam profile have been conducted. A side-view image setup is employed to measure the width and intensity distribution of the focused beam at each flow condition. The characterizations are carried out using a halogen white light source using a *short version* L-GRIN to achieve better image quality. Figure 2.6 shows a series of cross-sectional images of

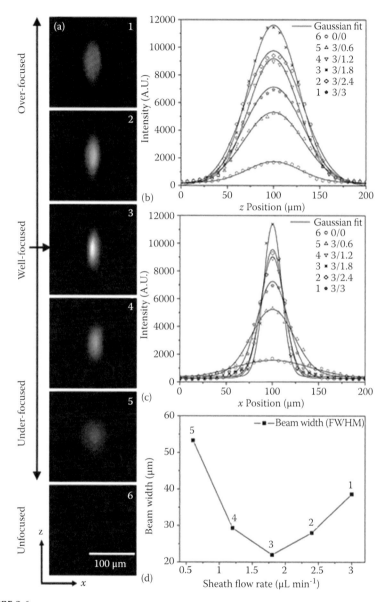

FIGURE 2.6

Characterization of the focused light beam profile in the translation mode. (a) Cross-sectional image of the focused light for different flow conditions (a1: 3/3, which represents $CaCl_2$ flow rates = 3 μL/min and H_2O flow rates = 3 μL/min; a2: 3/2.4; a3: 3/1.8; a4: 3/1.2; a5: 3/0.6; and a6: 0/0). (b and c) The intensity distributions of the focused light in z- and x-directions, respectively. The intensity readings were sampled at a 10 mm interval from the images in (a) and fitted with Gaussian curves. The flow conditions are indicated in the graphs (e.g., 3/3 represents $CaCl_2$ flow rate 3 μL/min and H_2O flow rate 3 μL/min). (d) Plot of beam width measured from (c) as a function of sheath flow rate. (Images are reproduced with permission from *Lab on a Chip,* Royal Society of Chemistry.)

the focused light beam. The well-focused position is first achieved under flow conditions of $F_{Ca} = 3$ µL/min and $F_H = 1.8$ µL/min. The position is then fixed as a reference for comparison with other flow conditions. As we established previously, changing flow rates of the water solution while fixing the $CaCl_2$ flow rates will lead to a tuning of the focal length, which can be perceived as defocusing from the fixed observing plane. The light spots will increase in size and intensity is reduced as the defocusing occurs. The extreme case of halting the injection of the fluids will create a uniform light distribution that covers the observation plane. Quantitative analysis on the intensity distribution along the z- and x-directions is shown in Figure 2.6b and c, while Figure 2.6d plots the beam width against sheath flow rate. A minimum focusing spot of approximately 22 µm can be achieved with the current setup with further improvement possible to achieve an even finer focusing effect for on-chip imaging and detection.

2.3.3.2 Swing Mode

The swing mode operates in a different flow condition than the translation mode. While the flow rate for $CaCl_2$ is kept constant at 3 µL/min as in all other cases, the flow rates of water solution are adjusted asymmetrically. To simplify our characterization of swing mode of the L-GRIN lens, the total flow rate of water solution from both sides is set as 6 µL/min. Based on this condition, several combinations are tested, including 5.4/0.6 (abbreviation for 5.4 µL/min for the left side and 0.6 µL/min for the right side), 4.2/1.8, 3/3, 1.8/4.2, and 0.6/5.4. The short version L-GRIN lens is used to achieve a large swing range.

Similar CFD and ray-tracing simulations are also employed in swing mode to estimate and optimize the L-GRIN lens performance. The CFD simulations clearly reveal the refractive index profile change across the cross section of the main channel. Figure 2.7a shows the refractive index distribution at a fluid combination of 4.2/1.8. As we can see, the refractive index profile tends to maintain its HS curves along the channel, though there is a shift in the center axis relative to that of all the translation modes. The refractive index at the entrance and exit, however, is no longer horizontal as in the translation modes, but changes gradually as diffusion occurs nonuniformly across the observing planes (Figure 2.7b). Additional tests on other flow rate combinations lead to similar refractive index profile distributions with various shifting distances, resulting in a shifting of the focusing point along the transverse direction (Figure 2.7c). The maximum swing modes in this case are achieved at flow rate conditions of 5.4/0.6 and 0.6/5.4, with the most unbalanced concentration distribution caused by the huge difference on flow rates. The ray-tracing simulations of swing modes are shown in Figure 2.7d.

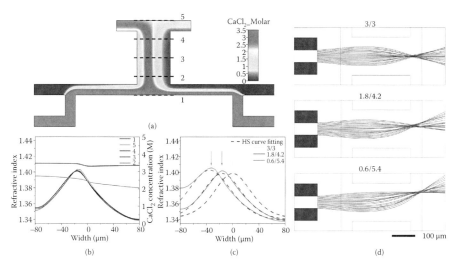

(a)

(b)

(c)

(d)

FIGURE 2.7
(See color insert.) Computational fluid dynamics (CFD) and ray-tracing simulation of the swing mode. (a) Simulated $CaCl_2$ concentration distribution ($CaCl_2$ flow rates = 3 μL/min on each side, and the H_2O flow rate on the left = 1.8 μL/min, and the H_2O flow rate on the right = 4.2 μL/min, abbreviated as 1.8/4.2 μL/min) in a short-version liquid gradient refractive index (L-GRIN) lens. The color bar represents the molar concentration of $CaCl_2$. (b) Cross-sectional $CaCl_2$ concentration/refractive index profiles at different locations (top to bottom: cross-sections 1, 5, 4, 3, and 2 defined in [a]). (c) Refractive index profile at the middle of the L-GRIN lens (cross-section 3) for different flow conditions (left to right: $CaCl_2$ solution flow rates were fixed at 3 μL/min for both sides, and the H_2O flow rates were 0.6/5.4, 1.8/4.2, and 3/3 μL/min, respectively). Dotted lines are hyperbolic secant curve fitting. (d) Ray-tracing simulation in a short-version L-GRIN lens using the parameters obtained from the CFD simulation. The flow conditions are indicated in the graph (i.e., 1.8/4.2 represents H_2O flow rate 1.8 μL/min on the left and 4.2 μL/min on the right). (Images are reproduced with permission from *Lab on a Chip*, Royal Society of Chemistry.)

Experimental verification on swing modes follows the exact flow rates used in CFD and ray-tracing simulations. The flow rates of the water solution are set as 5.4/0.6, 4.2/1.8, 3/3, 1.8/4.2, and 0.6/5.4. The optical images captured by the CCD camera are shown in Figure 2.8. A maximum swing angle of about 12 degrees is demonstrated with a larger swing angle possible through further optimization of the refractive index gradient and channel geometries.

2.4 Out-of-Plane Liquid Gradient Refractive Index Lens

The L-GRIN lens we discussed previously can be regarded as an in-plane L-GRIN lens. The light focusing and manipulation performed by this in-plane lens is restricted to the 2D x–y plane and lacks light-manipulating

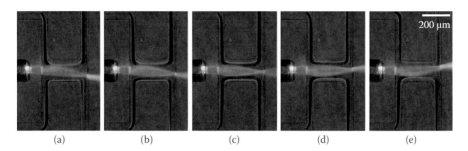

(a)   (b)   (c)   (d)   (e)

FIGURE 2.8
Ray-tracing experiments for the swing mode to characterize the variable output light direction at different flow conditions. The $CaCl_2$ flow rates were fixed at 3 µL/min on both sides, and the H_2O flow rates were (a) 5.4 µL/min on the left and 0.6 µL/min on the right (abbreviated as 5.4/0.6), (b) 4.2/1.8, (c) 3/3, (d) 1.8/4.2, and (e) 0.6/5.4 µL/min. (Images are reproduced with permission from *Lab on a Chip*, Royal Society of Chemistry.)

power in the z-direction. However, the design concepts of an in-plane L-GRIN lens can easily be extended to a three-dimensional (3D) space through the generation of a 2D L-GRIN created by a special microfluidics design. The design and working mechanism of such an out-of-plane L-GRIN lens is shown in Figure 2.9a [21]. The design is almost the same as that of a conventional L-GRIN lens (Figure 2.9b). An axis-symmetric 2D HS refractive index profile can be achieved with a microfluidic structure containing six inlets and two outlets (Figure 2.9c). Two of the six inlets are for $CaCl_2$ solutions, whereas the others are for deionized (DI) water. The two outlets allow the pressure balance inside the main chamber, reducing the adverse impact from the chamber on the concentration gradient. The concentration of the $CaCl_2$ solution in this case is 5 M, corresponding to a refractive index of 1.445. The refractive index of DI water is found to be around 1.335. Diffusion between these two solutions will create a concentration gradient at the interface of the two fluids, resulting in a refractive index gradient in the transverse direction of the diffusion interface, as shown in Figure 2.9d. The HS refractive index distribution again can be expressed using Equation 2.1, implying that it will be dependent on the flow rate ratio between the high-refractive-index fluid (in this case $CaCl_2$ solution) and low-refractive-index fluid (DI water).

The main diffusing chamber possesses similar parameters to a conventional L-GRIN lens, but is shared between two sets of channels to produce an HS refractive index profile in a 3D space. The difference between this setup and the previous one is in the action of the lens. The out-of-plane L-GRIN configuration works like a spherical lens with beam manipulation power in 3D space. The in-plane L-GRIN lens works like a cylindrical lens that only focuses light into a focusing line (lacking z-direction focusing), rather than a focusing point. The lens design of the out-of-plane L-GRIN is also different from the previous design, utilizing

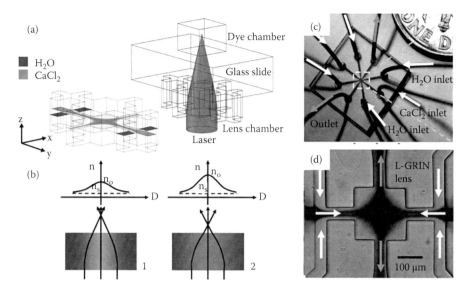

FIGURE 2.9
(See color insert.) Principle of the out-of-plane liquid gradient refractive index (L-GRIN) lens. (a) The out-of-plane L-GRIN lens structure is composed of an L-GRIN lens chamber, two inlets for CaCl$_2$ solution, four inlets for DI water, and two outlets. The diffusion of CaCl$_2$ inside the L-GRIN lens chamber results in a two-dimensional (2D) axis-symmetric hyperbolic secant (HS) refractive index profile in the x–y plane, which can be used to focus the light beam along the z-direction. The lens is bonded to a glass substrate and a dye chamber is bonded on the opposite side of the glass substrate for visualization purposes. The input light is generated via a laser diode aligned vertically to the device plane. (b) The side view of the refractive index distribution in the L-GRIN lens chamber. The focal point of the out-of-plane L-GRIN lens can be shifted along the z-axis (e.g., from b1 to b2 by adjusting the refractive index gradient within the liquid medium). (c) The fluid injection setup of the out-of-plane L-GRIN lens (dye chamber is not included in this image). (d) A microscopic image of the 2D diffusion pattern in the lens chamber. (Images are reproduced with permission from *Lab on a Chip,* Royal Society of Chemistry.)

a multilayered configuration stacked by a lens chamber (155 µm), a glass spacer (220 µm), and a dye chamber (155 µm). Both the lens chamber and dye chamber are designed as square channels with length 200 µm for all edges. The glass substrate is used to connect and support both chambers while considering the need for optical fiber insertion from the exposure side of the lens chamber. The widths of all the other injection channels are 50 µm. Rhodamine B with concentration of 10 µg/mL is used to fill the dye chamber. The whole setup is set on an inverted microscope (Nikon TE 2000U) and is characterized by a digital camera mounted onto the microscope (Nikon D50).

The CFD and ray-tracing simulation results of the out-of-plane L-GRIN lens configuration are shown in Figure 2.10. CaCl$_2$ solution flow rate

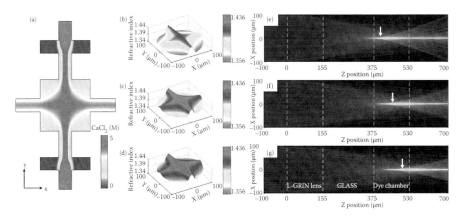

FIGURE 2.10

(See color insert.) Computational fluid dynamics (CFD) and ray-tracing simulation of the variable light focusing process. (a) Top view of the CFD simulation from mixing of CaCl$_2$ and DI water. (b–d) The CFD simulated two-dimensional hyperbolic secant refractive index profile within the lens chamber for different flow conditions (DI water flow rate is 8 µL/min, 3.2 µL/min, and 2.4 µL/min, respectively, and CaCl$_2$ solution flow rate is 8 µL/min in all three cases). (e–g) The simulated trajectories of the light beam during the focusing process with three different fluidic flow conditions shown in (b–d), respectively. The structure in the simulation is composed of three parts: the liquid gradient refractive index lens, the glass substrate, and the dye chamber. (Images are reproduced with permission from *Lab on a Chip*, Royal Society of Chemistry.)

is set as 8 µL/min. Star-shaped concentration diffusion profiles can be clearly observed in the center of the lens chamber. The refractive index profile will depend on the flow rate combinations of CaCl$_2$ and water, similar to the 2D L-GRIN lens. Ray-tracing simulations indicate tuning of the focal points in the dye chamber by adjusting flow conditions of DI water from 8 µL/min, 3.2 µL/min, and 2.4 µL/min. The experiments are optimized for characterization by confining all focusing points to the dye chambers. Agreement between simulations and experimental results were found after a series of L-GRIN tests. Side-view images of the experimental results are shown in Figure 2.11. A maximum tuning of the focal point at approximately 120 µm has been achieved. The transition between different translation modes can be completed within 2–3 seconds. Each mode can also be stabilized for several hours with constant flow injections.

From the experimental results, the maximum NA is measured to be 0.249. To improve the NA, one could increase the refractive index contrast or reduce the thickness of glass spacer. The beam-focusing qualities are also examined through top-view images showing the focusing shapes. The reference imaging plane is set at the position where the smallest focal point is located at a flow rate combination of $F_{Ca} = 8$ µL/min and $F_D = 3.2$ µL/min.

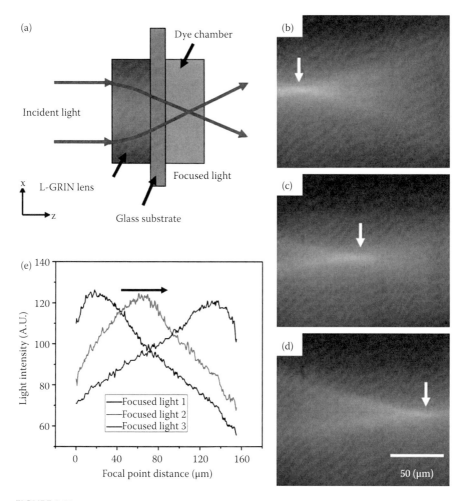

FIGURE 2.11

Experimental characterization of the variable light-focusing process. (a) The device includes three layers (i.e., the liquid gradient refractive index lens chamber, the glass substrate, and the dye chamber for visualization of the focused light beams). (b–d) Change of the light-focusing patterns and shifting of the focal point positions with different flow rate conditions (DI water flow rate is 8 μL/min, 3.2 μL/min, and 2.4 μL/min, respectively, and $CaCl_2$ solution flow rate is 8 μL/min in all three cases) are visualized in the dye chamber. (e) Light intensity plots along the z-axis, corresponding to the images in (b–d), indicate the shift of the focal point during the experiments. (Images are reproduced with permission from *Lab on a Chip*, Royal Society of Chemistry.)

A tight star-shaped focal point can be clearly observed with a minimum spot size around 20 μm. As the flow rate of DI water increases, the focal points deviate from the reference plane and expansion of the focal point size can be expected. Figure 2.12 shows the spot sizes of focal point at different DI water flow rates and their light intensity distribution.

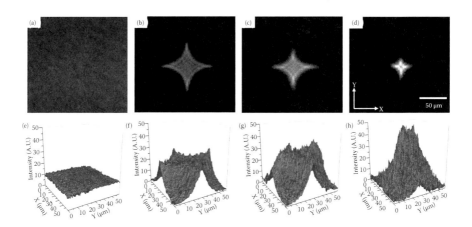

FIGURE 2.12

Top-view images of the light spots in the dye chamber after the light passing through the two-dimensional (2D) liquid gradient refractive index lens at different flow conditions. (a) Unfocused (stagnant flow); (b–d) from under-focused to well-focused light. DI water flow rate is 6.4 µL/min, 4.8 µL/min, and 3.2 µL/min, respectively; $CaCl_2$ solution flow rate is 8 µL/min. (e–h) The 2D light intensity distribution corresponding to (a–d). (Images are reproduced with permission from *Lab on a Chip,* Royal Society of Chemistry.)

2.5 Summary and Perspective

The L-GRIN lens, as a unique category of polymer-based microfluidic lens, has the following advantages over other microfluidic lenses: (1) It can work in different configurations. Both in-plane L-GRIN lenses and out-of-plane L-GRIN lenses can serve different purposes and fulfill different requirements in potential lab-on-a-chip applications. Almost all other microfluidic lenses can only work in one configuration and cannot be further adjusted after being built. Both of the discussed L-GRIN configurations can be tuned in real time to achieve dynamic lens performance easily in a short time. (2) L-GRIN lenses can have multiple operation modes. The light beam can not only be manipulated in translation modes, but also can be swung along the transverse direction to the channel axis. This is a unique feature that only the in-plane L-GRIN lens is capable of. Although it has not yet been demonstrated, swing modes are expected to work for out-of-plane L-GRIN as well. This provides a simple scanning mechanism, especially in comparison to most other scanning mechanisms, which require mechanical parts to work. (3) The spot size of the focal point and the NA of the L-GRIN lens can be further optimized to achieve even better performance. Preliminary results indicate a minimum spot size around 20 µm, which is significantly smaller as compared to the core size of multimode optical fibers. Further scaling down the spot size and increasing the NA are also quite possible, making the L-GRIN lens promising in on-chip optical imaging and manipulation.

(4) The stability of the L-GRIN lens is excellent. With constant flow injections, it will stand for days after use without malfunction. Most importantly, the L-GRIN lens is created through diffusion, rather than immiscible liquid interface generated from a very high flow rate. Therefore, it does not require high flow rate injection and is less likely to experience fluctuations from the mechanical injection system of pumps. (5) The slow flow rates required will help to reduce the consumption of solutions, which will reduce the necessary reservoir numbers and, ultimately, reduce the size and cost of the device for mass production.

To further demonstrate the capability of our L-GRIN lenses for on-chip excitation and detection, we performed a simple experiment to integrate the lens with a microfluidic flow cytometry device [53,54]. The results are shown in Figure 2.13. The L-GRIN lens has tight focusing and good selectivity, providing significant potential for on-chip imaging, detection, and manipulation applications. Further optimization of an out-of-plane L-GRIN lens and the use of computerized control in fluids injection could possibly be used to generate a beam scanning approach within a 3D chamber. With optimized lens performance, a tiny on-chip confocal microscope can be expected to image microscale bioassays, including single cells. The combination of the

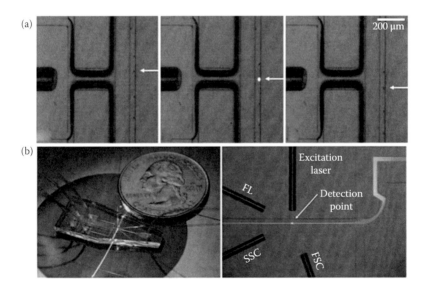

FIGURE 2.13
(a) Consecutive images showing the liquid gradient refractive index (L-GRIN) lens integrated with a microfluidic flow cytometry device. The traced particle (highlighted by a white arrow) only emitted fluorescence when entering the light-focusing region near the center axis of the L-GRIN lens. No fluorescent emission was observed from the neighboring particles that were outside the light-focusing region. (b) An example of a miniature flow cytometer. The optical detection part can be replaced with L-GRIN lens design to enrich its functionalities in the future. (Images are reproduced with permission from *Lab on a Chip,* Royal Society of Chemistry; *Biomicrofluidics,* American Institute of Physics.)

L-GRIN lens with recently developed lensless on-chip imaging devices could lead to an *all-in-one* on-chip system that has versatile functionalities including on-chip excitation, on-chip detection, and on-chip manipulation. We hope that these results can inspire other researchers to explore other possibilities wherein the uniqueness of the L-GRIN lens can be best utilized.

References

1. Jerjes, W. K. et al., The future of medical diagnostics: Review paper. *Head and Neck Oncology*, 2011. 3: 38.
2. Perelman, L. T., Optical diagnostic technology based on light scattering spectroscopy for early cancer detection. *Expert Review of Medical Devices*, 2006. 3(6): 787–803.
3. Balsam, J. et al., Low-cost technologies for medical diagnostics in low-resource settings. *Expert opinion on medical diagnostics*, 2013. 7(3): 243–55.
4. Myers, F. B. and L. P. Lee, Innovations in optical microfluidic technologies for point-of-care diagnostics. *Lab on a Chip*, 2008. 8(12): 2015–31.
5. Martinez, A. W. et al., Diagnostics for the developing world: Microfluidic paper-based analytical devices. *Analytical Chemistry*, 2010. 82(1): 3–10.
6. Mao, X. and T. J. Huang, Microfluidic diagnostics for the developing world. *Lab on a Chip*, 2012. 12(8): 1412–6.
7. Whitesides, G. M., The origins and the future of microfluidics. *Nature*, 2006. 442(7101): 368–73.
8. Arora, A. et al., Latest developments in micro total analysis systems. *Analytical Chemistry*, 2010. 82(12): 4830–47.
9. Neuzi, P. et al., Revisiting lab-on-a-chip technology for drug discovery. *Nature Reviews Drug Discovery*, 2012. 11(8): 620–32.
10. Horowitz, V. R., D. D. Awschalom, and S. Pennathur, Optofluidics: Field or technique? *Lab on a Chip*, 2008. 8(11): 1856–63.
11. Monat, C., P. Domachuk, and B. J. Eggleton, Integrated optofluidics: A new river of light. *Nature Photonics*, 2007. 1(2): 106–14.
12. Nguyen, N. T., Micro-optofluidic lenses: A review. *Biomicrofluidics*, 2010. 4(3). doi: 10.1063/1.3460392.
13. Mao, X. et al., Optofluidic tunable microlens by manipulating the liquid meniscus using a flared microfluidic structure. *Biomicrofluidics*, 2010. 4(4): 43007.
14. Mao, X. et al., Hydrodynamically tunable optofluidic cylindrical microlens. *Lab on a Chip*, 2007. 7(10): 1303–8.
15. Fei, P. et al., Discretely tunable optofluidic compound microlenses. *Lab on a Chip*, 2011. 11(17): 2835–41.
16. Huang, H. et al., Tunable two-dimensional liquid gradient refractive index (L-GRIN) lens for variable light focusing. *Lab on a Chip*, 2010. 10(18): 2387–93.
17. Tang, S. K. et al., A multi-color fast-switching microfluidic droplet dye laser. *Lab on a Chip*, 2009. 9(19): 2767–71.
18. Schmidt, H. and A. R. Hawkins, The photonic integration of non-solid media using optofluidics. *Nature Photonics*, 2011. 5(10): 598–604.

19. Li, Z. et al., Mechanically tunable optofluidic distributed feedback dye laser. *Optics Express*, 2006. 14(22): 10494.
20. Mao, X. L. et al., Tunable liquid gradient refractive index (L-GRIN) lens with two degrees of freedom. *Lab on a Chip*, 2009. 9(14): 2050–8.
21. Huang, H. et al., Tunable two-dimensional liquid gradient refractive index (L-GRIN) lens for variable light focusing. *Lab on a Chip*, 2010. 10(18): 2387–93.
22. Tang, S. K., C. A. Stan and G. M. Whitesides, Dynamically reconfigurable liquid-core liquid-cladding lens in a microfluidic channel. *Lab on a Chip*, 2008. 8(3): 395–401.
23. Song, W. and D. Psaltis, Pneumatically tunable optofluidic 2 x 2 switch for reconfigurable optical circuit. *Lab on a Chip*, 2011. 11(14): 2397–402.
24. Jo, B. H. et al., Three-dimensional micro-channel fabrication in polydimethyl-siloxane (PDMS) elastomer. *Journal of Microelectromechanical Systems*, 2000. 9(1): 76–81.
25. McDonald, J. C. and G. M. Whitesides, Poly (dimethylsiloxane) as a material for fabricating microfluidic devices. *Accounts of Chemical Research*, 2002. 35(7): 491–9.
26. Stephan, K. et al., Fast prototyping using a dry film photoresist: Microfabrication of soft-lithography masters for microfluidic structures. *Journal of Micromechanics and Microengineering*, 2007. 17(10): N69–N74.
27. Sundararajan, N., D. S. Kim, and A. A. Berlin, Microfluidic operations using deformable polymer membranes fabricated by single layer soft lithography. *Lab on a Chip*, 2005. 5(3): 350–4.
28. Love, J. C., J. R. Anderson, and G. M. Whitesides, Fabrication of three-dimensional microfluidic systems by soft lithography. *MRS Bulletin*, 2001. 26(7): 523–8.
29. Dimov, I. K. et al., Hybrid integrated PDMS microfluidics with a silica capillary. *Lab on a Chip*, 2010. 10(11): 1468–71.
30. Lee, K. S. and R. J. Ram, Plastic-PDMS bonding for high pressure hydrolytically stable active microfluidics. *Lab on a Chip*, 2009. 9(11): 1618–24.
31. Zhang, W. H. et al., PMMA/PDMS valves and pumps for disposable microfluidics. *Lab on a Chip*, 2009. 9(21): 3088–94.
32. Dodge, A. et al., PDMS-based microfluidics for proteomic analysis. *Analyst*, 2006. 131(10): 1122–8.
33. Nagarah, J. M. et al., Silicon chip-based patch-clamp electrodes integrated with PDMS microfluidics. *Biophysical Journal*, 2005. 88(1): 522A.
34. Pantoja, R. et al., Silicon chip-based patch-clamp electrodes integrated with PDMS microfluidics. *Biosensors & Bioelectronics*, 2004. 20(3): 509–17.
35. Sugiura, N. and S. Morita, Variable-focus liquid-filled optical lens. *Applied Optics*, 1993. 32(22): 4181–6.
36. Berge, B. and J. Peseux, Variable focal lens controlled by an external voltage: An application of electrowetting. *European Physical Journal E*, 2000. 3(2): 159–63.
37. Zhang, D. Y. et al., Fluidic adaptive lens with high focal length tunability. *Applied Physics Letters*, 2003. 82(19): 3171–2.
38. Kuiper, S. and B. H. W. Hendriks, Variable-focus liquid lens for miniature cameras. *Applied Physics Letters*, 2004. 85(7): 1128–30.
39. Werber, A. and H. Zappe, Tunable microfluidic microlenses. *Applied Optics*, 2005. 44(16): 3238–45.
40. Hong, K. S. et al., Tunable microfluidic optical devices with an integrated microlens array. *Journal of Micromechanics and Microengineering*, 2006. 16(8): 1660–6.

41. Dong, L. and H. Jiang, Tunable and movable liquid microlens in situ fabricated within microfluidic channels. *Applied Physics Letters*, 2007. 91(4): 041109.
42. Camou, S., H. Fujita, and T. Fujii, PDMS 2D optical lens integrated with microfluidic channels: Principle and characterization. *Lab on a Chip*, 2003. 3(1): 40–45.
43. Seo, J. and L. P. Lee, Disposable integrated microfluidics with self-aligned planar microlenses. *Sensors and Actuators B-Chemical*, 2004. 99(2–3): 615–22.
44. Wang, Z. et al., Measurements of scattered light on a microchip flow cytometer with integrated polymer-based optical elements. *Lab on a Chip*, 2004. 4(4): 372–7.
45. Godin, J., V. Lien, and Y. H. Lo, Demonstration of two-dimensional fluidic lens for integration into microfluidic flow cytometers. *Applied Physics Letters*, 2006. 89(6): 061106.
46. Fei, P. et al., Discretely tunable optofluidic compound microlenses. *Lab on a Chip*, 2011. 11(17): 2835–41.
47. Shi, J. J. et al., Tunable optofluidic microlens through active pressure control of an air-liquid interface. *Microfluidics and Nanofluidics*, 2010. 9(2–3): 313–18.
48. Mao, X. L., J. R. Waldeisen, and T. J. Huang, "Microfluidic drifting"—implementing three-dimensional hydrodynamic focusing with a single-layer planar microfluidic device. *Lab on a Chip*, 2007. 7(10): 1260–2.
49. Mao, X. L. et al., Hydrodynamically tunable optofluidic cylindrical microlens. *Lab on a Chip*, 2007. 7(10): 1303–8.
50. Krupenkin, T., S. Yang, and P. Mach, Tunable liquid microlens. *Applied Physics Letters*, 2003. 82(3): 316–8.
51. Li, H., T. N. Wong, and N. T. Nguyen. A tunable optofluidic lens based on combined effect of hydrodynamics and electroosmosis. *Microfluidics and Nanofluidics*, 2011. 10: 1033–1043.
52. Li, H., T. N. Wong, N. T. Nguyen, and J. C. Chai. Numerical modeling of tunable optofluidic lens based on combined effect of hydrodynamics and electroosmosis. *International Journal of Heat and Mass Transfer*, 2012. 55: 2647–2655.
53. Mao, X. L. et al., An integrated, multiparametric flow cytometry chip using "microfluidic drifting" based three-dimensional hydrodynamic focusing. *Biomicrofluidics*, 2012. 6(2): 24113–39.
54. Lapsley, M. I., L. Wang, and T. J. Huang, On-chip flow cytometry: Where is it now and where is it going? *Biomarkers in Medicine*, 2013. 7(1): 75–8.

3

Photorefractive Polymers for 3D Display Application

Pierre-Alexandre Blanche

CONTENTS

Abbreviations

    3D, three-dimensional

    DC, direct current

    Δn, refractive index modulation

    d, material thickness

    EA, electron affinity

    E_{ext}, externally applied electric field

E_{sc}, internal electric space-charge field

FOM, figure of merit

FWM, four-wave mixing

Γ, gain coefficient

Θ, grating phase shift in respect to the illumination pattern

θ, incidence angle

HOMO, highest occupied molecular orbital

I_p, ionization potential

K, grating vector

Λ, grating period

λ, wavelength

LUMO, lowest unoccupied molecular orbital

η, diffraction efficiency

OLED, organic light-emitting diode

OPV, organic photovoltaic

PR, photorefractive

QD, quantum dot

r_{eff}, effective electro-optic coefficient

RGB, red, green, blue

SLM, spatial light modulator

TBC, two-beam coupling

T_g, glass transition temperature

$\chi^{(n)}$, nth-order nonlinear susceptibility

ψ, grating "slant" angle: angle between the grating vector and the surface normal

3.1 Introduction

Photorefractive (PR) materials are a class of organic and inorganic compounds with the ability to change their index of refraction on illumination. The change in refractive index is both dynamic and reversible: dynamic because no external processing is required for the index modulation to be revealed, and reversible because the index change can be modified or suppressed by altering the illumination pattern. These properties make PR materials very attractive candidates for writing updatable holograms, and a key component in holographic 3D displays. Photorefractivity was initially discovered in inorganic crystals by Ashkin et al. in 1966 [1]. In 1990, Sutter et al. found the effect in an organic compound [2], and Ducharme et al. observed the phenomenon shortly afterward in polymers [3]. Since then, significant improvement in the material characteristics has enabled its use in applications as diverse as image restoration [4], correlation [5], beam conjugation [6], nondestructive testing [7], data storage [8], imaging trough scattering media [9], holographic imaging [10] and display [11], and many others [12].

The PR effect should not be confused with other mechanisms that also produce refractive index change, such as photochromism or molecular photo-orientation. As presented in Figure 3.1, the index change in PR material occurs via a multistep process. The incident light excites electrical charges that migrate in the bulk of the material and become trapped in the dark regions. This charge redistribution generates an internal space-charge electric field that ultimately induces an index change through the electro-optic effect and/or molecular reorientation.

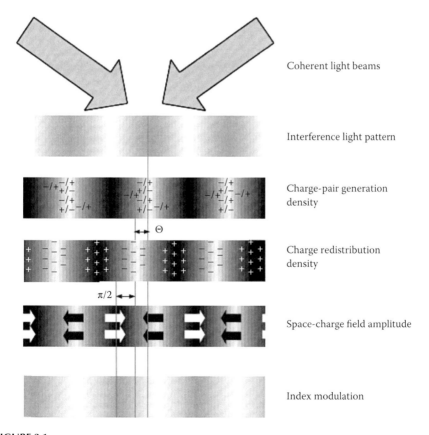

FIGURE 3.1
The different mechanisms involved in the photorefractive effect, transforming a light modulation (top) into refractive index change (bottom).

The PR effect has two specific properties that make it uniquely suitable for holographic displays and other applications. The mechanism is a reversible electronic process that does not induce any fatigue in the material, and there is a phase shift between the illumination pattern and the index modulation $(\Theta + \pi/2)$.

Different applications benefit from optimization of different aspects of the PR material. For example, coherent beam amplification needs a large index modulation and a phase shift as close as possible to $\pi/2$; data storage requires high sensitivity and long persistence. The electronic excitation wavelength should also be different if working within the infrared telecommunication domain, or for visible image processing and display. With inorganic crystals, these optimizations are performed by changing the doping levels of metallic impurities [13], but the range of adjustment is strongly limited by the stoichiometry and crystal structure. Polymer materials, on the other hand, can be tuned over large ranges through molecular engineering and incorporation of organic dopants. Other advantages of polymer films are their lower manufacturing cost, easier processability, and the ability to scale to large film sizes; these characteristics are of tremendous importance in the field of imaging and display. Since the discovery of the PR effect, inorganic crystals typically exhibited much better performances than organic compounds. However, recent advances in the field of nanocomposite and quantum dot (QD) dopants have dramatically improved the performance of polymer materials [14–17] to the point where the performance of polymer films often exceeds that of inorganic crystals. The field of PR polymers is also closely related to organic photovoltaic (OPV) materials and organic light-emitting diode (OLED), specifically with respect to the research into charge sensitizer molecules, and photoconduction polymers that occurs in these fields. Discoveries made in OPV and OLED technologies have been quickly adopted into PR applications and vice versa [18–21].

Thanks to those advances, it is now possible to synthesize organic PR compounds with extremely large index modulation [22], submillisecond response time [23], two-photon [24] or infrared sensitivity [25,26], and low-voltage operation [27]. Inorganic crystals [28] and hybrid devices [29] continue to hold the record for two-beam coupling (TBC) gain, but the polymers outperform inorganics in all other aspects.

Currently, there is renewed interest in the field of 3D display. Several factors have contributed to this boom: For the general public, the popularity of 3D movies at the theaters and the introduction of stereoscopic television played a large role. In the scientific world, the introduction of spatial light modulators (SLMs), and the emergence of pixelated light-emitting diode screens have accelerated the research and development of new holographic and autostereoscopy (glass-free 3D) techniques. In this chapter, we will show how PR polymers are making important contributions to the field of 3D display. We will first introduce the electrochemistry of PR polymers, the molecular structure of several commonly used compounds and dopants, as well as the device manufacturing techniques. We will then describe the application of PR materials to 3D displays using the specific properties of this unique material.

3.2 Mechanisms

As mentioned in Section 3.1, the PR effect is an electronic multistep process involving charge photogeneration, transport on a macroscopic distance, and trapping. The separation and delocalization of the charge leads to the creation of an electrical space-charge field inside the bulk of the material that ultimately generates an index modulation through the nonlinear electro-optics effect, and/or the molecular orientational birefringence (Figure 3.1). This section will describe the different aspects of the PR effect in polymer in further detail, and a review can be found in Lynn et al. [30].

3.2.1 Photogeneration

In a PR polymer, when an electron is excited by a photon, it moves from the highest occupied molecular orbital (HOMO) to the lowest unoccupied molecular orbital (LUMO). The energy level of the HOMO is characterized by the ionization potential (I_p) of the molecule, that is, the energy required to remove a loosely bound electron. The LUMO level is defined by the electron affinity, which is the energy involved as an electron is added to the molecule (Figure 3.2). This is a two-level electronic system as it can be found in semiconductor, with the distinction that, in a polymer, there is a statistical distribution of the energy levels due to different molecular interactions. In a

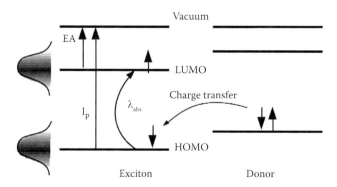

FIGURE 3.2
Highest occupied molecular orbital and lowest unoccupied molecular orbital molecular levels, and charge generation in photoconducting polymers. EA, electron affinity; I_p, ionization potential; λ_{abs}, absorption wavelength.

polymer, instead of having a well-defined absorption line, the peak is broadened around a central value.

Once an electron has been excited, it is still bound to the initial molecule (HOMO level), and there is a high probability that the electron-hole pair exciton will recombine within the same molecule via nonradiative energy transfer. The key step for charge transfer is the interaction between the excited molecule and a transport molecule. For the transfer to occur in a hole transport material, as is the case for most photoconducting polymers, the I_p of the transport molecule (donor) must be smaller than that of the excited molecule (acceptor). To assist the charge dissociation and further transport, the material is submitted to an external electric field applied through electrodes coated to the top and bottom faces of the polymer film. Such a device is usually made by pressing the PR compound between electrode-coated glass plates, and the electrode geometries will be discussed in further detail in Section 3.4.1.

The interaction between the acceptor and the donor molecule can lead to new absorption bands that are not present in the spectrum of either of the constituents. Sensitization to different parts of the spectrum can be achieved by this mechanism even though the component molecules do not manifest a strong absorption in that region. This is the case when polyvinylcarbazole (PVK) is mixed with trinitrofluorenone (TNF), creating an absorption peak around 900 nm. The theory governing the dissociation of ionic pairs in weak electrolyte under external electric field has been developed by Onsager [31], and works reasonably well for PR polymers. The model assumes that, after the formation of the exciton, an intermediate charge-transfer state is produced between the acceptor and the donor molecule in which the electron and the hole are thermalized and separated by an average distance r_0. One of the charge carriers (electron or hole) can escape from the other one with a probability proportional to the ratio between r_0 and the Coulomb radius r_c. The value of r_0 is calculated by best fit of the experimental data obtained by measuring the quantum efficiency according to the external electric field. The photogeneration efficiency is commonly measured using xerographic discharge technique [32] or estimated by DC field photoconductivity [33].

Most recent PR polymers use a C_{60} derivative as a sensitizer, such as [6,6]-phenyl-C61-butyric acid methyl ester (PCBM) described in Section 3.3. The advantages of this molecule are a large quantum efficiency, good charge transfer, and a relatively flat absorption spectrum over the entire visible region.

3.2.2 Transport and Trapping

Three different charge transport mechanisms have been identified in inorganic PR crystals: diffusion, photovoltaic, and drift. So far, only drift has been observed in organic materials. This distinction is important because in both the diffusion and the photovoltaic processes, the macroscopic charge transport occurs without the need of an external electric field. Drift, on the other hand, requires the assistance of such a field. All organic compounds that have been found to exhibit true photorefractivity require the application of an external electric field. The need to apply an electric field to the material has important implications on device fabrication and experimental geometry, as we will see in Section 3.4. Recently, TBC experiments on polymers have shown some asymmetric gain without the application of an external field, which is often taken as the macroscopic manifestation of photorefractivity [34,35]. However, other mechanisms may be causing the asymmetric gain [36], and further investigation is needed.

The microscopic explanation of the charge transport in polymeric compounds is based on the charge carrier hopping between molecular sites and delocalization through the conjugate part of the polymer chain [37]. Electron transport occurs among the LUMO levels, and hole transport takes place in the HOMO levels. In most PR polymers, both holes and electrons are mobile, but one of the species has a higher mobility and dominates the transport process. Charge carrier mobility can be measured by photocurrent time of flight [38], holographic time of flight [39], or estimated via the photoconductivity [40].

We can distinguish two types of charge trapping sites in photoconducting polymers: shallow and deep traps. Shallow traps hold the charge for a limited amount of time, but the carriers can be thermally re-excited, further contributing to the transport process. Shallow traps are responsible for the large mobility distribution in the material as seen in time-of-flight experiments. Deep traps, on the other hand, have an energy level at least 10 times deeper than shallow traps. There, the charge stops for an extended time and contributes to the space-charge field. Shallow traps are usually attributed to structural or conformational defects in the polymer, while deep traps are related to molecular species such as the chromophore. Figure 3.3 presents a sketch of charge generation, hole transport, trapping (shallow and deep), and recombination in organic PR materials.

This difference in mobility and trapping between the electrons and holes induces the initial phase shift Θ between the charge pair generation pattern, and the charge redistribution density presented in Figure 3.1. This phase shift depends on the material properties, but also on the experimental conditions such as the applied electrical field and grating frequency. In some compounds, bipolar transport can also be the genesis of dynamics competitive gratings, a case in which a faster but weaker grating is responsible for the initial diffraction, but is overcome by a second stronger grating with a different phase at a later time [41].

3.2.3 Space-Charge Field

The formation of the space-charge field is the outcome of the previous phenomena: charge generation, transport, and trapping. The field is the first derivative of the charge distribution, which is the source of the $\pi/2$ phase shift between a sinusoidal distribution and the field as presented in Figure 3.1.

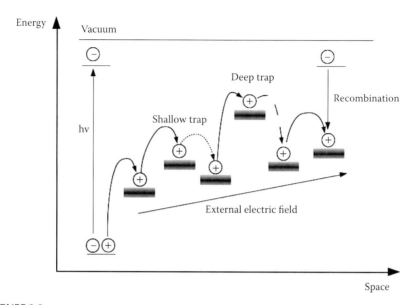

FIGURE 3.3
Hole transport mechanism in photoconducting polymers. An electron is photoexcited in a sensitizer molecule where dissociation of the exciton occurs through charge transfer to a donor molecule. Temporary charge trapping in the shallow trap is interspersed with the hopping process until trapped in a deep trap. Thermal re-excitation and eventual recombination with an electron complete the process.

The initial model for the formation of the space-charge field in PR inorganic crystals was developed by Kukhtarev et al. [42]. That model describes the equilibrium space-charge field by solving a set of rate equations for the charge transport in localized energetic bands. Though Kukhtarev's model is still the baseline model for polymers, many refinements have been added to take into account the particularities observed in those materials. One major contribution from Schildkraut et al. formalized the field dependency of the quantum efficiency and mobility [43]. Modeling the space-charge field is still a very active subject of research, and one of the most recent models has been developed by Ostroverkhova and Singer, in which thermally accessible deep traps as well as shallow traps have been included [32]. The introduction of differentiated traps into the model improved the prediction of the dynamics, especially the reduction of the speed due to trap filling and ionized acceptor growth.

3.2.4 Index Modulation

The generation of the refractive index modulation in polymer compounds was initially attributed to the nonlinear susceptibility $\chi^{(2)}$. In this representation, the molecular polarization in an electric field **E** is the sum of terms where the first one is given by the ground state dipole moment (μ), then the linear polarizability (α), followed by the higher hyperpolarizability terms β, γ, and so on, where all terms are tensors:

$$p = \mu + \alpha\,E + \beta\,E^2 + \gamma\,E^3$$

On a macroscopic level:

$$P = \chi\,E + \chi^{(2)}\,E^2 + \chi^{(3)}\,E^3$$

The oriented gas model presented by Burland et al. can predict the macroscopic polarizability from the orientational distribution of the molecules according to their microscopic properties, the electric field, and their density [44].

The second-order nonlinear susceptibility, $\chi^{(2)}$, is always zero in centrosymmetric media, which includes randomly oriented dipoles as found in chromophore-loaded polymers. To break the centrosymmetry, molecular orientation is required. This can be achieved by prepoling the material by heating it above the glass transition temperature (T_g), applying an electric field to orient the dipole molecules, and reducing the temperature before the field is switched off. This procedure leaves the dipoles oriented in the rigid matrix.

In a prepoled polymer compound where T_g is high above the ambient temperature, the index modulation is purely due to the electro-optic effect, as follows:

$$\Delta n(x) = \frac{-1}{2} n^3\,r_{\text{eff}} E_{\text{sc}}(x)$$

where r_{eff} is the effective electro-optic coefficient (proportional to $\chi^{(2)}$), and E_{sc} is the space-charge field.

In low glass transition temperature compounds, where the chromophore molecules have some rotational mobility inside the polymer matrix at ambient temperature, the chromophores become oriented by the space-charge field, and the linear polarizability (the birefringence of the molecule) becomes the dominant term in the index modulation. As presented in Figure 3.4, the electric field perceived by the molecules is the vector sum of the external field and the alternating space-charge field. From this overall modulating field results the periodic orientation of the chromophores. This orientational enhancement phenomenon was pointed out for the first time by Moerner et al. [45].

Because of the orientational enhancement, the total refractive index change is now a summation of the first- and second-order contributions, which leads to the definition of a new figure of merit for the chromophore molecules. This new figure of merit is not only based on the optimization of β, but also $\Delta\alpha$, the change in birefringence due to orientational enhancement:

$$FOM = \frac{2}{9kT}\mu\,\Delta\alpha + \mu\beta \tag{3.1}$$

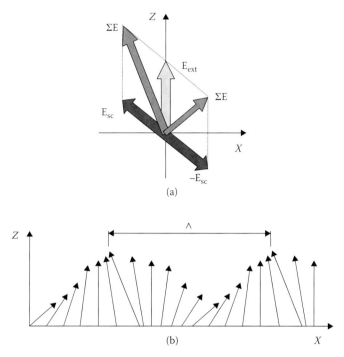

FIGURE 3.4
Orientational enhancement mechanism. (a) Sum of external and space-charge electric field vectors. (b) Field orientation along the grating vector K over a grating period Λ.

Due to the large index modulation that is achievable through orientational enhancement, most of the PR polymers that are developed today have a low glass transition temperature and chromophores, which are optimized for both their birefringence and first-order hyperpolarizability.

3.3 Material Composition

Recent research has focused on optimizing the different figure of merits of PR polymers for use in various applications. One of the most effective approaches has been the guest–host system, where the polymer is loaded with various organic molecules, each of which performs a different function required to complete the PR effect. The four basic components are the following:

- A polymer matrix that provides charge transport and structural integrity
- A sensitizer that ensures photogeneration of the charge carriers and the dissociation of the initial exciton
- A chromophore that generates the refractive index modulation

- A plasticizer that reduces the glass transition temperature of the polymer to allow orientation of the chromophores in response to the space-charge field

After purification, these different organic molecules are mixed in solution, dispersing them into the polymer matrix. The solvent is then evaporated to form a solid phase material. To prevent phase separation of the different species or crystallization of the polymer, some of the molecules (often the chromophore) can also be chemically grafted to the polymer as a side chain. Approaches other than the guest–host system that have been successfully implemented to obtain a PR organic material, including organic crystals [2], sol–gels [46,47], and liquid crystals [48,49].

3.3.1 Polymer Matrix

The polymer matrix plays the role of the photoconductor. To do so, it should be highly conjugated, that is, contain a large number of delocalized π electrons. The photoconduction not only depends on the structure of the polymer backbone itself, but also on that of the side chain(s) and the average length of the polymer (the molecular weight) [50]. Furthermore, since PR polymers are heavily loaded with chromophores, the photoconduction can also be influenced by these low molecular weight dopant molecules [51,52].

Among the various polymers tested to form PR compounds, two particular structures have been extensively used: poly (*N*-vinylcarbazole) and poly (acrylic tetraphenyldiaminobiphenol) (PATPD). Their chemical composition is presented in Figure 3.5.

3.3.2 Sensitizer

The sensitizer generates the initial exciton through photon absorption. To maximize the available carriers, the quantum efficiency of the sensitizer should be as high as possible for the wavelength of interest. The sensitizer should also be optimized to minimize the relative energy difference between the sensitizer and the polymer HOMO levels for an efficient charge transfer to the trap level for fast photoconduction [53,54]. Figure 3.6 presents the molecular structure of different sensitizers that have successfully been used. PCBM provides sensitization in the visible, (2,4,7-trinitro-9-fluorenylidene) malononitrile (TNFDM) in the near infrared (700–800 nm), and 2-[2-{5-[4-(di-n-butylamino)phenyl]-2,4-pentadienylidene}-1,1-dioxido-1-benzothien-3(2H)-ylidene]malononitrile (DBM) at even longer wavelengths (900–1000 nm).

Recently, researchers have shown some success in using nanoparticles and QDs as sensitizers. Cadmium selenide (CdSe) QDs treated with 4-methylbenzenethiol surfactant have shown efficient photoinduced charge generation in PR composites, leading to grating buildup times of 100 ms and below [14]. To sensitize the material in the infrared, where organic molecules are generally

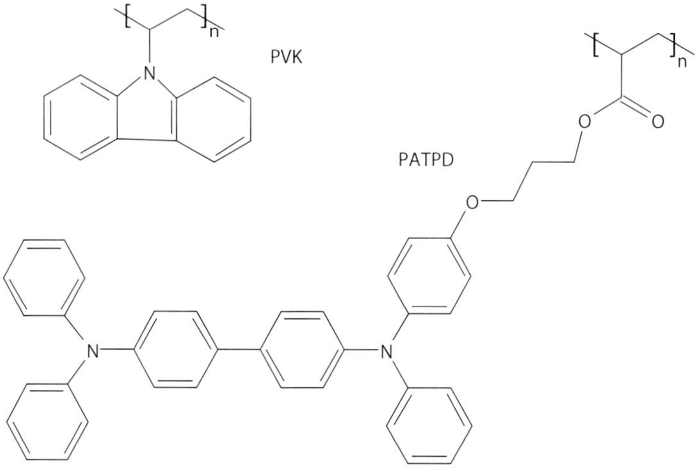

FIGURE 3.5
Molecular structure of photoconducting polymers polyvinylcarbazole (PVK) and poly (acrylic tetraphenyldiaminobiphenol) (PATPD).

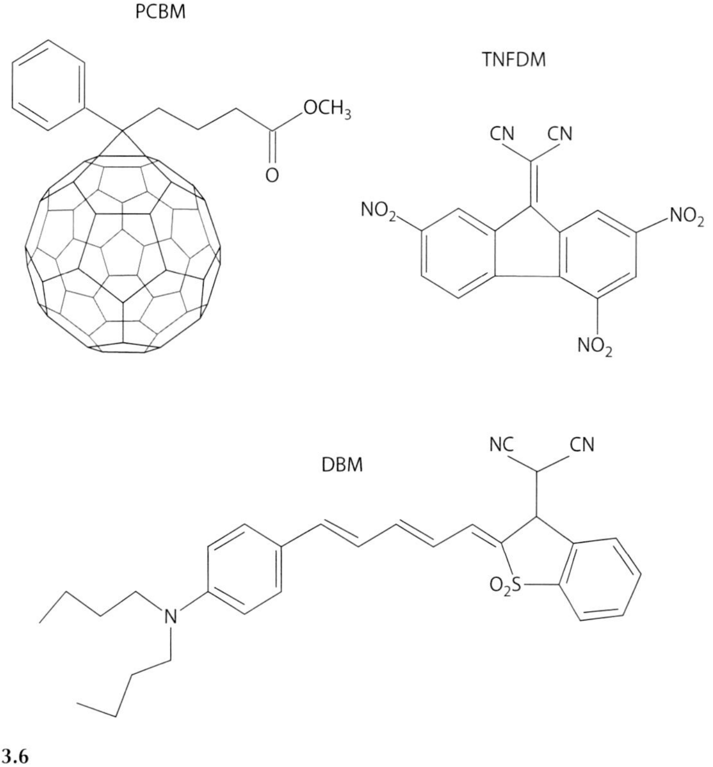

FIGURE 3.6
Molecular structure of sensitizer molecules [6,6]-phenyl-C61-butyric acid methyl ester (PCBM), (2,4,7-trinitro-9-fluorenylidene) malononitrile (TNFDM), and 2-[2-{5-[4-(di-n-butylamino)phenyl]-2,4-pentadienylidene}-1,1-dioxido-1-benzothien-3(2H)-ylidene]malononitrile (DBM).

not efficient, lead selenide (PbSe) QDs with spectral response at the optical communication wavelength of 1550 nm have been tested [15]. Different core/shell nanoparticles such as cadmium selenite/cadmium sulfide (CdSe/CdS) and cadmium selenite/zinc sulfide (CdSe/ZnS) have shown that the energy level of the outer layer (or shell) is important to prevent the recombination of the charge carriers [16]. Particularly, a type II QD such as a cadmium selenide/cadmium telluride (CdSe/CdTe) core/shell is appropriate to facilitate hole separation from the excitonic state to the polymer matrix. It has been shown that these QDs can double the photocurrents and improve both the TBC gain and the four-wave mixing (FWM) diffraction efficiency [55]. Gold nanoparticles have been found to enhance the TBC gain by affecting the phase shift (and not the index modulation), and acting as a trap center [56]. However, those successes are mitigated by the fact that nanoparticles or QD-loaded PR compounds have not yet achieved the performance demonstrated by all-organic materials. Most recently, graphene has been successfully tested as a sensitizer [57], and it is expected that its functionalization will lead to even better performances.

3.3.3 Chromophore

Because the chromophore molecules are ultimately responsible for the modulation of the refractive index, their optimization is of critical importance. Prior to the discovery of orientational enhancement phenomena in 1994 [45], optimization of the chromophore focused only on the improvement of the molecular hyperpolarizability (β) by increasing the conjugation length and the acceptor–donor strength found at each end of the molecule. For this reason, numerous azoic dyes have been used as PR chromophores because they have a large conjugated chain terminated with a donor and an acceptor. However, once it was recognized that the linear polarizability ($\Delta\alpha$) also played an important role, a new selection criteria was

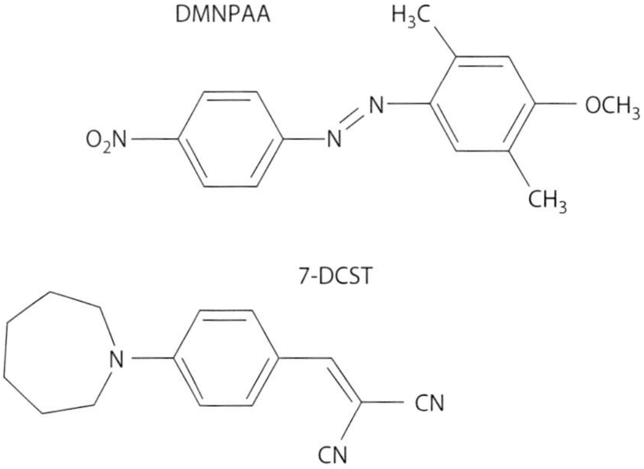

FIGURE 3.7
Molecular structure of chromophores 2,5-dimethyl-4′-nitrophenylazoanisole (DMNPAA) and 4-azacycloheptylbenzylidene-malonitrile (7-DCST).

introduced (see Equation 3.1) and new materials emerged [58,59]. Examples of both types of chromophores are given in Figure 3.7. 2,5-dimethyl-4′-nitrophenylazoanisole is an azo dye, and 4-azacycloheptylbenzylidene-malonitrile is of the high linear polarizability sort.

Increasing the chromophore concentration might also be viewed as a way to improve performance, but additional loading often leads to phase separation, causing the material to turn milky and highly diffuse. In most cases, the clarity can be restored by heating the material above the melting point, and then cooling it rapidly (quenching). The shelf life of PR polymers ranges from few hours to several years. It is possible to improve the stability of the material by using two (or more) chromophores in the same material [60], or grafting some chromophores on the polymer main chain, which increases the solubility [59]. Concentration as high as 40 wt% of chromophore is common in PR compounds. It should be noted that some applications require specific considerations in the chromophore design, as we will see in Section 3.5, where transparency in the visible must also be optimized.

3.3.4 Plasticizer

To facilitate the chromophore reorientation inside the polymer matrix, it is desirable to have a glass transition temperature (T_g) below the ambient temperature. If the original T_g of the material is too high, a plasticizer molecule can be added to soften the rigid structure of the polymer and facilitate the movement of the chromophore molecules.

The plasticizer could be the monomer of the main structural polymer, such as N-ethylcarbazone in the case of PVK, or a separate low molecular mass molecule like *n*-butyl benzyl phthalate; both are presented in Figure 3.8. The chromophores themselves are also acting as a plasticizers due to their low molecular mass and high concentration (up to 40 wt%). Sensitizer molecules, on the other hand, are generally used in such low concentration (0.5 wt%) that they have only a minor effect on the T_g. Care must be taken to identify plasticizers that do not accelerate phase separation of the various compounds and induce faster material crystallization.

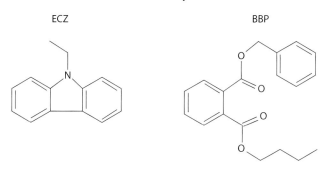

FIGURE 3.8
Molecular structure of plasticizers N-ethylcarbazone (ECZ) and *n*-butyl benzyl phthalate (BBP).

3.4 Holographic Characterization

As with any photonic material, there is a large battery of tests that can be done to chemically and optically characterize PR polymers, including molecule purity, nuclear magnetic resonance, absorption spectrum, scattering, photoconductivity, electro-induced birefringence, and so on. In this section, we will only focus on the holographic characterization methods that are specific to the PR effect. Two fundamental experiments will be detailed: TBC and FWM. Before describing the experimental techniques, a quick discussion of electrode geometries and device fabrication will provide a better understanding of the overall configuration requirements.

3.4.1 Device and Geometry Considerations

Most PR polymer devices are fabricated by pressing the material between two glass plates, whose inner faces are coated with transparent indium tin oxide to act as electrodes. In this case, the external electric field is orthogonal to the device surface. When illuminated by two coherent beams incident at the same angle from either side of the surface normal, as presented in Figure 3.9a, the interference fringes are parallel to the electric field. In this case, the electrical charges photogenerated inside the material travel along the bright fringes, pulled by the external field, and do not get trapped in the dark regions. No space-charge field is created, and no index modulation is observed. To ensure that the charges are forced into the dark regions, the interference fringes should be tilted with respect to the sample normal: the grating vector K must have a component along the field vector E. This is referred to as the tilted configuration, presented in Figure 3.9b.

To circumvent this limitation, different electrode geometries have been tested, such as interdigitated coplanar electrodes, where the anode and cathode have the shape of interleaving combs, and are coated on the same glass plate [61]. Authors have shown that these interdigitated electrodes provide similar performance to standard devices and geometries, but without the external slant angle that can pose difficulties in some applications. Other PR device manufacturing techniques include spin coating and injection molding [62].

Figure 3.9 shows two writing geometries where both writing beams are incident on the same side of the device. When such a diffraction grating is read, the reading beam goes *through* the sample and emerges to the other side. This configuration is referred to as *transmission geometry*. When the writing beams are incident from different sides to the device, the grating is said to be in *reflection geometry*, because the reading beam is diffracted *back* to the same side from which it is incident (see Figure 3.10) [63].

It should be noted that the grating spacing (Λ) between transmission and reflection geometries is very different at the angles usually employed in PR

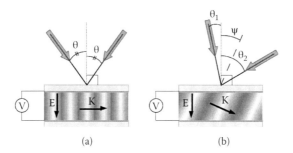

FIGURE 3.9
(See color insert.) Holographic grating transmission geometries. (a) Symmetric. (b) Tilted.

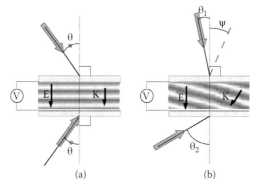

FIGURE 3.10
(See color insert.) Holographic grating reflection geometries. (a) Symmetric. (b) Tilted.

experiments. The grating spacing can be calculated using the diffraction grating equation:

$$\Lambda = \frac{2n\sin[(\theta_2 - \theta_1)/2]}{\lambda}$$

where n is the refractive index of the material, λ is the light wavelength, and θ_1 and θ_2 are the writing beam incidence angles.

In transmission, $\theta_1 = 50°$ and $\theta_2 = 70°$ in the air with a bisector tilted at $\psi = 60°$. $\Lambda = 400$ lp/mm. In reflection, $\theta_1 = \theta_2 = 72°$ (bisector $\psi = 0°$). $\Lambda = 4200$ lp/mm. Calculated with 532 nm writing wavelength, refractive index = 1.6.

This factor of 10 in the interfringe spacing has a large impact on the space-charge field amplitude because it requires a larger trap density in reflection than in transmission. For this reason, and in addition to the competition between the electro-optic effect and the molecular birefringence in reflection geometry, the diffraction efficiency is usually much weaker in reflection than in transmission [64]. Besides the geometry configuration, polarization of the beams also plays an important role, as we will see when discussing the FWM experiment.

3.4.2 Two-Beam Coupling

The TBC experiment can be summarized as interfering two beams in the material and monitoring the output intensity of those same beams (Figure 3.11). This simple setup is very powerful because it can be used to determine the gain coefficient of the material as well as the phase and the amplitude of the grating, three critical parameters necessary to understand the PR effect in a particular material.

Because of the phase shift between the intensity modulation and the index modulation in PR materials (Figure 3.1), the grating itself diffracts the writing beams, transferring energy from one beam to the other. This asymmetric energy transfer is measured by the gain coefficient (Γ), which depends on the amplitude of the index modulation (Δn), and the phase shift (Θ):

$$\Gamma = \frac{4\pi}{\lambda}(\hat{e}_1 \cdot \hat{e}_2^*)\Delta n \sin\Theta \tag{3.2}$$

with λ the wavelength, and $\hat{e}_x$ the polarization vectors of the writing beams.

The measure of Γ is given by

$$\Gamma = \frac{1}{d}[\cos\theta_1 \ln\gamma_1 - \cos\theta_2 \ln\gamma_2] \tag{3.3}$$

where d is the material thickness, θ_1 and θ_2 are the internal beam angles measured according to the sample normal, and γ_1 and γ_2 are the beam ratios defined as

$$\gamma_1 = \frac{I_1(I_2 \neq 0)}{I_1(I_2 = 0)}$$

$$\gamma_2 = \frac{I_2(I_1 \neq 0)}{I_2(I_1 = 0)}$$

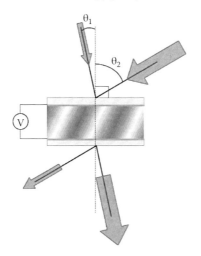

FIGURE 3.11
(See color insert.) Two-beam coupling measurement configuration.

The gain is given in per centimeter units, and net gain occurs when the gain value is larger than the absorption coefficient (also in per centimeter units) at the wavelength of the writing beams. As we can see from Equations 3.2 and 3.3, the measure of the gain alone cannot disambiguate the amplitude and the phase of the grating. To do so, an independent measurement should be performed. In the TBC setup, a phase shift is introduced to artificially move the illumination pattern relative to the grating. This can be done by changing the phase of one of the writing beams (the illumination pattern shifts), or moving the sample (the grating shifts). Doing so, one observes a sinusoidal modulation in the output intensity of the writing beams. The phase at the displacement origin is the natural (not induced) phase shift Θ [65].

Of course, in the phase-shifting experiment, one must make sure the induced shift is much faster than the dynamics of the grating, otherwise the intensity modulation will write a new pattern in the material, and the measurement will not be valid. A number of experimental parameters influence the phase and amplitude of the grating, for example, the writing beam polarization, the intensity ratio, the applied voltage, the inter-beam angle, and the slant angle (ψ). By going through different grating geometries, the different material constants (such as the trap density) can be determined through the model described in Yuan et al. [66].

3.4.3 Four-Wave Mixing

As depicted in Figure 3.12, the four beams present in the FWM experiment are the two writing beams, a probe beam that addresses the grating, and the diffracted probe beam. The probe beam can be of the same wavelength as the writing beams, called degenerate FWM, or the probe beam can be of another wavelength. In this later case, the incident angle should be corrected for the spectral dispersion according to Bragg's law:

$$\lambda = \frac{2n\sin[(\theta_2 - \theta_1)/2]}{\Lambda}$$

$$\text{if } \theta_2 = \theta_1 : \frac{\lambda_R}{\lambda_W} = \frac{\sin(\theta_R)}{\sin(\theta_W)}$$

where n is the refractive index of the material, λ is the light wavelength, θ_1 and θ_2 are the writing beam incidence angles, and Λ is the grating spacing (in number of fringes per unit of distance). The R and W subscripts stand for reading and writing, respectively.

The advantage of using different wavelengths for writing and reading the grating is that they can be carefully chosen such that the writing beams excite the sensitizer molecules, but the reading beam uses a transparency

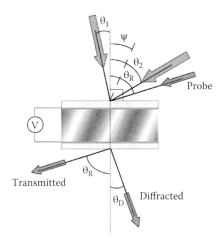

FIGURE 3.12
(See color insert.) Four-wave mixing measurement configuration.

window in the material, preventing perturbation of the trapped charges. When the same wavelength is used, the intensity of the probe beam must be significantly smaller than the writing beams to ensure that the grating is not disturbed.

In the FWM experiment, the choice of the probe beam polarization is very important. The index modulation experienced by the s- or p-polarizations is usually very different due to the rod-like shape of the chromophore molecules index ellipsoid. Figure 3.13 gives a visual representation. If the writing beams are in the YZ plane of the laboratory, the angle φ is close to $\pi/2$, and the index modulation is given by an oscillation along the θ angle. A reading beam that is s-polarized (orthogonal to the YZ plane of incidence) will intersect the ellipsoid along X and will only see the xy section of the ellipsoid, whereas the p-polarization (within the YZ plane of incidence) will see the z section of the ellipsoid, and experience a larger index modulation. This reasoning is also valid for the TBC experiment discussed in Section 3.4.2, and explains why the coupling coefficient is larger with p-polarized beams than with s-polarized.

The dynamics of the PR grating can be observed with FWM experiments. Initially, the writing beams are off and there are no gratings in the material. The probe beam goes through without being diffracted. When the writing beams are turned on, the grating begins to form and the diffraction increases until it reaches a steady state value. When the writing beams are turned off again, the grating naturally disappears (dark decay) due to charge recombination, and the diffraction decreases to a minimum (see Figure 3.14, solid line).

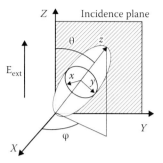

FIGURE 3.13
Geometry of the chromophore refractive index ellipsoid relative to the external electric field
and incident writing beam plane in a holographic recording experiment.

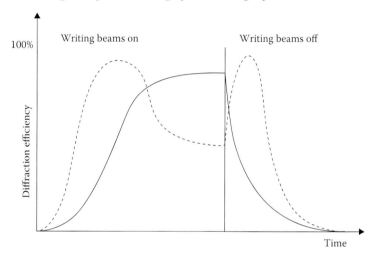

FIGURE 3.14
Dynamic four-wave mixing: diffraction efficiency according to time. Typical behaviors of pho-
torefractive polymers. Solid line: external voltage set below or equal to maximum modulation.
Dotted line: external voltage set above maximum modulation.

In transmission geometry, the diffraction efficiency (η) is proportional to
the sine squared of the index modulation (Δn):

$$\eta = \sin^2\left[\frac{\pi \Delta n\, d}{\lambda\sqrt{\cos\theta_1 \cos\theta_2}}\cos(\theta_1 - \theta_2)\right]$$

where d is the material thickness, λ is the wavelength, and θ_1 and θ_2 are the
incidence and diffracted angle measured inside the material, respectively.
The index modulation increases until the argument of the sine exceeds $\pi/2$,
after which the efficiency starts decreasing. This is represented by the dotted
line in Figure 3.14.

The dynamic behavior of PR polymers can be modeled by a double exponential function, or by a stretched exponential:

$$\Delta n(t) = A[1 - m\exp(-t / \tau_1) - (1 - m)\exp(-t / \tau_2)]$$
$$\Delta n(t) = A\left\{1 - \exp\left[-(t / \tau)^\beta\right]\right\}$$

where A is an amplitude factor, m is the proportion of each exponential, τ_1 and τ_2 are the time constants, and β is the stretch factor for which $0 < \beta < 1$.

A second important experiment can be run with the FWM setup. With both writing beams on, the external voltage can be increased slowly, and the steady state diffraction value for each increment can be recorded. The diffraction efficiency follows as a power law of the field:

$$\eta \propto \sin^2\left(E^p\right)$$

with $p \sim 1.9\text{–}2$ [27].

This measurement helps to determine to optimum working voltage for the device, where the efficiency is maximum. If the voltage is further increased, the efficiency decreases (because of the sine relation), and the grating is said to be overmodulated (Figure 3.15). Overmodulation voltage can be used in some types of applications to improve the response time. This type of external field manipulation is called voltage kickoff [67].

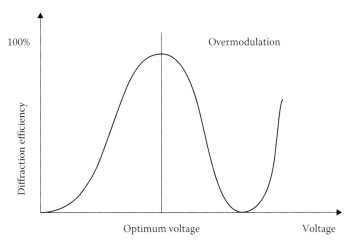

FIGURE 3.15
Diffraction efficiency according to external voltage. Typical behaviors of photorefractive polymers.

3.5 3D Display

Many applications take advantage of the unique properties of PR material in that they can be used to form a holographic recording device that is both dynamic and refreshable. In contrast, other materials such as silver halide emulsion, photopolymers, or dichromated gelatin need a postprocessing treatment to reveal the grating and are permanently recorded. These materials are well suited to single-use holography applications but cannot be updated. Photochromic materials, such as azo dyes or doped glasses, are also dynamic and refreshable, but the process is generally orders of magnitude slower and less sensitive than PR materials, which can react in submilliseconds, and/or with a single-nanosecond laser pulse [11,23]. While an external electric field needs to be applied to the PR polymer, far from being an impediment, it can be used to improve the response of the device, or provide a longer dark decay [67].

We should distinguish here first between imaging and display applications. In imaging applications, the material is used as a relay to record an image that is already formed elsewhere in the setup. This is the case when one of the writing beams is scattered from an object, or structured via an SLM. Figure 3.16 presents these two scenarios, with recording and reading differentiated in the case of a real object, and a dynamic recording-reading using an SLM. In this latter case, when the image on the SLM is updated, thanks to the dynamic nature of the PR effect, the diffracted image also changes [68]. However, the diffracted image is similar in nature to the SLM image: 2D, and not 3D. This is quite analogous to the FWM experiment presented in Figure 3.12.

Though imaging experiments are useful to determine the characteristics of the recording material, in the framework of a display, it is certainly more convenient to directly look at the first image (the real object, or the SLM), instead of the diffracted beam. By contrast, a holographic display system uses the diffraction from the device to structure the light, and directly presents the information, not simply reimaging it. In such a system, there is no other place where a viewer could see the image than in the diffracted beam.

Holographic displays can be made by calculating the diffraction pattern, loading it to an SLM [69] or an acousto-optic modulator [70], and ultimately diffracting a reading light beam to produce the 3D image. Such a system is referred to as electroholographic 3D display, and does not require a recording material such as a PR device. While this type of approach might lead to the ultimate 3D display, the computational demand to realize a large area and large field of view display, is beyond current capabilities, and the reproduction of textures, which provide important visual cues, has never been achieved.

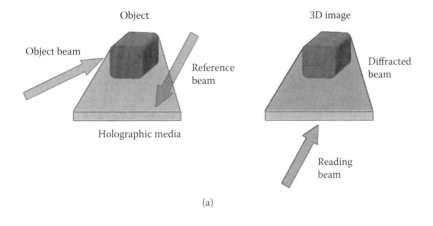

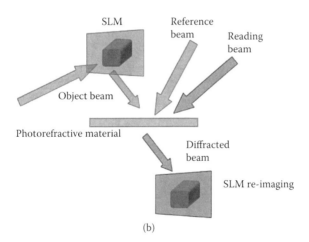

FIGURE 3.16
Imaging configuration. (a) Recording a real object and reading the hologram afterward. The diffracted beam reconstructs a 3D image of the object. (b) Dynamic recording and reading using a spatial light modulator (SLM) in the object beam. The diffracted beam reconstructs the 2D SLM screen.

Meanwhile, there exists a class of holograms that have demonstrated excellent reproduction of volume, exceptional color rendering, and the possibility to reproduce real or computer-generated objects: stereographic holograms. Presented for the first time by DeBitetto in 1969 [71], and improved numerous times [72,73], the recording principle of stereographic holograms is called holographic printing and is depicted in Figure 3.17. In this technique, the object beam is structured by an SLM, and is then focused by a lens into a holographic pixel (hogel) on the recording material. There, the object beam interferes with an unstructured reference beam, and the holographic pattern

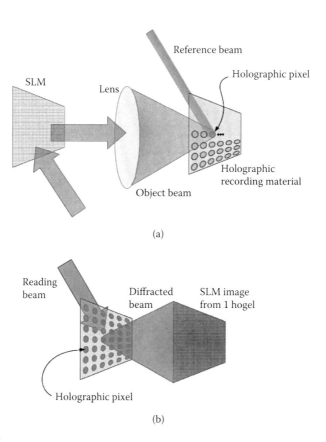

FIGURE 3.17
Holographic stereogram recording and reading configurations. (a) The object beam is reflected from a spatial light modulator (SLM) and focused into a spot on the recording material where it interferes with the reference beam, recording one hogel. Hogels are recorded to fill the entire display area. (b) Each hogel diffracts the reading beam and reconstructs its own SLM image. All the SLM images form the 3D picture.

is written. The hologram is recorded hogel after hogel to fill the entire material surface. Reading is accomplished by directing a simple monochromatic light source, such as an LED, at the Bragg angle to the hologram. The reading light color can be different from the recording wavelength.

The important point is that in focusing the object beam, a structured cone of light is generated that reproduces the different pixels that were on the SLM during recording (see Figure 3.17b). This cone of light is reconstructed by the reading beam and provides the viewing angles in the display. The collection of individual hogels reconstructs an entire 3D image that changes as the viewer changes position. This configuration generates different images that intercept the left and right eye, reproducing not only the stereo disparity but also motion parallax.

Holographic stereograms do not reproduce the wave front as do electro-holographic displays; they only reproduce the light field. This distinction means that, similar to all the other light field reproduction techniques [74], stereograms have a limited depth of field [75]. This is the trade-off that can be made to simplify the image calculation algorithm, and reduce the volume of data (phase is discarded). Notwithstanding, a technique was introduced by Smithwick et al. in 2011 to partially restore the phase, and thus the wave front, in holographic stereograms, at the cost of additional computation [76].

Since its inception, holographic printing has been refined to achieve near perfection [77]. 3D images produced with this technique are among the most realistic in terms of depth rendering, color, and resolution [78]. However, until the development of PR polymers, the applications of holographic printing were limited to static 3D displays. Their technology was missing a refreshable holographic material that would be capable of moving holographic printing from static 3D imaging to an actual dynamic display.

PR polymer properties fit the requirements for a dynamic holographic stereographic display. The diffraction efficiency is large enough to achieve the required screen brightness, the buildup speed can be tuned to different refresh rates (even video-rate), their sensitivity places them within reach of commercial laser systems, and as shown in Figure 3.18, large devices can be made at reasonable price.

In 2008, Tay et al. demonstrated the first refreshable holographic 3D display using PR polymers [79]. The material was composed of a copolymer with a polyacrylic backbone to attach pendant groups, tetraphenyldiamino-biphenyl-type (TPD) and carbaldehyde aniline (CAAN) through an alkoxy linker in a ratio of 10:1, The chromophore was a fluorinated dicyanostyrene (FDCST) and 9-ethyl carbazole (ECZ) was added as plasticizer. Relative weight concentration was PATPD-CAAN:FDCST:ECZ (50:30:20 wt%). In this

FIGURE 3.18
Large photorefractive device.

setup, a continuous wave laser recorded the hologram in a few minutes. By increasing the applied voltage during recording and reducing it during reading (voltage kickoff), the persistence was lengthened to few hours. In 2010, Blanche et al. used a nanosecond pulsed laser with a 50 Hz repetition rate to achieve a recording rate of 2 seconds [11]. This was also enabled by the use of PCBM sensitizer, causing the material to better respond to pulse illumination: PATPD/CAAN:FDCST:ECZ:PCBM (49.5:30:20:0.5 wt%). Colored holograms were presented where each color channel was independently recorded by angular multiplexing. This technique takes advantage of the angular

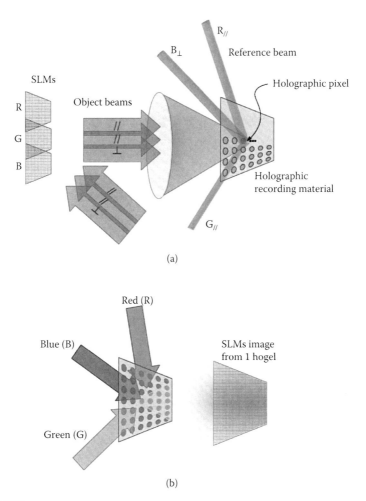

FIGURE 3.19
Color stereogram angular multiplexing, recording, and reading configurations. (a) Three object beams reflected by three spatial light modulators are focused into one single pixel where they interfere with three reference beams at different angles. Note the different polarizations. (b) Reading with three different beams (RGB) at different angles of incidence.

selectivity of thick holograms, which diffracts the reading light only when it is incident at the Bragg angle. The principle of angular multiplexing recording and reading is presented in Figure 3.19; three different SLMs, one for each color channel, reflect three collinear object beams (R,G,B), whereas three different reference beams are incident at different angles. To interfere the correct object and reference beam pairs together (RR, GG, BB) and not the dissimilar (RB, BG), orthogonal polarizations are used for (RG) and B. However, R and G could still interfere because they both have the same polarization (there is not a third orthogonal polarization). Therefore the incidence angle of the G reference beam is selected to be dramatically different than the R beam so the diffraction will be as far apart as possible, and eventually not seen by the viewer.

Note that in the transmission angular multiplexing technique, all the writing beams come from the same laser, have the same wavelength, and are incident at the same location at the same time. The only differences are the angle of incidence and polarization. To read the hologram, three different light sources with narrow spectrum (LED) are incident at the correct Bragg angle. The diffraction from those three sources is directed toward the viewer's eye where the colors are integrated. Figure 3.20 shows a picture of a full-color hologram recoded on PR polymer by angular multiplexing. Other possible configurations are discussed in Blanche [80].

FIGURE 3.20
Picture of a full-color holographic stereogram printed by angular multiplexing on photorefractive polymer. The three primary color (RGB) mixing allows for a large color gamut reproduction.

Future efforts on the PR-based holographic 3D display should be aimed at demonstrating a video-rate ($\geq$30 Hz) refresh rate. Generating the full hologram at that pace requires the recording of hundreds (or even thousands) of hogels in a fraction of a second (0.03 s). That would necessitate massive spatial multiplexing (recording several hogels with one laser pulse) and kHz repetition rate laser. The mechanical scanning system that was previously used in Blanche et al. [11] and Tay et al. [79] is not extendable to 30 Hz operation; a different solution should be implemented. However, it has already been proven in FWM configuration that PR polymers can sustain video-rate [27], making them very well adapted for such a system.

Since their discovery, PR polymers have graduated from laboratory curiosities to an entire class of optical materials. Spanning the entire visible spectrum and sensitive deep into the infrared telecommunication wavelengths, the different merits of PR polymers are now worthy of close attention for real-life applications. This section was focused on 3D displays, but it is obvious that PR polymers have a role to play in the photonics revolution we are seeing every day.

Acknowledgments

The author is grateful to Dr. Lloyd LaComb and Brittany Lynn for carefully reviewing the manuscript.

References

1. A. Ashkin, G. D. Boyd, J. M. Dziedzic, R. G. Smith, A. A. Ballman, J. J. Levinstein, and K. Nassau, "Optically-induced refractive index inhomogeneities in LiNbO3 and LiTaO3," *Applied Physics Letters*, 9, 72 (1966).
2. K. Sutter and P. Gunter, "Photorefractive gratings in the organic crystal 2-cyclooctylamino-5-nitropyridine doped with 7,7,8,8-tetracyanoquinodimethane," *JOSA B*, 7(12), 2274–2278 (1990).
3. S. Ducharme, J. C. Scott, R. J. Twieg, and W. E. Moerner, "Observation of the photorefractive effect in a polymer," *Physical Review Letters*, 66(14), 1846–1849 (1991).
4. J. G. Winiarz and F. Ghebremichael, "Beam cleanup and image restoration with a photorefractive polymeric composite," *Applied Optics*, 43, 15, 3166–3170 (2004).
5. B. L. Volodin, B. Kippelen, K. Meerholz, B. Javidi, and N. Peyghambarian, "A polymeric optical pattern-recognition system for security verification," *Nature*, 383, 58–60 (1996).
6. A. Grunnet-Jepsen, C. L. Thompson, and W. E. Moerner, "Spontaneous oscillation and self-pumped phase conjugation in a photorefractive polymer optical amplifier," *Science*, 227(5325), 549–552 (1997).

7. M. P. Georges, V. S. Scauflaire, and P. C. Lemaire, "Compact and portable holographic camera using photorefractive crystals. Application in various metrological problems," *Applied Physics B*, 72, 761–765 (2001).

8. L. Hesselink, S. S. Orlov, A. Liu, A. Akella, D. Lande, and R. R. Neurgaonkar, "Photorefractive materials for nonvolatile volume holographic data storage," *Science*, 282, 5391, 1089–1094 (1998).

9. M. Salvador, J. Prauzner, S. Köber, K. Meerholz, J. J. Turek, K. Jeong, and D. D. Nolte, "Three-dimensional holographic imaging of living tissue using a highly sensitive photorefractive polymer device," *Optics Express*, 17(14), 11834–11849 (2009).

10. B. P. Ketchel, C. A. Heid, G. L. Wood, M. J. Miller, A. G. Mott, R. J. Anderson, and G. J. Salamo, "Three-dimensional color holographic display," *Applied Optics*, 38(29), 6159 (1999).

11. P.-A. Blanche, A. Bablumian, R. Voorakaranam, C. Christenson, W. Lin, T. Gu, D. Flores, et al., "Holographic three-dimensional telepresence using large-area photorefractive polymer," *Nature*, 468, 80–83 (2010).

12. P. Günter and J.-P. Huignard, eds, *Phototrefractive Materials and Their Applications 3*, Springer Series in Optical Sciences, volume 115, Springer, New York (2007).

13. T. Woike, U. Dörfler, L. Tsankov, G. Weckwerth, D. Wolf, M. Wöhlecke, T. Granzow, R. Pankrath, M. Imlau, and W. Kleemann, "Photorefractive properties of Cr-doped Sr0.61Ba0.39Nb2O6 related to crystal purity and doping concentration," *Applied Physics B*, 72(6), 661–666 (2001).

14. C. Fuentes-Hernandez, D. J. Suh, B. Kippelen, and S. R. Marder, "High-performance photorefractive polymers sensitized by cadmium selenide nanoparticles," *Applied Physics Letters*, 85(4), 534 (2004).

15. K. R. Choudhury, Y. Sahoo, and P. N. Prasad, "Hybrid quantum-dot–polymer nanocomposites for infrared photorefractivity at an optical communication wavelength," *Advanced Materials*, 17(23), 2877–2881 (2005).

16. F. Aslam, D. J. Binks, M. D. Rahn, D. P. West, P. O'Brien, N. Pickett, and S. Daniels, "Photorefractive performance of a CdSe/ZnS core/shell nanoparticle-sensitized polymer," *Journal of Chemical Physics*, 122(18), 184713 (2005).

17. X. Li, C. Bullen, J. W. M. Chon, R. A. Evans, and M. Gu, "Two-photon-induced three-dimensional optical data storage in CdS quantum-dot doped photopolymer," *Applied Physics Letters*, 90, 161116 (2007).

18. Q. Wei, Y. Liu, Z. Chen, M. Huang, J. Zhang, Q. Gong, X. Chen, and Q. Zhou, "Improvement in photorefractivity of a polymeric composite doped with the electron-injecting material Alq3," *Journal of Optics A: Pure and Applied Optics*, 6(9), 890 (2004).

19. W. You, L. Wang, Q. Wang, and L. Yu, "Synthesis and structure/property correlation of fully functionalized photorefractive polymers," *Macromolecules*, 35, 12, 4636–4645 (2002).

20. N. M. Kronenberg, M. Deppisch, F. Würthner, H. W. A. Lademann, K. Deing, and K. Meerholz, "Bulk heterojunction organic solar cells based on merocyanine colorants," *Chemical Communications*, 48, 6489–6491 (2008).

21. Wai-Yeung Wong and Cheuk-Lam Ho, "Functional metallophosphors for effective charge carrier injection/transport: New robust OLED materials with emerging applications," *Journal of Materials Chemistry*, 19, 4457–4482 (2009).

22. C. Fuentes-Hernandez, D. J. Suh, B. Kippelen, and S. R. Marder, "High-performance photorefractive polymers sensitized by cadmium selenide nanoparticles," *Applied Physics Letters*, 85, 4 (2004).

23. M. Eralp, J. Thomas, S. Tay, G. Li, A. Schülzgen, R. A. Norwood, M. Yamamoto, and N. Peyghambarian, "Submillisecond response of a photorefractive polymer under single nanosecond pulse exposure," *Applied Physics Letters*, 89, 114105 (2006).

24. P.-A. Blanche, B. Kippelen, A. Schülzgen, C. Fuentes-Hernandez, G. Ramos-Ortiz, J. F. Wang, E. Hendrickx, N. Peyghambarian, and S. R. Marder, "Photorefractive polymers sensitized by two-photon absorption," *Optics Letters*, 27(1), 19–21 (2002).

25. S. Tay, J. Thomas, M. Eralp, G. Li, R. A. Norwood, A. Schülzgen, M. Yamamoto, S. Barlow, G. A. Walker, S. R. Marder, and N. Peyghambarian, "High-performance photorefractive polymer operating at 1550 nm with near-video-rate response time," *Applied Physics Letters*, 87, 171101 (2005).

26. L. Licea-Jiménez, A. D. Grishina, L. Ya. Pereshivko, T. V. Krivenko, V. V. Savelyev, R. W. Rychwalski, A. V. Vannikov, "Near-infrared photorefractive polymer composites based on carbon nanotubes," *Carbon*, 44(1), 113–120 (2006).

27. M. Eralp, J. Thomas, G. Li, S. Tay, A. Schülzgen, R. A. Norwood, N. Peyghambarian, and M. Yamamoto, "Photorefractive polymer device with video-rate response time operating at low voltages," *Optics Letters*, 31(10), 1408–1410 (2006).

28. G. Cook, C. J. Finnan, D. C. Jones, "High optical gain using counterpropagating beams in iron and terbium-doped photorefractive lithium niobate," *Applied Physics B*, 68, 911–916 (1999).

29. D. R. Evans and G. Cook, "Bragg-matched photorefractive two-beam coupling in organic–inorganic hybrids," *Journal of Nonlinear Optical Physics & Materials*, 16(3), 271–280 (2007).

30. B. Lynn, P.-A. Blanche, and N. Peyghambarian, "Photorefractive polymer for holography," review article, *Journal of Polymer Science Part B: Polymer Physics*, 2014, 52, 193–231 (November 2013).

31. L. Onsager, "Initial recombination of ions," *Physical Review*, 54, 554 (1938).

32. O. Ostroverkhova and K. D. Singer, "Space-charge dynamics in photorefractive polymers," *Journal of Applied Physics*, 92(4), 1727 (2002).

33. E. Hendrickx, Y. D. Zhang, K. B. Ferrio, J. A. Herlocker, J. Anderson, N. R. Armstrong, E. A. Mash, A. P. Persoons, N. Peyghambarian, B. Kippelen, "Photoconductive properties of PVK-based photorefractive polymer composites doped with fluorinated styrene chromophores," *Journal of Materials Chemistry*, 9, 2251–2258 (1999).

34. N. Tsutsumi and Y. Shimizu, "Asymmetric two-beam coupling with high optical gain and high beam diffraction in external-electric-field-free polymer composites," *Japanese Journal of Applied Physics*, 43, 6A, 3466 (2004).

35. A. Tanaka, J. Nishide, and H. Sasabe, "Asymmetric energy transfer in photorefractive polymer composites under non-electric field," *Molecular Crystals and Liquid Crystals*, 504(1), 44–51 (2009).

36. F. Gallego-Gómez, F. del Monte, and K. Meerholz, "Optical gain by a simple photoisomerization process," *Nature*, 7, 490–497 (2008).

37. P. W. M. Blom and M. Vissenberg, "Charge transport in poly (p-phenylene vinylene) light-emitting diodes," *Materials Science and Engineering*, 27, 53–94 (2000).

38. T. Jakob, S. Schloter, U. Hofmann, M. Grasruck, A. Schreiber, and D. Haarer, "Influence of the dispersivity of charge transport on the holographic properties of organic photorefractive materials," *Journal of Chemical Physics*, 111, 23 (1999).

39. S. J Zilker, M. Grasruck, J. Wolff, S. Schloter, A. Leopold, M. A. Kol'chenko, U. Hofmann, A. Schreiber, P. Strohriegl, C. Hohle, D. Haarer, "Characterization of

charge generation and transport in a photorefractive organic glass: Comparison between conventional and holographic time-of-flight experiments," *Chemical Physics Letters*, 306(5–6), 285–290 (1999).

40. L. Kulikovsky, D. Neher, E. Mecher, K. Meerholz, H. H. Hörhold, and O. Ostroverkhova, "Photocurrent dynamics in a poly (phenylene vinylene)-based photorefractive composite," *Physical Review B*, 69, 125216 (2004).

41. C. W. Christenson, J. Thomas, P.-A. Blanche, R. Voorakaranam, R. A. Norwood, M. Yamamoto, and N. Peyghambarian, "Grating dynamics in a photorefractive polymer with Alq3 electron traps," *Optics Express*, 18(9) 9358–9365 (2010).

42. N. V. Kukhtarev, V. B. Markov, S. G. Odulov, M. S. Soskin, and V. L. Vinetskii, "Holographic storage in electrooptic crystals," *Ferroelectric*, 22(1), 949–960 (1979).

43. J. S. Schildkraut and A. V. Buettner, "Theory and simulation of the formation and erasure of space charge gratings in photoconductive polymers," *Journal of Applied Physics*, 72(5), 1888 (1992).

44. D. M. Burland, R. D. Miller, and C. A. Walsh, "Second-order nonlinearity in poled-polymer systems," *Chemical Review*, 94(1), 31–75 (1994).

45. W. E. Moerner, S. M. Silence, F. Hache, and G. C. Bjorklund, "Orientationally enhanced photorefractive effect in polymers," *Journal of the Optical Society of America B*, 11(2), 320–330 (1994).

46. P. Cheben, F. del Monte, D. J. Worsfold, D. J. Carlsson, C. P. Grover, and J. D. Mackenzie, "A photorefractive organically modified silica glass with high optical gain," *Nature*, 408, 64–67 (2000).

47. O. Ostroverkhova, W. E. Moerner, M. He, and R. J. Twieg, "High-performance photorefractive organic glass with near-infrared sensitivity," *Applied Physics Letters*, 82(21), 3602–3604 (2003).

48. M. Talarico and A. Golemme, "Optical control of orientational bistability in photorefractive liquid crystals," *Nature Materials*, 5, 185–188 (2006).

49. G. P. Wiederrecht, "Photorefractive liquid crystals," *Annual Review of Materials Research*, 31, 139–169 (2001).

50. N. Tsutsumi and H. Kasaba, "Effect of molecular weight of poly (N-vinyl carbazole) on photorefractive performances," *Journal of Applied Physics*, 104(7), 73102–73105 (2008).

51. J. C. Scott, L. Th. Pautmeier, and W. E. Moerner, "Photoconductivity studies of photorefractive polymers," *Journal of the Optical Society of America B*, 9(11), 2059–2064 (1992).

52. P. Pingel, A. Zen, R. D. Abello, F. C. Grozema, L. D. A. Siebbeles, and D. Neher, "Temperature-resolved local and macroscopic charge carrier transport in thin P3HT layers," *Advanced Functional Materials*, 20(14), 2286–2295 (2010).

53. J. Thomas, C. Fuentes-Hernandez, M. Yamamoto, K. Cammack, K. Matsumoto, G. A. Walker, S. Barlow, B. Kippelen, G. Meredith, S. R. Marder, and N. Peyghambarian, "Bistriarylamine polymer-based composites for photorefractive applications," *Advanced Materials*, 16(22), 2032–2036 (2004).

54. S. Köber, F. Gallego-Gomez, M. Salvador, F. B. Kooistra, J. C. Hummelen, K. Aleman, S. Mansurova, and K. Meerholz, "Influence of the sensitizer reduction potential on the sensitivity of photorefractive polymer composites," *Journal of Materials Chemistry*, 20(29), 6170–6175 (2010).

55. X. Li, J. Van Embden, R. A. Evans, and M. Gu, "Type-II core/shell nanoparticle induced photorefractivity," *Applied Physics Letters*, 98(23), 231107 (2011).

56. F. Wang, Z. Chen, B. Zhang, Q. Gong, K. Wu, X. Wang, B. Zhang, and F. Tang, "Nanometer-Au-particle-enhanced photorefractivity in a polymer composite," *Applied Physics Letters*, 75(21), 3243 (1999).

57. P. Chantharasupawong, C. Christenson, R. Philip, L. Tetard, L. Zhai, J. Winiarz, M. Yamamoto, R. Raveendran Nair, and J. Thomas, "Photorefractive performances of a graphene-doped PATPD/7-DCST/ECZ composite," *Journal of Materials Chemistry C*, Accepted (2014).

58. R. Wortmann, C. Glania, P. Krämer, K. Lukaszuk, R. Matschiner, R. J. Twieg, and F. You, "Highly transparent and birefringent chromophores for organic photore-fractive materials," *Chemical Physics*, 245(1–3), 107–120 (1999).

59. F. Würthner, R. Wortmann, K. Meerholz, "Chromophore design for photorefrac-tive organic materials," *Chemphyschem*, 3(1), 17–31 (2002).

60. J. Thomas, M. Eralp, S. Tay, G. Li, S. R. Marder, G. Meredith, A. Schülzgen, R. A. Norwood, and N. Peyghambarian, "Near-infrared photorefractive polymer composites with high diffraction efficiency and fast response time," in *Organic Holographic Materials and Applications II*, K. Meerholz, Ed., SPIE proceedings 5521, 2004.

61. C. W. Christenson, C. Greenlee, B. Lynn, J. Thomas, P.-A. Blanche, R. Vooraka-ranam, P. St. Hilaire, L. J. LaComb, Jr, R. A. Norwood, M. Yamamoto, and N. Peyghambarian, "Interdigitated coplanar electrodes for enhanced sensitivity in a photorefractive polymer," *Optics Letters*, 36(17), 3377–3379 (2011).

62. J. A. Herlocker, C. Fuentes-Hernandez, J. F. Wang, N. Peyghambarian, B. Kip-pelen, Q. Zhang, and S. R. Marder, "Photorefractive polymer composites fabri-cated by injection molding," *Applied Physics Letters*, 80(7), 1156 (2002).

63. P.-A. *Blanche, Field Guide to Holography*, SPIE Field Guides, volume FG31, SPIE Press, Bellingham, WA (2014).

64. F. Gallego-Gomez, M. Salvador, S. Köber, and K. Meerholz, "High-performance reflection gratings in photorefractive polymers," *Applied Physics Letters*, 90(25), 261113 (2007).

65. C. A. Walsh and W. E. Moerner, "Two-beam coupling measurements of grating phase in a photorefractive polymer," *Journal of the Optical Society of America B*, 9(9), 1642–1647 (1992).

66. B. Yuan, X. Sun, Z. Zhou, Y. Li, Y. Jiang, and C. Hou, "Theory of space-charge field with a moving fringe in photorefractive polymers," *Journal of Applied Physics*, 89(11), 5881 (2001).

67. P.-A. Blanche, S. Tay, R. Voorakaranam, P. Saint-Hilaire, C. Christenson, T. Gu, W. Lin, et al., "An updatable holographic display for 3D visualization," *Journal of Display Technology*, 4, 4 (2008).

68. N. Tsutsumi, K. Kinashi, W. Sakai, J. Nishide, Y. Kawabe, and H. Sasabe, "Real-time three-dimensional holographic display using a monolithic organic compound dispersed film," *Optical Materials Express*, 2(8), 1003–1010 (2012).

69. C. Slinger, C. Cameron, and M. Stanley, "Computer-generated holography as a generic display technology," *Computer*, 38(8), 46–53 (2005).

70. D. E. Smalley, Q. Y. J. Smithwick, V. M. Bove, Jr., "Holographic video display based on guided-wave acousto-optic devices," in *Practical Holography XXI: Materials and Applications*, R. A. Lessard, H. I. Bjelkhagen, Ed., 6488, Proceedings of SPIE—the International Society for Optical Engineering, San Jose, CA, 2007.

71. D. J. DeBitetto, "Holographic panoramic stereograms synthesized from white light recordings," *Applied Optics*, 8(8), 1740, 1741 (1969).

72. M. W. Halle, S. A. Benton, M. A. Klug, J. S. Underkoffler, "Ultragram: A generalized holographic stereogram," in *Practical Holography V*, S. A. Benton, Ed., SPIE Proceedings 1461, 1991.

73. M. Yamaguchi, N. Ohyama, and T. Honda, "Holographic three-dimensional printer: New method," *Applied Optics*, 31(2), 217–222 (1992).

74. R. Yang, X. Huang, S. Li, and C. Jaynes, "Toward the light field display: Autostereoscopic rendering via a cluster of projectors," *IEEE Transactions on Visualization and Computer Graphics*, 14(1), 84–96 (2008).

75. P. St. Hilaire, P. Blanche, C. Christenson, R. Voorakaranam, L. LaComb, B. Lynn, and N. Peyghambarian, "Are stereograms holograms? A human perception analysis of sampled perspective holography," *Journal of Physics: Conference Series* 415, Conference 1, 2013.

76. Q. Y. J. Smithwick, J. Barabas, D. E. Smalley, and V. M. Bove, Jr., "Interactive holographic stereograms with accommodation cues," in *Practical Holography XXIV: Materials and Applications*, H. I. Bjelkhagen, R. K. Kostuk, Ed., SPIE Proceedings 7619, 2010.

77. H. Bjelkhagen and D. Brotherton-Ratcliffe, *Ultra-Realistic Imaging: Advanced Techniques in Analogue and Digital Colour Holography*, CRC Press, Taylor and Francis Group, Boca Raton, FL (2013).

78. Zebra Imaging, RabbitHoles, Geola, Available at http://www.zebraimaging.com and http://www.geola.lt

79. S. Tay, P.-A. Blanche, R. Voorakaranam, A. V. Tunç, W. Lin, S. Rokutanda, T. Gu, et al., "An updateable holographic 3D display," *Nature*, 451, 694–698 (2008).

80. P.-A. Blanche, "Holographic visualization of 3D data," in *Optical and Digital Image Processing: Fundamentals and Applications*, G. Cristobal, P. Schelkens, and H. Thienpont, Eds., Wiley-VCH, Weinheim, Germany 2011.

4

Optical Gene Detection Using Conjugated Polymers

Jadranka Travas-Sejdic and Anupama Rao Gulur Srinivas

CONTENTS

Abbreviations

A, adenine

C, cytosine

CapODN, capture probe ODN

CP, conjugated polymer

DNA, deoxyribonucleic acid

ds-DNA, double-stranded probe DNA

EET, electronic energy transfer

F, fluorophore

FRET, fluorescence resonance energy transfer

G, guanine

IFC, imaging flow cytometry

MB, molecular beacon

ODN, oligonucleotides

PCR, polymerase chain reaction

PCT, photoinduced energy transfer

PDA, polydiacetylene

PEG, poly(ethylene glycol)

PFP, poly(fluorene-co-phenylene)

PNA, peptide nucleic acid

PPE, poly(p-phenylene ethynylene)

PPP, poly(p-phenylene)

PPV, poly(p-phenylene vinylene)

PPV-MagSi, PPV-coated MagSi beads

PRU, polymer repeat units

PT, polythiophene

Q, quencher

RET, resonance energy transfer

SERS, surface-enhanced Raman spectroscopy

SP1, signaling probe 1

SP2, signaling probe 2

SPR, surface plasmon resonance

ss-DNA, single-stranded probe DNA

T, thymine

4.1 Introduction

There is enormous demand for the commercial use of biosensors in a diverse range of fields including health care, environmental monitoring, food industry, pharmaceuticals, military, and veterinary applications. A biosensor is defined as an analytical device that provides information about its environment via interactions with an analyte [1,2]. Interactions are then converted into measurable signals such as optical, piezoelectric, and electrochemical signals. Research has been focused toward developing biosensors that are sensitive, are selective toward the target, and have fast response times [3].

Deoxyribonucleic acid (DNA) sensors are those whose probe–analyte interactions comprise hybridization of complementary DNA sequences (Figure 4.1). This process can then be converted into signals by various readout methodologies [4]. Broadly speaking, DNA sensors can be classified into three categories depending on the signals produced to monitor hybridization: optical, piezoelectric, and electrochemical.

DNA microarrays (commonly called gene chips, DNA chips, or biochips) allow parallel detection and analysis of multiple genes in a single experiment [5]. The current commercial, conventional microarray technology is based on detection of fluorescence emission from the hybridization of dye-labeled

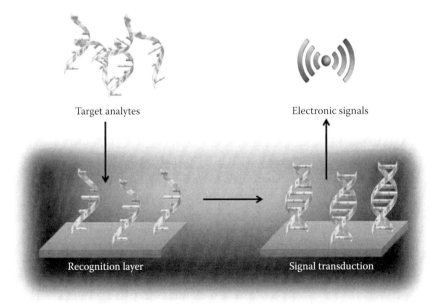

Target analytes   Electronic signals

Recognition layer   Signal transduction

FIGURE 4.1
The components involved in designing a DNA sensor based on hybridization.

DNA targets with immobilized probes [6]. However, there is room for improvement. Disadvantages of the current technology include low sensitivity (fluorescence signal is weak when target DNA concentration is low) and utilization of amplifying techniques such as polymerase chain reaction (PCR) and long analysis times. [6] Although DNA microarrays utilizing photoluminescent polymers have been recently developed to improve sensitivity, they suffer from not being easily amenable to detecting multiple target detections [6]. Conjugated polymers (CPs) with their exceptional optical and electrical properties have found extensive use in functioning as both a passive support for entrapment of DNA and a transducer of the hybridization event into an observable and measurable electrical [4,7–10] or optical [11–14] signal.

This chapter focusses on DNA sensors that utilize optical properties of CPs. First, the different types of DNA sensors will be described, followed by brief descriptions of the underlying physics of fluorescence in photoluminescent polymers and energy and electron transfer mechanisms. We have endeavored to cover major scientific publications up to 2013 while not attempting to cover the patent literature. This account is not intended to be exhaustive; rather, it is intended to provide examples of specific sensing system designs that illustrate the main ideas and best results achieved to date with the use of photoluminescent CPs. Readers wishing to obtain full details of the gene sensing designs are directed to the primary literature.

4.1.1 DNA and DNA Sensors

4.1.1.1 Structure of DNA

Since the first description of the DNA structure by Watson and Crick [15] in 1953, it has been known that genetic information is contained within specific sequences involving two types of aromatic planar bases, namely, purines (guanine [G] and adenine [A]) and pyrimidines (cytosine [C] and thymine [T]) (Figure 4.2) [2,16].

A single-stranded DNA sequence consists of these bases linked together by a backbone of alternating carbon sugars, β-D-2-deoxyribose, and phosphate molecules in the form of 3′-5′ phosphatediester bonds [17–19,16]. The secondary structure of DNA is a right-handed double helix consisting of two linear strands of DNA that run antiparallel in such a way that the negatively charged phosphatediesters are positioned outside while the bases are stacked in the middle of the helix, held in place by hydrophobic interactions and hydrogen bonding (Figure 4.2) [20] in a process known as hybridization [17–19].

Recently, peptide nucleic acids (PNAs) have been developed as synthetic analogues of DNA with the major difference of having a neutral backbone (containing N-[2-aminoethyl] glycine molecules) instead of the charged phosphate backbone of DNA [21]. Although PNAs are expensive, due to lack of charges on the backbone, they have the advantages of being more selective [21].

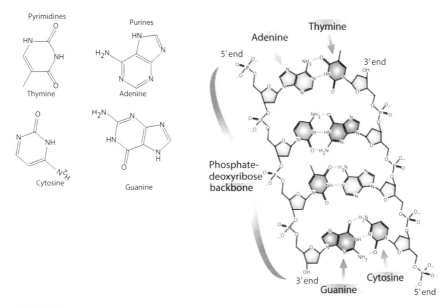

FIGURE 4.2
The structures of purines and pyrimidines and the secondary double helix structure of DNA.

4.1.1.2 DNA Sensors and Types of Readout Methodologies Employed

Single strands of DNA probes are usually immobilized on the receptor surface of a sensor to hybridize with complementary DNA strands in the analyte [3,18]. Conditions such as temperature, solvent ionic strength, and pH affect the hybridization process. Hybridization can be done in either homogenous or heterogeneous media whereby the probes are either in solution or immobilized on a solid support and in contact with the target in solution phase [18].

4.1.1.3 Piezoelectric DNA Sensors

Piezoelectric methods of DNA sensing are extremely sensitive toward changes in mass, which are monitored using microgravimetric detectors [18]. Quartz crystals are the most common transducers used for DNA detection. The change in mass accompanied by hybridization can be analyzed using the Sauerbrey equation [18]. These methods are promising due to advantages such as low cost, chemical inertness, durability, and potential for quantitative analysis [18]. An added advantage is the reduction of pretreatment and posttreatment steps compared to methods that rely on labeling the probe with enzymes, fluorescent molecules, and antibodies [22].

4.1.1.4 Electrochemical DNA Sensors

Novel electrochemical DNA sensing methods are based on monitoring hybridization-induced changes in current and conductivity. Electrochemical techniques used include cyclic voltammetry, differential pulse voltammetry, square-wave voltammetry, and stripping voltammetry [3,18,23,24]. Electrochemical DNA sensors can be categorized according to the type of detection, where hybridization is measured either directly or indirectly [3,18,24]. Direct methods involving label-free detection of DNA monitor hybridization altering the interfacial properties of molecules and inherent properties of electroactivity of DNA bases (guanine and adenosine) [3]. The measured current is directly proportional to the relative amounts of guanidine and adenosine present in the double helix [3]. Another type of direct detection, based on oxidation of the sugar backbone of DNA, has been developed, thus opening up new opportunities for sensitive indicator-free detection [3,4].

Indirect electrochemical DNA sensing methods are based on measuring changes in current after the binding of a sequence-selective redox indicator [3,22,25]. Reusable sensors have been obtained by utilizing electroactive molecules such as metal complexes, nanoparticles, quantum dots, [26] planar cationic molecules, and enzymes [3,18,24,23]. Rather than using the redox indicators on their own, applications including them on interfaces such as conducting polymers [27] have been developed to augment the sensitivity and detection limit [3]. Novel electrochemical DNA sensors based on utilizing the inherent conducting properties of polythiophenes, polypyrroles, and polyaniline [3,4,18,22,27–39] to enhance the electrochemical readout on hybridization have been reported. Such sensors have the additional advantages of low cost and compatibility with micromanufacturing technology [5]. Readers interested in the exhaustive list of published piezoelectric and electrochemical DNA sensors should refer to the literature [4,7,10,40–43].

4.1.1.5 Optical DNA Sensors

Fluorescence, surface plasmon resonance (SPR), Raman spectroscopy, chemiluminescence, colorimetry, and interferometry are some of the most common techniques used for optical gene sensing [5]. More recently demonstrated, novel optical methods for DNA sensing include the use of SPR and Raman spectroscopy [18,44]. Raman spectroscopy [2] has been used for the detection of bacterial and viral genes as it has excellent chemical group identification capabilities [1]. Problems involving low sensitivity were improved in recent years by using surface-enhanced Raman spectroscopy, which utilizes molecules that have absorption bands near or on metal surfaces [1,2,18]. Although such designs are highly sensitive, drawbacks of expensive instrumentation and long analysis times hinder developments toward commercialization [18]. SPR-based [5] DNA sensors monitor hybridization by change in refractive index at a metal-coated prism surface, which is proportional to the mass of biomolecules bound to the surface. Although incredibly low concentrations of label-free DNA can be detected by SPR, this technique has limitations in detecting short-chain DNA molecules (due to slight shifts in resonance angle), which can be remedied by using fluorescence (fluorescence tagging) [5].

Classical optical monitoring of DNA hybridization using radioactive labeled methods and nonradioactive labeled methods has been phased out due to high levels of toxicity, and steric effects encountered during labeling that hinder hybridization [18,19]. Furthermore, designs involving fiber optics have also been developed [30]. These techniques use quartz optical fibers immobilized with probes and fluorescent intercalating agents whose fluorescence response is proportional to the amount of complementary DNA present in the solution [30]. More recently, novel materials such as CPs, quantum dots, and metallic nanoparticles have replaced radioactive and organic dye-labeled methods [14,45,46].

4.1.2 Conjugated Polymers

Photoluminescent CPs are an attractive alternative to dyes because the color of emission can be tuned. Other desirable properties include ease of processing, low costs, and high environmental stability. The optical properties of CPs arise due to absorption of photons, which induces creation of electron–hole pairs (excitons) by transitions from π-π^* states delocalized along the backbone of the polymer [25]. Efficiency of luminescence is dependent on the polarizability and delocalization of the excitons along the backbone of CPs. [47]. Vibronic transitions also play a role, in particular with polythiophenes where changes in their planar or nonplanar conformations affect the optical properties [47]. In addition, CPs can amplify the emission intensities as a result of ultrafast migration of excitons along the polymer chains in both intrachain and interchain manners [14,48–51].

Some of the promising applications of CPs include chemical [47] and biological sensors, [7,12,49] printable electronics, [52] flexible displays, and photovoltaics, [49] which may require suitable processing of CPs. [49] A severe limitation of unfunctionalized CPs is their large dielectric constants, limiting their solubility in common organic solvents [53]. This can be overcome by functionalization to render CPs soluble. As a result, water solubility, which is critical to the use of these materials in biological and biomedical applications, is a rare find in CPs. On the basis of the chemical structure of the polymer backbone, photoluminescent CPs can be classified into the following types (Figure 4.3): poly(fluorene-co-phenylene) (PFP), poly(*p*-phenylene) (PPP), poly(*p*-phenylenevinylene) (PPV), poly(*p*-phenyleneethynylene) (PPE), polydiacetylene (PDA), and poly(thiophene) (PT).

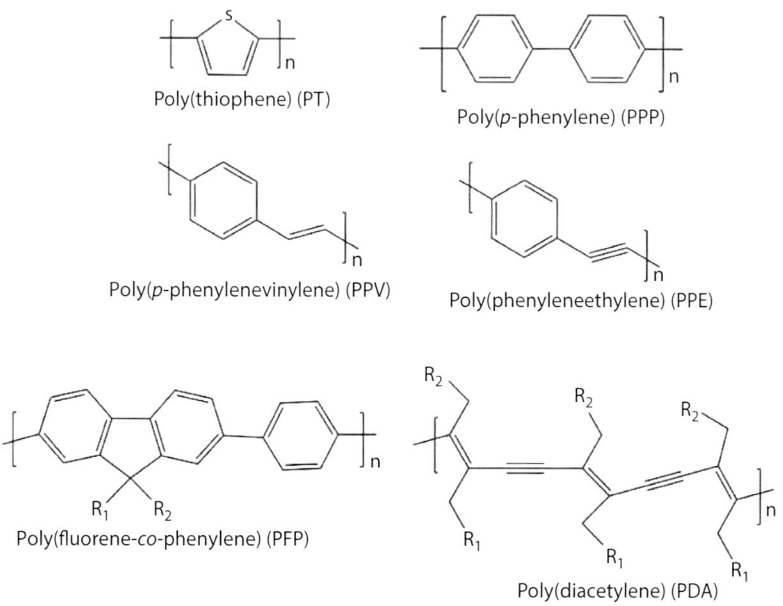

FIGURE 4.3
Basic structures of photoluminescent conjugated polymers.

Broadly speaking, Grignard, $FeCl_3$ oxidative polymerization reactions (for PT), Suzuki, Yamamoto coupling (for PPP and PFP), Wittig–Horner, Heck, Gilch, Wessling reactions (for PPV), Sonogashira coupling (for PPE), and 1,4-photopolymerization (for PDA) have been employed for the synthesis of photoluminescent CPs. Readers interested in the exhaustive list of published synthesis procedures should refer to the literature [49,50,54–56].

4.2 Optical Properties of Photoluminescent CPs

Fluorescence from a molecule is caused by the excitation of electrons from a ground state to an excited energy state, which is followed by radiative decay back to the ground state (from the lowest vibrational level of the excited state) [57] (Figure 4.4). The efficiency of a molecule to exhibit fluorescence is defined by the quantum yield, Φ, which is defined as the amount of photons emitted per photon absorbed [57].

Understanding the underlying physics and dynamics of photoluminescence in CPs is the key to designing, optimizing, and fine-tuning properties for optoelectronic applications [58]. Luminescence studies of CPs give information about excited-state dynamics [58]. Typical photoluminescence characteristics of CPs include high quantum yields in solution, large apparent

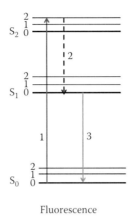

Fluorescence

FIGURE 4.4
Schematic diagram of the mechanism of fluorescence.

Stokes' shifts (due to intramolecular geometrical reorganization energy associated with excitation to the excited state causing planarization, followed by energy migration toward the lowest energy trapping sites), and broad absorption bands [58]. The optical properties of CPs are highly dependent on chain conformation, conformational disorder (due to the low energy barrier around σ bonds in the conjugated backbone), and π-system conjugation lengths. Typical conjugation lengths of polymers are estimated to be between 5 and 10 subunits, whereas breaks in conjugation are thought to occur most commonly from chemical defects and configurational imperfections [58].

When a chromophore in a polymer chain is excited, the energy is funneled to lower-energy sites on the chain by electronic energy transfer (EET) prior to photoluminescence emission. Such EET can occur along the chain (intrachain) or between subunits that are nearby (interchain). Mechanistically, EET in CPs has been explained by the well-known Forster (energy transfer is mediated via long-range dipole–dipole interactions) and Dexter (energy transfer is brought about by short-range spatial overlap of wave functions of molecules) energy transfer models [58]. Thus, CPs can be thought of as a series of chromophores of altering size and energy that couple electronically, determining the overall photophysical properties. As this chapter focuses on the applications of photoluminescent CPs in gene sensing, readers interested in the exhaustive list of published studies toward understanding the optical properties and underlying physics are referred to the literature [58–60].

4.2.1 Optical Readout Methodologies

There are reports on promising sensitive, optical DNA sensors based on CPs that have utilized CPs for two main purposes: immobilization of the DNA probe and transducing the hybridization event [14,29,35,1,61,62]. Traditionally, optical detection using analytical techniques such as fluorescence resonance energy transfer (FRET) [63] or superquenching [64] has been utilized due to rapid, simple, and low-cost assays and instrumentation. Dynamic quenching, static quenching, FRET, photoinduced electron transfer, [57] and photoinduced photon transfer are some of the processes that occur after a molecule has been electronically excited [57].

4.2.1.1 Fluorescence Resonance Energy Transfer

FRET is characterized by long-range energy transfer from an excited molecule (donor) to another molecule (acceptor) on coupling of their dipole–dipole interactions and followed by relaxation of the donor (Figure 4.5). FRET rate (k_t) is highly dependent on the proximity of donor and acceptor, orientation factor, and overlap integral (J) of the emission and absorption spectra of the donor and acceptor, respectively [65,66]. In the presence of the acceptor, the donor transfers the energy ($h\gamma_{em}{}^D$) to the acceptor, which then returns to the ground state by the emission of photons ($h\gamma_{em}{}^A$) [66].

4.2.1.2 Superquenching

Fluorescence quenching is a competing process with fluorescence, where chromophores can interact with quenchers resulting in decreased fluorescence intensity and lifetime. Such interactions can occur in either the excited state or the ground state of the chromophores via energy transfer,

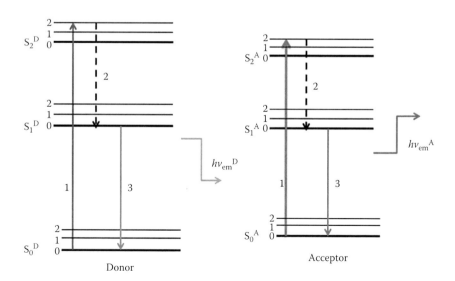

FIGURE 4.5
Schematic simplified Jablonski diagram of the mechanism of fluorescence resonance energy transfer. (1) Absorption of a photon, (2) loss of vibrational energy, and (3) emission of a photon.

electron transfer, and dynamic (collision) and static (complex) quenching [57]. Dynamic (collisions of the quencher molecule Q with the excited fluorophore F*, resulting in the deactivation of F* to the ground state without the emission of a photon) quenching and static (formation of a nonemissive F–Q complex) quenching are the most common mechanisms encountered [57]. It should be noted that in most quenching systems these mechanisms occur concurrently and can be distinguished in a number of ways: (1) change in fluorescence lifetime of the chromophore in the presence of the quencher, (2) temperature dependence on quenching efficiency, and (3) static complexes exhibiting different absorption compared to chromophores [67].

CPs exhibit amplified quenching abilities for certain moieties, which is termed *superquenching* [64,67]. Superquenching in polymers is thought to occur due to efficient electron/energy transfer from the excited photoluminescent CP to the quenchers [68]. Electrostatic attractions between the CP and the quencher are thought to contribute to efficient superquenching [67]. When a polymer is excited, the energy/electron transfer is delocalized over a certain number of polymer repeat units (PRUs) along the backbone of the polymer, which is then effectively quenched by quencher molecules in the vicinity [64]. The sensitivity toward superquenching by a given quencher is measured using the quenching constant K_{sv} and the number of PRUs [64,67]. The larger the numbers for K_{sv} and PRU, the higher the sensitivity toward the quenchers as lower concentrations of quencher molecules are required to quench the fluorescence of the CPs [67].

4.3 Gene Sensor Designs Using Photoluminescent CPs

The gene sensor designs based on photoluminescent CPs can be broadly classified into two categories: (1) designs utilizing the conformational change or aggregation of CPs on hybridization and (2) energy/electron transfer–based designs. The polymers PDA and PT have been used for the former design, whereas polymers based on PFP, PPP, PPV, and PPE have been utilized in the latter designs. Both types of design are discussed in Sections 4.3.1 and 4.3.2.

4.3.1 Gene Sensing Exploiting Change in Configuration/Aggregation of CPs

4.3.1.1 Poly(thiophene)-Based Sensors

The majority of optical DNA sensing designs that utilize PT transduce hybridization as a colorimetric change due to change in conformation of the backbone [69,70]. PTs have demonstrated interesting chromic properties in the presence of different stimuli, which have been attributed to the change from the nonplanar (deaggregated) structure to the planar (aggregated) structure [70]. Such drastic changes in the chromic properties of PT have been explained by the modification in the delocalization of π electrons along the backbone in different conformations [70]. Leclerc's group [70] pioneered the detection of DNA using such chromic properties of PT where label-free optical PT DNA sensors utilized water-soluble methylimidazol [69] and cationic trimethyl ammonium [71]–substituted alkoxy PT. The mechanism of sensing is through the formation of a duplex between cationic PT and single-stranded probe DNA (ss-DNA), which alters the conformation of the PT to a planar, aggregated red-colored polymer solution. In the presence of target DNA, the PT switches back to the nonaggregated, nonplanar yellow-colored polymer due to the formation of a triplex with the double-stranded helix (ds-DNA) (Figure 4.6) [72].

This type of homogenous sensing design has been extended toward heterogeneous sensors, which use surfaces such as glass slides or microbeads [73,74]. A detection limit of 2.5×10^{-13} M and single-base mismatch were achieved, but the heterogeneous sensors suffered from low fluorescence yields and increased cost due to expensive PNA.

Recently, PT was used by Brouard et al. [75] for developing a novel resonant energy transfer (RET)-based DNA sensor, which detects hybridization on dye-doped, multilayer, and metal core-silica beads (Figure 4.7). Such nanoparticles have other functions, such as enhancement of excitation efficiency and emission rates, while also reducing the lifetime of excited states and self-quenching due to plasmonic enhancement [75]. The complexation of a cationic polymer with negatively charged probe-grafted beads forms aggregates due to partial neutralization of charges, which leads to enhanced plasmonic coupling and enhanced emission from the polymer and eosin dye. The authors used imaging flow cytometry to detect the fluorescence of

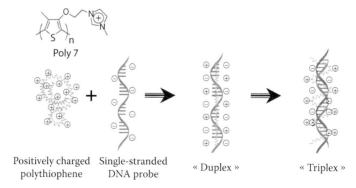

FIGURE 4.6
(See color insert.) Schematic description of the formation of polythiophene/single-stranded probe deoxyribonucleic acid (DNA) duplex and polythiophene/double-stranded probe DNA triplex complex. (Reprinted from *Tetrahedron*, 60, Bera-Aberem et al., Functional polythiophenes as optical chemo- and biosensors, 11169, Copyright 2004, with permission from Elsevier.)

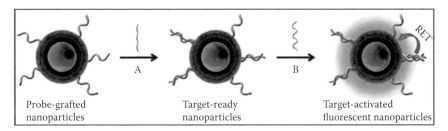

| Probe-grafted nanoparticles | Target-ready nanoparticles | Target-activated fluorescent nanoparticles |

FIGURE 4.7
Schematic description of the principle of DNA detection on fluorescent multilayer core-shell nanoparticles. (Reprinted with permission from Brouard et al., 1888. Copyright 2011 American Chemical Society.)

the dye in the presence of complementary and mismatched sequences and reported a very low detection limit of 10^{-16} M. In spite of high sensitivity and selectivity, the sensor design suffers from the relatively high hybridization temperature (60°C), high polymer concentration (10^{-6} M) relative to the DNA probe, and false positive signals obtained due to nonspecific interactions at high polymer concentrations.

4.3.1.2 Polydiacetylene-Based Sensors

PDA-based designs utilize the ability of the blue-phase nonfluorescent PDAs [76] (absorption at 640 nm) to switch to the red-phase fluorescent PDAs [76] (absorption at 550 nm) on environmental simulation [76]. This type of chromism is thought to arise from the conformational change of the PDA backbone from planar to nonplanar structures. Optical gene sensing designs employ aggregation induced by target DNA to simulate PDAs to the fluorescent red phase. An improved design utilizes nonfluorescent positively charged PDA liposomes responding to negatively charged DNA, which results in the production of red-phase fluorescence in liposomes. A limitation was observed where the control PDA liposomes reacted with PCR buffer as well [76].

4.3.2 Gene Sensing with Photoluminescent CPs Utilizing Energy Transfer/Electron Transfer

The second major class of optical gene sensing designs involving CPs (PPV, PPE, PPP, and PF) exploits the characteristic of efficient excitation throughout the conjugated backbone, which can be transferred to a dye resulting in FRET, or to a quencher causing superquenching.

4.3.2.1 FRET-Based Designs Using Cationic CPs

The ability of light-harvesting cationic CPs to interact with the anionic phosphate backbone of DNA has been used extensively in homogenous- and heterogeneous-based designs. A donor cationic CP is brought in close proximity to an acceptor dye during hybridization, which is transduced as a FRET signal when the CP is selectively excited [13]. Thus, dyes have been introduced into the sensing system by labeling the target DNA sequence, labeling the probe DNA sequence, or intercalation of dye molecules after hybridization (Figure 4.8).

The earlier designs developed by Bazan et al. [13,77–80] used a dye-labeled target to monitor hybridization through FRET when it hybridizes to a CP-probe complex. Although these novel designs demonstrated high sensitivity, target-labeled designs have been phased out due to the additional step of labeling the target. Designs utilizing labeled probe DNA and cationic PFP, PPP, and PPV have since been developed by our group [14] and others [13,77–83]. The detection mechanism [14,78] is such that there is a change in FRET efficiency as target concentration increases. The proposed mechanism takes into account electrostatic and hydrophobic interactions that bring the

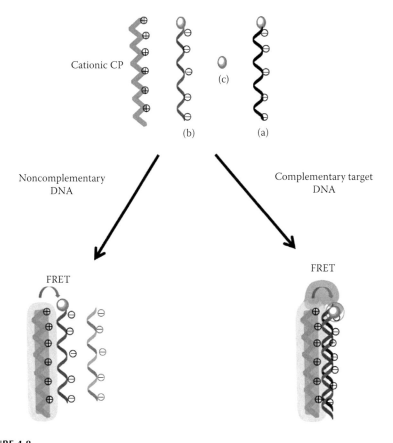

FIGURE 4.8
(See color insert.) The components involved in cationic conjugated-polymer-based fluorescence resonance energy transfer a DNA sensor with (a) labeled target, (b) labeled probe, and (c) intercalating dye.

cationic CP and the labeled probe DNA in close vicinity to trigger efficient FRET [14,78]. As the target concentration increases, the CP switches from a flexible random coil conformation (in the presence of probe) to a rigid structure (in the presence of the stiffer target-probe hybrid), thereby changing the distance between the CP and the dye [78]. The major limiting factor of such sensors is nonspecific electrostatic interactions with noncomplementary DNA, thereby reducing sensitivity and selectivity of the DNA sensors.

Consequently, Bazan and coworkers have extensively developed [80,82,83] similar sensing designs and studied the different factors that affect FRET efficiency in PFP-based sensing designs. Their sensing strategy [13,79] utilized expensive labeled PNA instead of the anionic probe, thus reducing nonspecific interactions. The cationic CP has the chance to electrostatically interact with the labeled PNA probe only in the presence of the target [79]. Further improvements were done to reduce the following: (1) inefficient overlap between the donor–acceptor pairs resulting in direct excitation of the acceptor and (2) self-interactions of the polymer and dye resulting in lower quantum yields. Efforts such as improving the shape of the polymer, spectral overlap, and side-chain length have also been reported [13,80]. Tuning molecular orbital energy levels by substitution of phenylene with electron-withdrawing fluorine and electron-donating alkoxy functional groups has also been reported [80].

Oligomers with six polyfluorene units and tetrahedral polyfluorenes have shown improved selectivity between hybrid and probe compared to linear polymers [80]. Studies were also conducted for tuning FRET efficiency by changing the size of counterions, as chances of FRET increase after the distance between FRET pairs is increased [84]. Diminished interactions between the CP and the dye have been reported by incorporating cosolvents such as Tetrahydrofuran in aqueous buffers, dilutions of probe with unlabeled single strands, and introduction of surfactants [85,86] while also implementing the same design on surfaces [13,79,87,88]. Surfactants have been reported to enhance polymer quantum yields in aqueous solutions as they induce breakup of polymer aggregates [86] and reduce fluorescence quenching by water (as hydrophobic polymer backbones are incorporated into surfactant micelles), [86] thereby increasing the FRET efficiency in these sensing designs [86].

Recently, optical label-free sensors utilizing FRET processes to monitor hybridization from cationic PFP polymers to intercalating dyes such as pico green, thiazole orange, and ethidium bromide have been developed to circumvent the labeling of probes [89]. FRET is instigated when hybridization brings together cationic CP, intercalating dye, unmodified probe, and target DNA. Advantages of a high detection limit and single-base mismatch detection were observed but have been overshadowed by nonspecific interactions of DNA with increased concentrations of the dye [89,81].

Taking advantage of the optical amplification of cationic CPs, Wang and Liu [90] reported a labeled FRET-based DNA sensor that makes use of probe-immobilized silica beads electrostatically interacting with the fluorescein-labeled target and the cationic CP [90]. A detection limit as low as 10 pM was observed but with disadvantages of decreased selectivity between two-base mismatch and noncomplementary sequences. A similar design with large magnetic beads derivatized with streptavidin was used in sandwich-type hybridization between biotin-labeled capturing probe, fluorescein-labeled signaling probe, and unlabeled target. An excellent detection limit of 100 fM was estimated by the authors. As the sensing was done off the bead, the drawbacks of this sensing design were introduction of time-consuming steps of hybridization–dehybridization using NaOH and dilution of sample by the *capture and release* scheme [91].

Inspired by the design developed by Wang and Liu [90], Wang et al. [92] reported a similar DNA sensor based on silica beads, but without labeling of the target, which was able to detect single-nucleotide polymorphisms. The sensing mechanism involved energy transfer between cationic tetrahedral-fluorene and intercalator ethidium bromide, signaling that hybridization had taken place (Figure 4.9) [92]. Although the additional step of labeling of probes is eliminated, this design suffers from relatively large concentrations of ethidium bromide (3 µM) and cationic tetrahedralfluorene (1.4 µM) while also being limited by the occurrence of false signals from nonspecific interactions of the intercalator with DNA [92].

Liu and coworkers [93] have recently reported an improved FRET-based DNA microchip with an excellent detection limit of 10^{-17} M, which utilizes electrostatic interactions between Cy5-dye-labeled target, self-assembled polystyrene beads functionalized with PNA probe, and cationic CP [93]. The CP had a reasonable fluorescence quantum yield of 31% while emitting

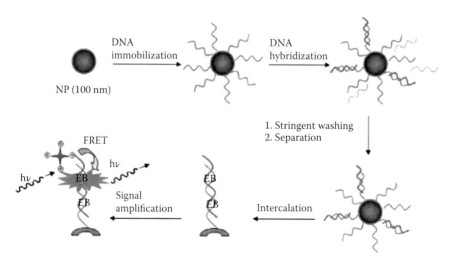

FIGURE 4.9

(See color insert.) Schematic cationic conjugated-polymer-assisted nanoparticle-based DNA assay. (Reprinted with permission from Wang et al., 7214. Copyright 2007 American Chemical Society.)

at a maximum wavelength of 480 nm. The silicon wafer is modified with polyethylene glycol, making it hydrophilic, which minimizes nonspecific interactions, while monitoring of the FRET signal is done by confocal microscopy. The authors have also replaced the Cy5-dye-labeled target with an unmodified target to test a label-free DNA sensor [93]. This sensing design involved monitoring of the fluorescence of CP due to increased electrostatic interactions with the hybridized DNA-modified polystyrene bead compared to the probe-modified bead. A detection limit of 10^{-16} M was observed [93].

A novel sensing platform with labeled G-quadruplex DNA involving FRET from CPs to fluorescein and ethidium bromide dyes was reported by Feng et al. [48] The mechanism exploits the affinity of K^+ toward the formation of negatively charged quadruplexes, which bind electrostatically to the cationic CP. Thus, efficient FRET occurs between CP and fluorescein-dye-labeled DNA strands on addition of the KCl salt. This quadruplex can be forced to form a double-stranded DNA hybrid once complementary ethidium-bromide-labeled DNA target is introduced. This causes two FRET processes—FRET from CP to fluorescein followed by FRET from fluorescein to ethidium bromide—to occur, which can be used to detect and measure the number of base mismatches. The capability of the single-base extension reaction of DNA polymerase was used to detect single-nucleotide polymorphisms and DNA methylation.

An unfortunate feature often observed in the proposed sensing designs based on the optical properties of CPs is the occurrence of nonspecific interactions, particularly between anionic DNA and cationic CPs, as electrostatic complexes can form with the free DNA present in solution. CP–CP interaction can also occur when more than one CP is bound to the DNA/DNA–PNA hybrid [86]. Hydrophobic interactions between CP and PNA have also been reported to further reduce the selectivity of the sensors [86]. A low quantum yield of CPs in aqueous solutions is an additional obstacle that is often encountered, which reduces the possibilities of optimizing the designs toward ultralow sensitivities [86]. Use of intercalating molecules to eliminate the labeling of DNA has additional problems of low specificity, along with the requirement of high concentration of intercalators.

4.3.2.2 FRET-Based Designs Using Neutral/Anionic CPs

A labeled DNA sensor on solid surfaces was tested by using a copolymer with fluorene, oxadiazole, and phenylene units covalently bound to the glass surface with probes while hybridization was carried out with HEX-dye-labeled target. Although a good detection limit (10^{-10} M) was obtained, the sensor's use was limited by the presence of a labeled probe [94,95]. Instead of covalent attachment of the polymer, Langmuir–Blodgett films were formed on hexamethyldisilazane-coated glass (Figure 4.10) [96]. A water-soluble PPE derivative with carboxylic acids was used for covalent attachment of DNA probes, whereas dye-modified targets were used for detection based on

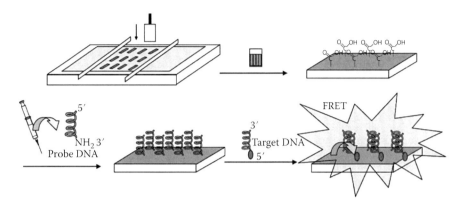

FIGURE 4.10
Schematic representation of the fabrication and amplified detection of a labeled solid-state DNA sensor. (Reprinted with permission from Pun et al., 7461. Copyright 2006 American Chemical Society.)

FRET. Although the sensor design suffers from the disadvantage of labeling of the target, it offers advantages such as reusability, high quantum yield, and fluorescence enhancement.

Another label-free DNA sensor using a different copolymer with fluorene, oxadiazole, and phenylene units grafted to the glass surface functionalized with probes was tested using FRET from an intercalating SYBR green 1 dye [97]. A FRET signal is produced when the asymmetric dye is intercalated by selectively binding to the major groove of double-stranded DNA. Although a good detection limit of 10^{-10} M and good selectivity were obtained, large amounts of dye were needed to obtain a reproducible signal (50 nM).

We have recently demonstrated a novel switch on/off sensor design [12] that utilizes either FRET (switch-on) or superquenching (switch-off) to detect DNA. The uniqueness and flexibility of this design is that two alternative readout modes were generated in the same sensing design depending on the type of signaling probe used. The detection mechanism involved sandwich hybridization between an anionic PPV-bound capture probe coated onto magnetic beads, target, and signaling probe. FRET from the donor CP to the acceptor dye occurs when capture probe oligonucleotides (ODNs) (CapODN) attach to PPV-coated magnetic beads (PPV-MagSi) and 5′-end dye-labeled signaling probes (SP1) hybridize with the target ODN (Figure 4.11, Scheme 1A).

FRET was observed when a polymer was selectively excited. Replacing SP1 with the 3′-end dye-labeled signaling probes (SP2) produced quenching of the fluorescence of PPV (Figure 4.11, Scheme 1B). As a result, detection of the target was easily visualized in the sensing system due to the increase in fluorescence of the dye in the FRET (switch-on) design, and the quenching of fluorescence in the switch-off design. A good selectivity as well as a low detection limit of 240 fmol in 2 mL for the switch-on FRET sensor was achieved.

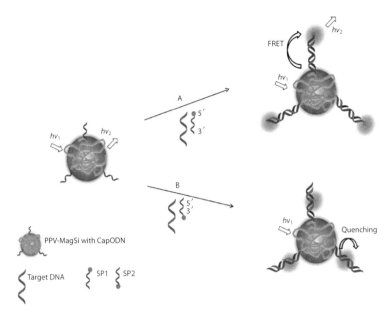

FIGURE 4.11
Principles of DNA detection systems based on two different routes, namely, fluorescence resonance energy transfer (route A) and superquenching (route B). Poly(*p*-phenylene vinylene) (PPV)-modified magnetic beads (PPV-MagSi) immobilized with CapODNs form sandwich assays with SP1/SP2. (Reprinted from *Biosensors & Bioelectronics*, 35, Gulur Srinivas et al., Switch on or switch off: An optical DNA sensor based on poly(p-phenylenevinylene) grafted magnetic beads, 498, Copyright 2012, with permission from Elsevier.)

4.3.2.3 Superquenching-Based Designs Using CPs

Turn-off optical DNA sensors have been developed using the superquenching properties of CPs as they provide amplification of fluorescence and efficient quenching by small molecules. The mechanism is termed *turn-off* as the fluorescence of the CP is quenched in the presence of a quencher. A design by Li and coworkers [98] using an anionic PFP polymer whose fluorescence is switched off by iron porphyrin containing hemins was demonstrated (Figure 4.12). As hemins are known to be efficiently oxidized and have well-defined absorption wavelengths between 415 and 430 nm, quenching of fluorescence (turn-off design) occurred via electron transfer (from the polymer to the iron) on hybridization. Furthermore, a FRET process (turn-on design) also occurred between the polymer (donor) and the hemin (acceptor). The latter process uses the inability of hemin to bind to hairpin aptamers. Hybridization forms a duplex that includes an aptamer/hemin complex while the fluorescence of the polymer is unquenched. Single-mismatch detection, high fluorescence quantum yield, minimization of nonspecific interactions with DNA, and reasonable detection limits (1.4 and 0.6 nM, respectively) were the advantages of using such sensors. It should be noted

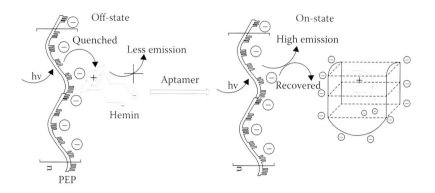

FIGURE 4.12
(See color insert.) Schematics for the principle of fluorescent switch combining poly(fluorene-co-phenylene), hemin, and hemin aptamer. (Reprinted with permission from Li et al., 3544. Copyright 2009 American Chemical Society.)

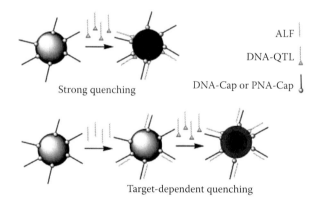

FIGURE 4.13
Superquenching-based DNA sensing system with target-dependent quenching. (Reprinted with permission from Kushon et al., 6456. Copyright 2003 American Chemical Society.)

that this design suffers from harsh conditions such as high salt concentrations, while purification of the target is also needed to eliminate false positives.

A robust platform for DNA sensing comprising biotin and biotin-binding proteins has been explored by Kushon and coworkers [21,64,91]. Superquenching of anionic biotin-functionalized PPE polymer bound to the surface of polystyrene microspheres was explored [21,64] where the sensing was based on target-concentration-dependent quenching, involving two steps (Figure 4.13) of hybridization of the unlabeled target followed by hybridization with the quencher-labeled target. This resulted in a dynamic range of quenching of 50% of the total fluorescence of the PPE-coated microspheres [21,64].

PNA in place of anionic DNA was also used to improve affinity and selectivity of the sensor to detect single-base mismatches at 500 fmol concentration within 20 minutes of hybridization [21,64]. Although the sensing system is selective and sensitive, a nonspecific background of 15% was generated by nonspecific responses of the quencher-labeled target on the positively charged microspheres [21].

4.3.2.4 Molecular Beacon Gene Sensing Designs Using CPs

Label-free DNA sensors based on CPs and molecular beacons (MBs) have been recently developed by a number of groups. MBs are hairpin structures containing recognition loops and stem structures with fluorescent dyes and quenchers at the 5′ and 3′ end of the long sequences [99]. They offer advantages of elimination of target labeling, single-base mismatch detection, high sensitivity, and selectivity while also minimizing nonspecific interactions. The stable hairpin formation on partial hybridization of the stem results in FRET quenching of the fluorophore [99]. The introduction of perfectly matching targets brings about the formation of thermodynamically stable hybrids [99] while fluorescence of the fluorophore is detected. Such sensors offer an additional advantage of amplifying the superquenching signal.

Novel PPE–DNA conjugates covalently attached during either the polymerization process or heterogeneous on-chip DNA synthesis have been used as derivatized MBs, whereby PPE acts as the fluorophore [95,100,101].

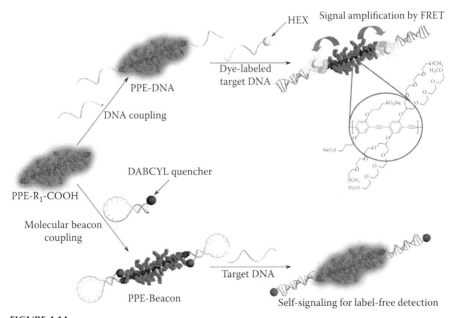

FIGURE 4.14
(See color insert.) Homogenous molecular-beacon-based DNA sensing system with signal transduction afforded by the poly(*p*-phenylene ethynylene). (Lee et al.: *Advanced Functional Materials*. 2007. 17. 2580. Copyright Wiley-VCH Verlag GmbH & Co. KGaA. Reproduced with permission.)

The sensing system has been extended to a turn-off sensor demonstrated by Lee and coworkers (Figure 4.14) using FRET between a fluorescent dye (fluorescein or Hex) and PPE. FRET emission of the dye was recorded only in the hybridized open form.

4.4 Conclusions and Future Outlook

Optical, electrochemical, or gravimetric (piezoelectric) methods have been the major approaches used for DNA detection. Extremely low concentrations of DNA can be detected by optical methods and especially by fluorescence-based detection, which also has the benefits of rapidity, nondestructiveness, and comparatively inexpensive analysis. On the other hand, electrochemical methods of detecting hybridization afford advantages of portable analyzers but generally suffer from low sensitivity. Consequently, by assessing the pros and cons of different categories of DNA sensing, fluorescence-based optical DNA sensing stands out as one of the most powerful, versatile, and economically efficient approaches.

Different strategies used to develop sensitive and selective optical DNA sensors based on optical properties of CPs have been outlined and discussed in this chapter. Labeled DNA sensors have very limited commercial applications as they are time-consuming, complex, and expensive. Label-free DNA sensing, on the other hand, either in homogenous assays or on solid substrates, has been preferred over labeled DNA sensing due to advantages of having rapid assay times, low-cost procedures, and simplicity in detecting DNA. Thus, a majority of the recent research focusses on developing novel approaches to label-free DNA sensors. Furthermore, turn-off DNA sensors are generally less preferred over turn-on sensors due to false positives in the assay from background signals [88,102].

Significant effort has been made in enhancing the selectivity and sensitivity of DNA sensors by several groups using new design concepts. This emerging field of sensing platforms using CPs can be envisioned to have applications in many areas, including DNA diagnostics, gene analysis, biodefense, forensics, and food contamination testing. Although optical DNA sensors based on CPs appear highly efficient with enormous potential, major limitations such as low quantum yield in water, biocompatibility, self-quenching due to aggregation, and nonspecific adsorption of biomaterials have to be overcome before commercialization can be undertaken. Because of the versatility of synthesis procedures, CPs can be tailor-made to have improved water solubility, quantum yields, and stability. Designer CPs, whose optoelectronic properties can be controlled through molecular design, will continue to command research not only in DNA detection but also in wider optoelectronic fields.

References

1. Bilitewski, U. and Turner, A. P. F. *Biosensors for Environmental Monitoring*, Harwood Academic Publishers: Amsterdam, the Netherlands, 2000.
2. Hall, E. A. H. *Biosensors*, Open University Press: London, United Kingdom, 1990.
3. Wang, J. *Chemistry-A European Journal*, 1999, *5*, 1681.
4. Mikkelsen, S. R. *Electroanalysis*, 1996, *8*, 15.
5. Sassolas, A., Leca-Bouvier, B. D., and Blum, L. J. *Chemical Reviews (Washington, DC, United States)*, 2008, *108*, 109.
6. Lv, F., Liu, L., and Wang, S. *Current Organic Chemistry*, 2011, *15*, 548.
7. Peng, H., Zhang, L., Soeller, C., and Travas-Sejdic J. *Biomaterials*, 2009, *30*, 2132.
8. Kannan, B., Williams, D. E., Booth, M. A., and Travas-Sejdic, J. *Analytical Chemistry*. American Chemical Society: Washington, DC, United States, 2011, *83*, 3415.
9. Cosnier, S. *Analytical Letters* 2007, *40*, 1260.
10. Drummond, T. G., Hill, M. G., and Barton, J. K. *Nature Biotechnology*, 2003, *21*, 1192.
11. Peng, H., Zhang, L., Kjaellman, T. H. M., Soeller, C., and Travas-Sejdic, J. *Journal of the American Chemical Society*, 2007, *129*, 3048.
12. Gulur Srinivas, A. R., Travas-Sejdic, J., Barker, D., and Peng, H. *Biosensors & Bioelectronics*, 2012, *35*, 498.
13. Liu, B. and Bazan, G. C. *Chemistry of Materials*, 2004, *16*, 4467.
14. Peng, H., Soeller, C., and Travas-Sejdic, J. *Chemical Communications*. Royal Society of Chemistry: Cambridge, United Kingdom, 2006, 3735.
15. Watson, J. D. and Crick, F. H. C. *Nature*, 1953, *171*, 737.
16. Watson, J. D. and Crick, F. H. C. *Nature*, 1953, *171*, 964.
17. Richard, S. R. *DNA Structure and Function*, Academic Press: Houston, Texas, 1994.
18. Junhui, Z., Hong, C., and Ruifu, Y. *Biological Advances*, 1997, *15*, 43.
19. Keller, G. H. and Manak, M. M. *DNA Probes*, Macmillan Publishers, Ltd.: London, United Kingdom, 1989.
20. Tenover, F. C. *Clinical Microbiological Reviews*, 1988, *1*, 82.
21. Kushon, S. A., Bradford, K., Marin, V., Suhrada, C., Armitage, B. A., McBranch, D., and Whitten, D. *Langmuir*, 2003, *19*, 6456.
22. Downs, M. E. A., Kobayashi, S., and Karube, I. *Analytical Letters*, 1987, *20*, 1897.
23. Arora, K., Prabhakar, N., Chandb, S., and Malhotra, B. D. *Biosensors and Bioelectronics*, 2007, *23*, 613.
24. Kagan, K., Masaaki, K., and Eiichi, T. *Measurement Science and Technology*, 2004, *15*, R1.
25. Gavrilenko, A. V., Matos, T. D., Bonner, C. E., Sun, S. S., Zhang, C., and Gavrilenko, V. I. *Journal of Physical Chemistry C*, 2008, *112*, 7908.
26. Kjällman, T., Peng, H., Travas-Sejdic, J., and Soeller, C. *Biomedical Applications of Micro- and Nanoengineering III*, 2006, *6416*, 641602.
27. Floch, F. L., Ho, H. A., Harding-Lepage, P., Bedard, M., Neagu-Plesu, R., and Leclerc, M. *Advanced Materials*. 2005, *17*, 1251.
28. Cha, J., Han, J. I., Choi, Y., Yoon, D. S., Oh, K. W., and Lim, G. *Biosensors and Bioelectronics*, 2003, *18*, 1241.
29. Cosnier, S. *Analytical Letters*, 2007, *40*, 1260.
30. Floch, F. L., Ho, H. A., and Leclerc, M. *Analytical Chemistry*, American Chemical Society: Washington, DC, United States, 2006, *78*, 4727.

31. Garnier, F., Korri-Youssoufi, H., Srivastava, P., Mandrand, B., and Delair, T. *Synthetic Metals*, 1999, *100*, 89.
32. Gaylord, B. S., Heeger, A. J., and Bazan, G. C. *PNAS*, 2002, *99*, 10954.
33. Geetha, S., Chepuri, R. K. R., Vijayan, M., And Trivedi, D. C. *Analytica Chimica Acta*, 2006, *568*, 119.
34. Heeger, P. S. and Heeger, A. J. *PNAS*, 1999, *96*, 12219.
35. Malhotra, B. D., Chaubey, A., and Singh, S. P. *Analytica Chimica Acta*, 2006, *578*, 59.
36. Peng, H., Soeller, C., and Travas-Sejdic, J. *Macromolecules*, 2007, *40*, 909.
37. Peng, H., Soeller, C., Vigar, N., Kilmartin, P. A., Cannell, M. B., Bowmaker, G. A., Cooney, R. P., and Travas-Sejdic, J. *Biosensors and Bioelectronics*. 2005, *20*, 1821.
38. Peng, H., Zhang, L., Soeller, C., and Travas-Sejdic, J. *Polymer*, 2007, *48*, 3413.
39. Jean-Michel, P. and Reynolds, J. R. *Journal of Physical Chemistry B*, 2000, *104*, 4080.
40. Lange, U., Roznyatovskaya, N. V., and Mirsky, V. M. *Analytica Chimica Acta*, 2008, *614*, 1.
41. Guodong, L., Jun, W., Yuehe, L., and Joseph, W. *Electrochemical Sensors, Biosensors, and Their Biomedical Applications*. Academic Press, Inc.: Oxford, United Kingdom, 2008.
42. Fawcett, N. C., Evans, J. A., Chien, L. C., and Flowers, N. *Analytical Letters*, 1988, *21*, 1099.
43. Zhai, J., Hong, C., and Yang, R. *Biotechnology Advances*, 1997, *15*, 43.
44. Piunno, P. A. E. and Krull, U. J. *Analytical Bioanalytical Chemistry*, 2005, *381*, 1004.
45. Epstein, J. R., Biran, I., and Walt, D. R. *Analytica Chimica Acta*, 2002, *469*, 3.
46. Park, H. G., Song, J. Y., Park, K. H., and Kim, M. H. *Chemical Engineering Science*, 2005, *61*, 954.
47. McQuade, D. T., Pullen, A. E., and Swager, T. M. *Chemical Reviews (Washington, DC)*, 2000, *100*, 2537.
48. Feng, F., He, F., An, L., Wang, S., Li, and Y. Zhu, D. *Advanced Materials (Weinheim, Germany)*, 2008, *20*, 2959.
49. Jiang, H., Taranekar, P., Reynolds, J. R., and Schanze, K. S. *Angewandte Chemie, International Edition*, 2009, *48*, 4300.
50. Pinto, M. R. and Schanze, K. S. *Synthesis*, 2002, 1293.
51. Heeger, P. S. and Heeger, A. J. *Proceedings of the National Academy of Sciences of the United States of America*, 1999, *96*, 12219.
52. Singh, R. P. and Choi, J. W. *Sensors & Transducers Journal*, 2009, *104*, 1.
53. Gettinger, C. L. and Heeger, A. J. *Journal of Chemical Physics*, 1994, *101*, 1673.
54. Bunz, U. H. F. *Chemical Reviews (Washington, DC)*, 2000, *100*, 1605.
55. Gruber, J., Li, R. W. C., and Hiimmelgen, I. A. *Handbook of Advanced Electronic and Photonic Materials and Devices*, Vol. 8., Academic Press: London, United Kingdom, 2001.
56. Grimsdale, A. C., Chan, K. L., Martin, R. E., Jokisz, P. G., and Holmes, A. B. *Chemical Reviews*, 2009, *109*, 897.
57. Lakowicz, J. R. *Principles of Fluorescence Spectroscopy*. Springer: New York, 2006.
58. Laquai, F. D. R., Park, Y. S., Kim, J. J., Basché, T. *Macromolecular Rapid Communications*, 2009, *30*, 1203.
59. Lupton, J. M. *ChemPhysChem*, 2012, *13*, 901.
60. Hwang, I. and Scholes, G. D. *Chemistry of Materials*, 2011, *23*, 610.
61. Piunno, P. A. E. and Krull, U. J. *Analytical and Bioanalytical Chemistry*, 2005, *381*, 1004.

62. Rodriguez-Mozaz, S., Alda, M. J. L. D., and Barcelo, D. *Analytical Bioanalytical Chemistry*, 2006, *386*, 1025.
63. Downs, M. E. A., Kobayashi, S., and Karube, I. *Analytical Letters*, 1987, *20*, 1897.
64. Achyuthan, K. E., Bergstedt, T. S., Chen, L., Jones, R. M., Kumaraswamy, S., Kushon, S. A., Ley, K. D. et al. *Journal of Materials Chemistry*, 2005, *15*, 2648.
65. Knopf, G. K. and Bassi, A. S. *Smart Biosensor Technology*. CRC Press: Boca Raton, Florida, 2007.
66. Lilley, D. M. J. and Wilson, T. J. *Current Opinion in Chemical Biology*, 2000, *4*, 507.
67. Tan, C., Atas, E., Mueller, J. G., Pinto, M. R., Kleiman, V. D., and Schanze, K. S. *Journal of the American Chemical Society*, 2004, *126*, 13685.
68. Lopez Arbeloa, F., Ruiz Ojeda, P., and Lopez Arbeloa, I. *Journal of Photochemistry and Photobiology, A: Chemistry*, 1988, *45*, 313.
69. Ho, H. A., Doré, K., Boissinot, M., Bergeron, M. G., Tanguay, R. M., Boudreau, D., and Leclerc, M. *Journal of the American Chemical Society*, 2005, *127*, 12673–12676.
70. Ho, H. A., Bera-Aberem, M., and Leclerc, M. *Chemistry—A European Journal*, 2005, *11*, 1718.
71. Massey, M., Piunno, P. A. E., and Krull, U. J. *Springer Series on Chemical Sensors and Biosensors*, 2005, *3*, 227.
72. Bera-Aberem, M., Ho, H. A., and Leclerc, M. *Tetrahedron*, 2004, *60*, 11169.
73. Raymond, F. R., Ho, H. A., Peytavi, R., Bissonnette, L., Boissinot, M., Picard, F. J., Leclerc, M., and Bergeron, M. G. *BMC Biotechnology*, 2005, *5*.
74. Yang, X., Zhao, X., Zuo, X., Wang, K., Wen, J., and Zhang, H. *Talanta*, 2009, *77*, 1027.
75. Brouard, D., Viger, M. L., Bracamonte, A. G., and Boudreau, D. *ACS Nano*, 2011, *5*, 1888.
76. Chen, X., Zhou, G., Peng, X., and Yoon, J. *Chemical Society Reviews*, 2012, *41*, 4610.
77. Aidee Duarte, K.-Y. P., Liu, B., and Bazan, G. C. *Chemistry of Materials*, 2011, *23*, 501.
78. Gaylord, B. S., Heeger, A. J., and Bazan, G. C. *Journal of the American Chemical Society*, 2003, *125*, 896.
79. Liu, B. and Bazan, G. C. *Proceedings of the National Academy of Sciences of the United States of America*, 2005, *102*, 589.
80. Liu, B. and Bazan, G. C. *Journal of the American Chemical Society*, 2006, *128*, 1188.
81. Liu, B. and Bazan, G. C. *Macromolecular Rapid Communications*, 2007, *28*, 1804.
82. Woo, H. Y., Vak, D., and Bazan, G. C. *Materials Research Society Symposium Proceedings*, 2007, *965E*.
83. Woo, H. Y., Vak, D., Korystov, D., Mikhailovsky, A., Bazan, G. C., and Kim, D. Y. *Advanced Functional Materials*, 2007, *17*, 290.
84. Kang, M., Nag, O. K., Nayak, R. R., Hwang, S., Suh, H., and Woo, H. Y. *Macromolecules (Washington, DC, United States)*, 2009, *42*, 2708.
85. Nayak, R. R., Nag, O. K., Kang, M., Jin, Y., Suh, H., Lee, K., and Woo, H. Y. *Macromolecular Rapid Communications*, 2009, *30*, 633.
86. Attar, H. A. A. and Monkman, A. P. *Advanced Functional Materials*, 2008, *18*, 2498.
87. Pu, K. Y., Pan, S. Y. H., and Liu, B. *Journal of Physical Chemistry B*, 2008, *112*, 9295.
88. Pu, K. Y. and Liu, B. *Advanced Functional Materials*, 2009, *19*, 277.
89. Ren, X. and Xu, Q. H. *Langmuir*, 2009, *25*, 43.
90. Wang, Y. and Liu, B. *Chemical Communications*. Royal Society of Chemistry: Cambridge, United Kingdom, 2007, 3553.

91. Xu, H., Wu, H., Huang, F., Song, S., Li, W., Cao, Y., and Fan, C. *Nucleic Acids Research*, 2005, *33*, e83/1.
92. Wang, Y. and Liu, B. *Analytical Chemistry.* American Chemical Society: Washington, DC, United States, 2007, *79*, 7214.
93. Wang, C., Zhan, R., Pu, K. Y., and Liu, B. *Advanced Functional Materials*, 2010, *20*, 2597.
94. Lee, K., Rouillard, J. M., Pham, T., Gulari, E., and Kim, J. *Angewandte Chemie, International Edition*, 2007, *46*, 4667.
95. Lee, K., Povlich, L. K., and Kim, J. *Advanced Functional Materials*, 2007, *17*, 2580.
96. Pun, C. C., Lee, K., Kim, H. J., and Kim, J. *Macromolecules*, 2006, *39*, 7461.
97. Lee, K., Maisel, K., Rouillard, J. M., Gulari, E., and Kim, J. *Chemistry of Materials*, 2008, *20*, 2848.
98. Li, B., Qin, C., Li, T., Wang, L., and Dong, S. *Analytical Chemistry.* American Chemical Society: Washington, DC, United States, 2009, *81*, 3544.
99. Satterfield B. C., West J. A. A., and Caplan M. R. *Nucleic Acids Res*, 2007, *35*, e76.
100. Yang Chaoyong, J., Pinto, M., Schanze, K., and Tan, W. *Angewandte Chemie (International Edition in English)*, 2005, *44*, 2572.
101. Lee, K., Rouillard, J. M., Kim, B. G., Gulari, E., and Kim, J. *Advanced Functional Materials*, 2009, *19*, 3317.
102. Pu, K. Y. and Liu, B. *Biosensors & Bioelectronics*, 2009, *24*, 1067.

5

Polymer Ion Sensors Based on Intramolecular Charge-Transfer Interactions

Tsuyoshi Michinobu and Tsuyoshi Hyakutake

CONTENTS

5.1 Introduction

Chemical sensing of specific metal ions and anions is an important event in living organisms [1,2]. It is thus reasonable to mimic biological systems when artificial chemical sensors are designed. Biological systems are composed of assembled macromolecules with well-defined molecular weights and shape persistency, and the recognition events of guest molecules and ions are often governed by unique environments. For example, control of chiral molecular recognition has been the focus of supramolecular and macromolecular chemistries. To realize an artificial chiral molecular recognition, one-handed helical macromolecules and chiral molecular assemblies have been developed as host platforms, and their recognition behaviors toward enantiomeric guests suggested that an amplified chiral recognition can be achieved by supramolecular assemblies [3–6].

A similar amplification of recognition signals can also be obtained by polymer effects. Cooperative interactions between the side-chain chromophores of polymer sensors results in a high sensitivity due to the effect that is analogous to the allosteric regulation of enzymes or other proteins. Thus, side-chain-type polymer sensors are a versatile molecular design for sensing a small amount of specific ions in solution. They can be used to detect toxic heavy metal ions and some anions. In addition, the selective extraction of precious metal ions from industrial wastes enables the reuse of these resources in commercial products. Both applications are currently desired due to the social demand.

This chapter begins with the introduction of conventional polymer ion sensors, which are based on optical sensors of conjugated polymers. This is followed by a discussion on the recent fluorescent polymer ion sensors based on poly(triazole) building blocks prepared by Cu(I)-catalyzed azide–alkyne cycloaddition (CuAAC), known as a typical click chemistry reaction. Finally, the colorimetric visual detection of ion recognition behaviors is focused on using nonplanar donor–acceptor chromophores prepared by new types of click chemistry reactions.

5.2 Conjugated Polymer Sensors

The ion-sensitive color change of conjugated polymers is a useful recognition signal of polymeric chemosensors. To efficiently capture specific metal ions, crown ether units have often been used. The size of the crown ether units determines the selectivity of the alkali metal ions. A pioneering study was reported by Swager and coworkers in 1993 [7]. They used crown ether units for size-dependent metal ion recognition and combined these units with a colored polythiophene backbone (Figure 5.1). The precursor polythiophene showed the longest absorption maximum (λ_{max}) at 497 nm in CH_3CN, reflecting the lower steric hindrance between the adjacent thiophene rings. The

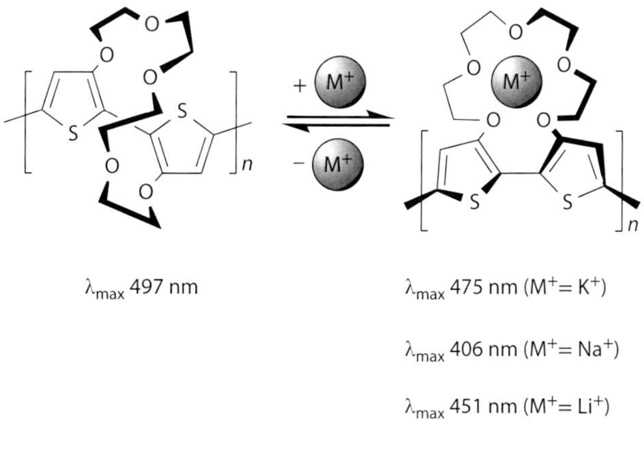

λ_{max} 497 nm

λ_{max} 475 nm (M$^+$= K$^+$)

λ_{max} 406 nm (M$^+$= Na$^+$)

λ_{max} 451 nm (M$^+$= Li$^+$)

FIGURE 5.1
Conformational changes of polythiophene containing crown ether units in response to metal ions. The longest absorptions in CH_3CN are shown. (Adapted with permission from Marsella and Swager, 12214–12215. Copyright 1997 American Chemical Society.)

addition of metal ions to this solution resulted in a hypsochromic shift in the λ_{max} values, suggesting a conformational change in the polythiophene backbone by metal ion complexation. The Na^+ ion showed the largest shift in λ_{max} (91 nm) compared to the Li^+ ion (46 nm) and the K^+ ion (22 nm). This order is consistent with the binding constants for the employed crown ether unit.

Bäuerle and coworkers reported similar polythiophene-based sensors functionalized with crown ethers or oligo(ethylene oxide) side chains (Figure 5.2) [8,9]. Due to the aliphatic spacers between the polythiophene backbone and the ether-based side-chain units, metal ion recognition led to negligible changes in the visible absorption originating from the polythiophene backbone. Accordingly, the responses to metal ions were analyzed by electrochemical techniques. The redox activity of polythiophenes bearing oligo(ethylene oxide) side chains was reduced to 35%–58% of the original level when alkali metal ions were added. This was because the oligo(ethylene oxide) unit displayed a similar degree of affinity to Li^+, Na^+, and K^+ ions. In contrast, the 18-crown-6-ether-appended polythiophene showed selectivity for K^+ ions due to the perfect match with cavity size. The addition of Li^+ and Na^+ ions to this polythiophene solution did not produce any changes in the electrochemical behavior.

Another important heteroatom for the recognition of metal ions is the nitrogen atom. It is well known that nitrogen-containing heteroaromatic ring units, such as bipyridyl and terpyridyl, show an excellent recognition behavior of the di- and trivalent main group and transition metal ions. This metal ion recognition could be visually detected when the heteroaromatic ring units were integrated into highly colored conjugated polymers. For example, Wang and Wasielewski reported poly(arylenevinylene)s containing the 2,2'-bipyridyl structure in the main chain repeat unit (Figure 5.3) [10]. This poly(arylenevinylene) displayed a λ_{max} at 455 nm in $CHCl_3$. A comparison with other poly(arylenevinylene)s suggested that the 2,2'-bipyridyl moiety of the poly(arylenevinylene) was twisted with the estimated dihedral angle of 20°. However, the coordination to metal ions resulted in the planarization of the bipyridyl unit, leading to the extended conjugation and thereby longer wavelength absorption. Among the many metal ions,

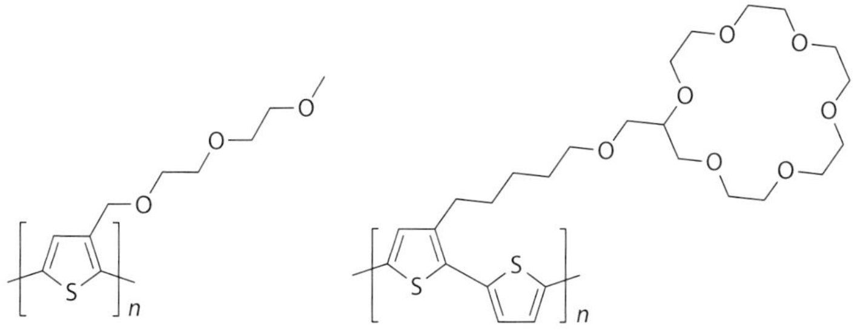

FIGURE 5.2
Chemical structures of metal ion–sensitive polythiophene derivatives.

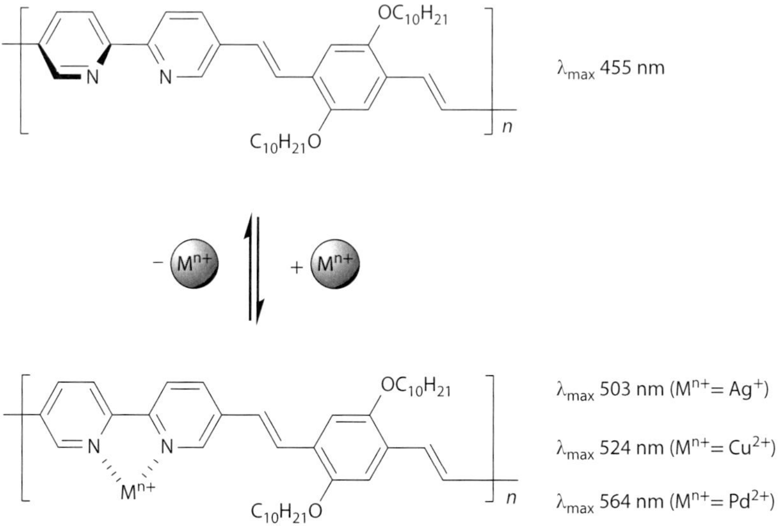

FIGURE 5.3
Conformational changes of the poly(arylenevinylene) containing the 2,2'-bipyridyl units in response to metal ions. The longest absorptions in CHCl$_3$ are shown. (Adapted with permission from Wang and Wasielewski, 12–21. Copyright 1997 American Chemical Society.)

Ag$^+$, Cu^{2+}, and Pd^{2+} ions induced the well-defined bathochromic shift in λ_{max}. The λ_{max} of the poly(arylenevinylene) was detected at 503 nm in the presence of Ag$^+$ ions. The extent of the bathochromic shift became more significant when Cu^{2+} or Pd^{2+} ions were added. Collectively, it was found that this polymer has the highest affinity toward the Pd^{2+} ions.

It should be noted that conjugated polymers including poly(arylenevinylene)s are usually highly fluorescent and this feature can be used as an effective sensory signal to detect various analytes. Compared to small conjugated molecule-based sensors, the use of conjugated polymers often enables significant amplification of the recognition signals [11]. The research groups of Kimura and Jones independently reported poly(phenylenevinylene) and poly(aryleneethynylene) substituted by terpyridyl segments, respectively (Figure 5.4). These polymers were sensitive to a wide variety of metal ions, leading to well-defined changes in their absorption and fluorescence spectra. The λ_{max} of poly(phenylenevinylene) in a CH$_2$Cl$_2$–methanol mixture shifted from 450 to 569 nm on the addition of Fe^{2+} ions [12]. The fluorescence of this polymer was detected at 524 nm, but this peak was completely quenched by the formation of the bistridentate-type Fe^{2+} complex due to an electron or energy transfer between the metal complex and the polymer backbone. Similar to this result, the poly(aryleneethynylene) substituted by terpyridyl units displayed chemosensor quenching responses [13]. The addition of Ni^{2+}, Cr^{6+}, Co^{2+}, Cu^{2+}, and Mn^{2+} ions to the polymer solution of tetrahydrofuran (THF) lowered the original emission intensity to 508 nm. This quenching

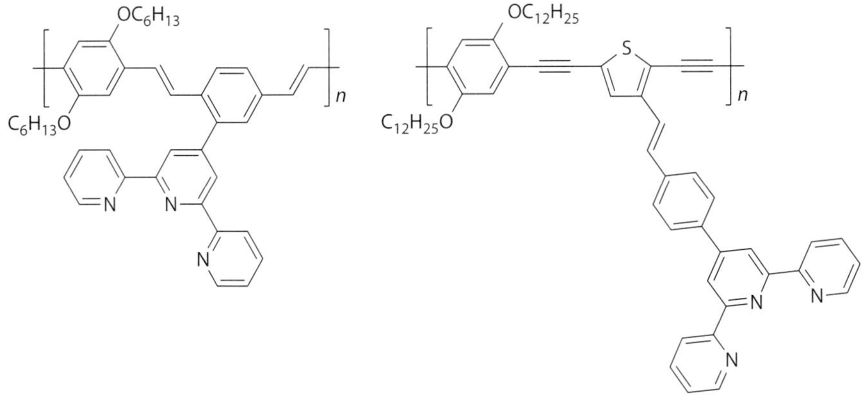

FIGURE 5.4
Chemical structures of the ion-sensitive poly(phenylenevinylene) and poly(aryleneethynylene) substituted by terpyridyl segments.

response was the most significant when Ni^{2+} ions were added. Thus, the emission quenching behavior on Ni^{2+} ion addition was analyzed by Stern–Volmer plots. In contrast, the optical spectra of poly(aryleneethynylene) did not change on the addition of Na^+, Ca^{2+}, Pb^{2+}, and Hg^{2+} ions, ensuring the selectivity of the metal ions.

5.3 Oligo- and Poly(triazole) Ion Sensors

As seen in Section 5.2, chemosensors have relied on conjugated polymers due to their sensitive absorption and fluorescence spectra. Recently, attention was also focused on an efficient synthetic method of producing chemosensor components. In 2001, Sharpless and coworkers proposed the new concept of click chemistry [14]. This concept is represented by highly efficient addition reactions featuring high product yields, mild reaction conditions, absence of by-products, and excellent selectivity of functional groups. The most common click reaction is CuAAC, producing 1,2,3-triazole derivatives. This reaction is a powerful tool in producing a wide variety of functional polymers in a straightforward way.

Click polymerization was conducted by the polyaddition between two complementary bifunctional monomers, namely, dialkyne monomers and diazide comonomers. Tang and coworkers optimized the polymerization conditions of CuAAC and eventually found that $Cu(PPh_3)_3Br$ is a good homogeneous catalyst in organic solvents [15,16]. The detection of explosives, such as picric acid, was shown by utilizing the aggregation-induced emission characteristics of the tetraphenylethene moiety as a sensing unit (Figure 5.5). They recently reported a metal-free click polymerization using

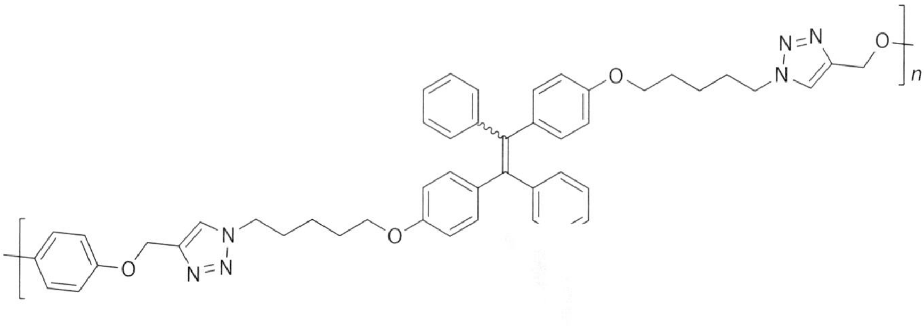

Aggregation-induced emission

FIGURE 5.5
(See color insert.) Chemical structure of the poly(triazole) sensor containing aggregation-induced emission chromophores.

the combination of activated aroyl acetylenes and azide functionalities, and the resulting poly(triazole)s exhibited a sensing behavior similar to those prepared by the conventional CuAAC polymerization [17,18].

The 1,2,3-triazole rings formed by the CuAAC click reaction do not have a visible absorption, but they show a weak fluorescence. In addition, the rings contain three nitrogen atoms for the recognition of metal ions in terms of ion-dipole interactions. Accordingly, they can become fluorophores for ion sensors when appropriately designed. Bunz and coworkers reported that the 4-(2-pyridyl)-1,2,3-triazole derivatives, prepared from 2-ethynylpyridine and azide compounds by CuAAC, are some of the best structures for metal ion recognition [19–21]. Thus, these sensor units were successfully substituted into the polymer main chains. Weng and coworkers designed the 2,6-bis(1,2,3-triazol-4-yl)pyridine unit, and the polyurethane containing this unit was prepared (Figure 5.6) [22]. Interestingly, this polymer showed selective recognition of Zn^{2+} and Eu^{3+} ions, which was indicated by colorimetric fluorescent color changes. For Zn^{2+} ion recognition, the polymer displayed a blue fluorescence (Figure 5.6). In contrast, the complexation with Eu^{3+} ions resulted in a red fluorescence. The addition of an equimolar amount of Zn^{2+} and Eu^{3+} ions produced intermediate emission colors. Furthermore, the recognition events of these metal ions were also suggested by the formation of supramolecular organogels in $CHCl_3$. The formed gels exhibited a repeatable autonomic self-healing ability.

Linear aromatic polymers containing the 1,2,3-triazole unit in the main chain were also prepared, and their metal ion sensing behaviors were comprehensively investigated. The aromatic polymer composed of alternating *p*-phenylene and 1,2,3-triazole rings was prepared by the click polymerization of the diethynylbenzene and diazidebenzene derivatives (Figure 5.7) [23]. The thermal analysis of the resulting polymer suggested the onset decomposition temperature of 240°C due to the high content of the aromatic moieties. This aromatic polymer showed selective recognition of Hg^{2+} ions by fluorescence

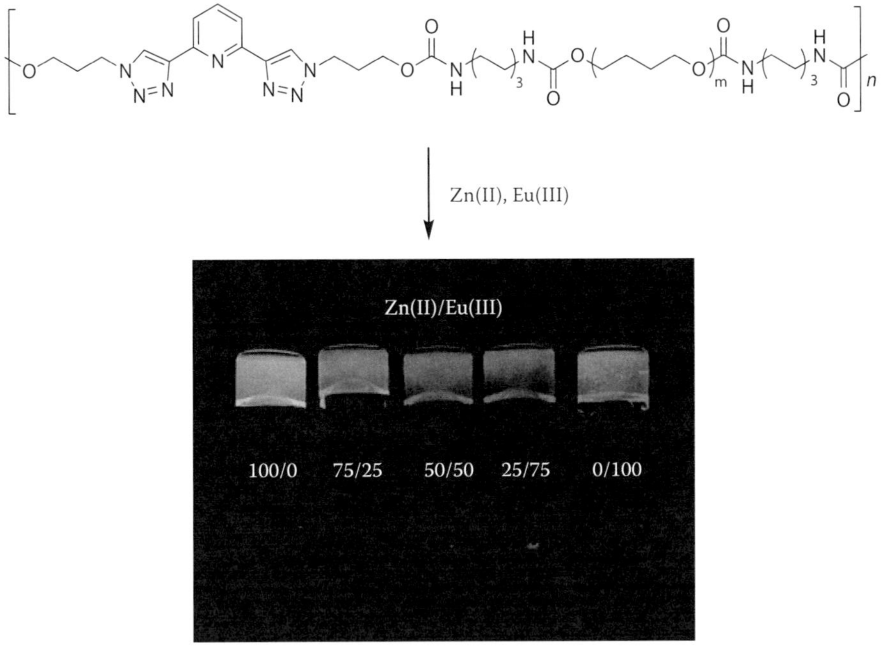

FIGURE 5.6
(See color insert.) Multicolored organogels of poly(triazole) with Zn^{2+} and Eu^{3+} ions. (Adapted with permission from Yuan et al., 11515–11522, Copyright 2012 The Royal Chemical Society.)

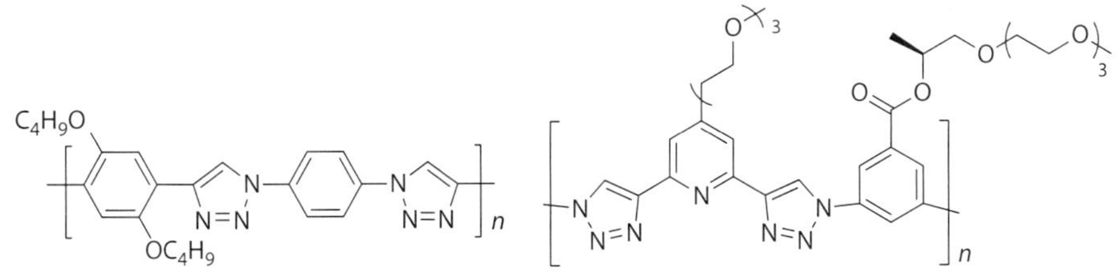

FIGURE 5.7
Chemical structures of the metal ion–sensitive aromatic poly(triazole)s.

quenching. Similar to this polymer, the aromatic polymer composed of alternating *m*-phenylene and 2,6-pyridine rings was prepared by click polymerization. Interestingly, this polymer adopted a one-handed helical backbone conformation due to the chiral side chains (Figure 5.7) [24]. The addition of Zn^{2+}, Fe^{2+}, and Eu^{3+} ions led to the smooth formation of organogels.

Side-chain-type polymers are a versatile molecular design of polymer ion sensors. Conventional side-chain-type polymer sensors have been synthesized by the preparation of vinyl monomers bearing chromophores (or fluorophores) followed by polymerization. However, recent advances in click chemistry have opened the door to a more simplified synthetic approach to preparing side-chain-type polymer sensors. The noticeable features of click chemistry are rapid reaction rates, high yields, and simple purification processes. All these features are suitable for clean transformation of polymer side chains. It should be noted that undesired products caused by side reactions cannot be separated in the case of polymer reactions. Thus, the quantitative yield is a key to the success of this approach. Also, the tedious purification processes, such as chromatography and distillation under high vacuum, increase the production costs.

Addition reactions yielding no by-products are the most suitable reaction types. The 1,2,3-triazole ring itself does not show any remarkable optical spectral changes in response to ion recognition, but the conjugated substitution at the 1- and/or 4-positions enables fluorophore formation. In this context, the 4-(2-pyridyl)-1,2,3-triazole moiety was attached to the side chain of polymethacrylate, which was prepared by copolymerization with methyl methacrylate using the reversible addition–fragmentation transfer polymerization technique (Figure 5.8) [25]. This side-chain-type polymethacrylate underwent efficient complexation with Fe^{2+} and Co^{2+} ions accompanied by well-defined changes in ultraviolet–visible absorption spectra and solution viscosities. Recently, highly cross-linked poly(triazole) gels were reported (Figure 5.8) [26]. The tetrakis(triazolylmethyl)ethylenediamine unit was prepared as the multidentate ligand of metal ions by quadruple CuAAC reaction. This ligand moiety exhibited a temperature-dependent Cd^{2+} ion extraction when *N*-isopropylacryl amide gels were formed.

Another noticeable ion recognition feature of poly(triazole)s is anion sensing. Craig and coworkers prepared the *m*-phenylene-linked oligo(triazole) derivative, and its foldamer formation in response to size-dependent anion recognition through ionic hydrogen bonding interactions was demonstrated (Figure 5.9) [27]. Among the many anions investigated, the Cl⁻ ion showed

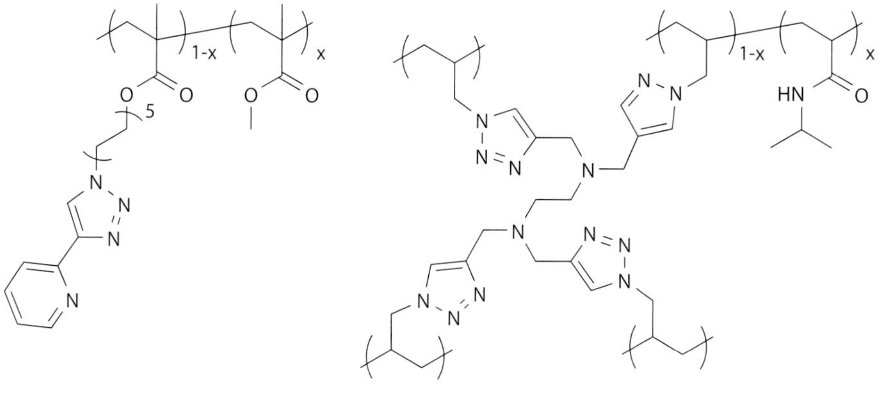

FIGURE 5.8
Chemical structures of side-chain-type poly(triazole) and highly cross-linked poly(triazole) gel with the recognition abilities of metal ions.

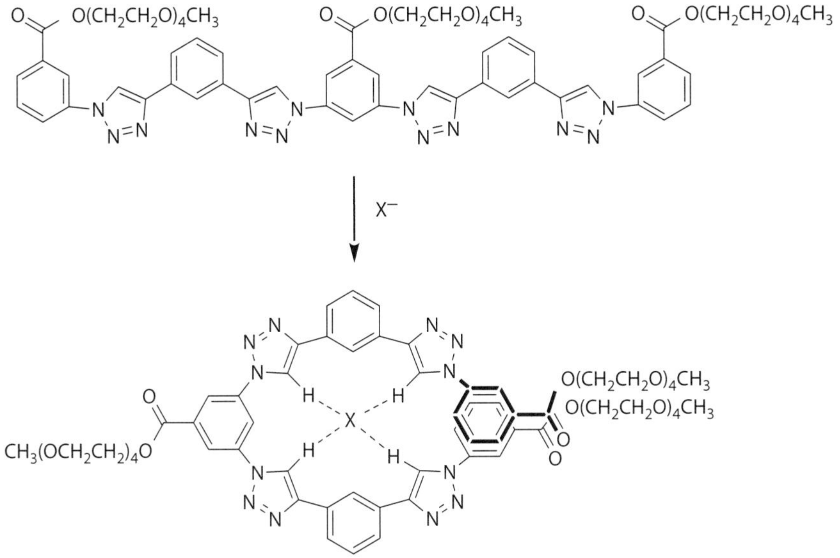

FIGURE 5.9
Size-dependent anion recognition of oligo(triazole). (Adapted with permission from Juwarker et al., 8924–8934. Copyright 2009 American Chemical Society.)

the highest binding constant due to the best match with the preorganized rigid macrocyclic receptor in addition to supramolecular interaction with electropositive 1,2,3-triazole CH protons.

Thiol-ene or thiol-yne reactions and Diels–Alder reactions are classified as second-generation click reactions. However, there are, to our surprise, few reports on the use of these click reactions for the preparation of polymer ion sensors. This is probably due to the inert characteristics of the new covalent bonds, that is, the S–C bond for the thiol-ene or thiol-yne reactions and the C–C bond for the Diels–Alder reactions, formed through these click reactions.

5.4 Nonplanar Donor–Acceptor Chromophores

Recently, a completely new type of click reaction was proposed as a third-generation type. These reactions are based on a [2+2] cycloaddition followed by electrocycloreversion between alkynes activated by electron-donating units and electron-deficient olefins of tetracyanoethylene (TCNE) or 7,7,8,8-tetracyanoquinodimethane (TCNQ). Diederich and coworkers systematically investigated the reactivity of alkyne molecules and found that aromatic amines, as represented by *N,N*-dialkylaniline, are one of the best substituents for the effective activation of the alkyne moiety [28]. Later, the pursuit of other donors revealed that thiophene, tetrathiafulvalene, porphyrin, azulene, and some organometallic substances are also effective for the activation of the alkyne moiety (Figure 5.10) [29]. Furthermore, it was also shown that substituted TCNQ derivatives, such as 2,3,5,6-tetrafluoro-TCNQ and 2,5-dialkoxy-TCNQ, can be employed as additional effective acceptors of these reactions [30]. The survey using the combination of small alkyne molecules and TCNE suggested that these reactions proceed under mild conditions in quantitative yields, thus ensuring the click reactions. A noticeable feature of these click

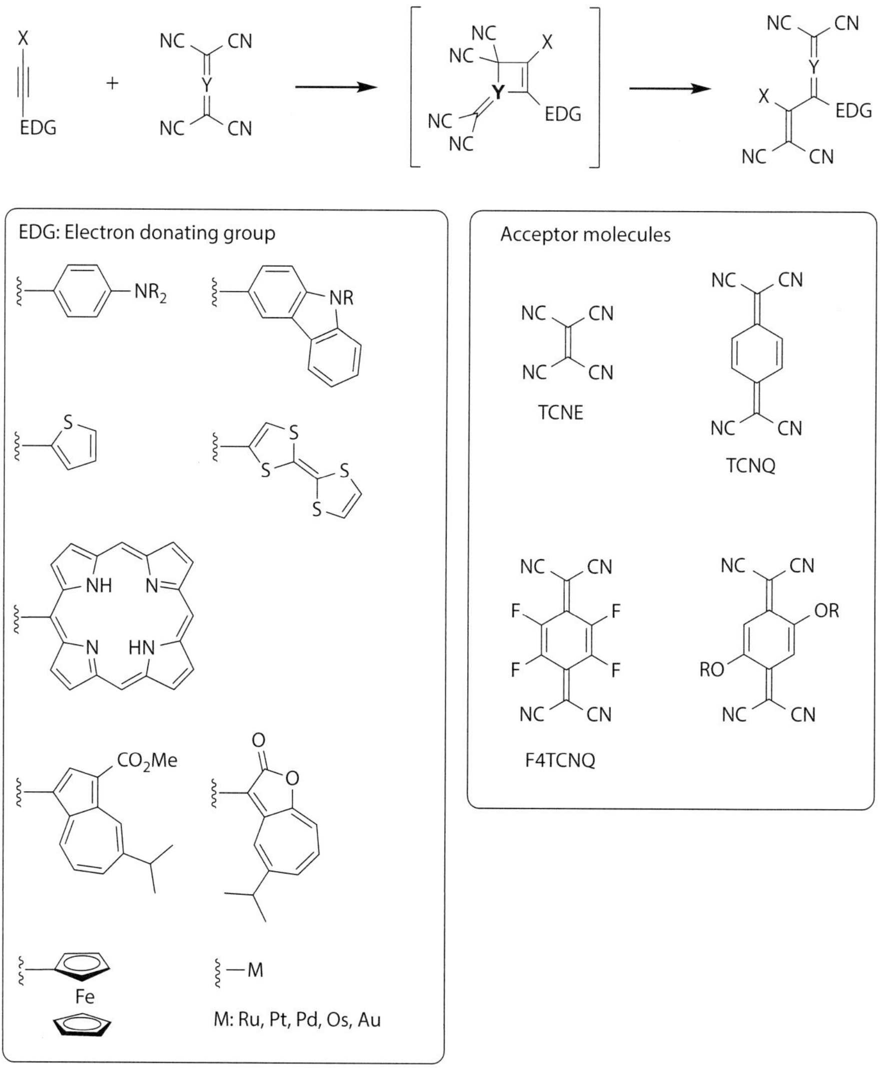

FIGURE 5.10
New types of click reactions based on cycloaddition–electrocycloreversion between electron donor–activated alkynes and electron-deficient olefins of acceptor units.

reactions, referred to as alkyne-acceptor click reactions, is the visible charge-transfer (CT) bands of the resulting donor–acceptor type products.

It should be noted that other conventional click reactions simply connect two different components through covalent bonds, and the linker moieties do not produce a well-defined visible absorption. In contrast, the *N,N*-dialkylaniline-substituted 1,1,4,4-tetracyanobutadienes (TCBDs) (TCNE adducts) and cyclohexa-2,5-diene-1,4-diylidene-expanded TCBDs (TCNQ adducts) are usually red- and green-colored compounds, respectively. The nitrogen atoms of these donor–acceptor chromophores are expected to serve as effective recognition sites of some specific ions. For example, the metal ion recognition through ion-dipole interactions affects the extent of the intramolecular CT interactions and, accordingly, the visual detection of the recognition events is realized. In Sections 5.5 and 5.6, examples of the visual detection of some metal ions and anions are introduced.

5.5 Colorimetric Polymer Sensors of Metal Ions

Michinobu and coworkers successfully applied the new alkyne-acceptor click reactions to the postfunctionalization of polymers [31]. A typical example is polystyrene derivatives bearing donor–acceptor side-chain chromophores, prepared by the radical polymerization of styrene monomers followed by the click addition of TCNE or TCNQ (Figure 5.11) [32,33]. Proof of the quantitative acceptor addition was achieved by combined analytical techniques including ^{1}H nuclear magnetic resonance and absorption spectra as well as microanalytical data. For example, the precursor polystyrene has a λ_{max} of 329 nm in 1,2-dichlorobenzene, whereas the TCNE and TCNQ adducts displayed bathochromically shifted λ_{max} values at 483 and 727 nm, respectively. The intensities of these bathochromically shifted absorption bands linearly increased as more acceptors were added, but they eventually saturated when a stoichiometric amount of acceptors was added.

The formed donor–acceptor moieties contain two different nitrogen atoms, namely, hard basic aniline nitrogen and soft basic cyano nitrogen atoms. It is generally known that hard bases prefer to bind with hard acids (metal ions), whereas soft bases tend to interact with soft acids (metal ions). Thus, the donor–acceptor units formed through the alkyne-acceptor click reactions would have two different modes of metal ion recognition. When hard metal ions were added to the solution of donor–acceptor chromophore-appended polystyrenes in CHCl$_3$, most of them unfortunately showed negligible spectral changes. However, some hard metal ions, that is, Fe^{3+}, Sn^{2+}, and Fe^{2+} ions for the TCNE-adducted polystyrene and Fe^{3+}, Cu^{2+}, Ti^{4+}, and Sc^{3+} ions for the TCNQ-adducted polystyrene, led to a dramatic decrease in the visible CT bands (Figure 5.11). Considering the fact that the spectral changes were similar to the protonation event of the aniline nitrogen moiety by the addition of

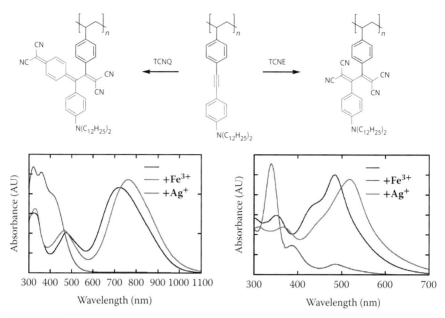

FIGURE 5.11
(See color insert.) Click postfunctionalization of polystyrene derivative by TCNE and TCNQ and their spectral changes in response to metal ion recognition. (Adapted with permission from Li et al., 1996–2005, Copyright 2012 The Royal Chemical Society.)

protonic acids, it is postulated that these metal ions were coordinated to the aniline nitrogen of the polystyrene side chains. Among the detected metal ions, Fe^{3+} ion produced the most significant spectral change, indicating the largest binding constant, K. The K values estimated for small-molecular-weight counter units indeed supported this result.

The detection mode of soft metal ions was different from that of the hard metal ions, such as the Fe^{3+} ion. Among the examined soft metal ions, only the Ag^+ ion displayed a spectral change. Due to the soft acidity of this metal ion, the recognition occurred at the softly basic cyano nitrogen atoms. As a consequence, a well-defined bathochromic shift in the CT bands occurred because the extent of intramolecular CT was altered by the Ag^+ coordination to the cyano groups. These spectral changes correspond to the color change from orange to pink for the TCNE-adducted polystyrene and from green to yellowish green for the TCNQ-adducted polystyrene (Figure 5.11). An additional investigation using small-molecular-weight counter units revealed that the Ag^+ ion recognition by the cyano groups is assisted by the cooperative polymer effect. In other words, multivalent coordination of the side-chain cyano groups to the Ag^+ ion enabled selective and rapid ion recognition. It should be noted that the precursor polystyrene did not show any metal ion recognition events despite the presence of an aniline nitrogen atom. Thus, the clicked polystyrenes are a readily available clever molecular design to distinguish soft and hard metal ions by visual color changes.

There is another design approach to prepare polymer ion sensors by these click reactions. A precursor polymer containing the TCNQ moiety in the main chain, namely, the TCNQ polyester, was selected as a precursor polymer (Figure 5.12). The TCNQ moieties were quantitatively converted into the donor–acceptor chromophore by the addition of the small alkyne molecule $(H_3C)_2NPhC\equiv CPhN(CH_3)_2$ [34]. Despite the possibilities of E- and Z-isomer products, all the spectroscopic characterization and computational calculations suggested the significantly stabilized Z-form as the single product structure. The λ_{max} of the precursor TCNQ polyester was 412 nm in 1,2-dichlorobenzene, whereas the postfunctionalized TCNQ polyester exhibited the longer wavelength λ_{max} at 623 nm due to the intramolecular CT interaction. The addition of Fe^{3+} and Ag^+ ions to the solution of the precursor TCNQ polyester did not change the absorption spectrum due to the absence of recognition abilities of both cyano and aniline nitrogen atoms. However, metal ion sensitivity similar to that of side-chain-type polymer ion sensors occurred when the TCNQ polyester was functionalized by the alkyne-acceptor addition reaction. The CT band intensity at 623 nm decreased upon addition of Fe^{3+} ions, whereas the band bathochromically shifted upon inter-action with Ag^+ ions [35]. These ion-specific spectral changes were similar to those observed for the TCNQ-adducted polystyrene derivative. Previously, the redox activities of TCNQ polymer films were used to detect metal ions in solutions [36]. However, the repeated use of the polymer films was hindered by the limited chemical stability of the reduced TCNQ moieties. In contrast to this result, all the spectral changes of the clicked polymer sensors fully

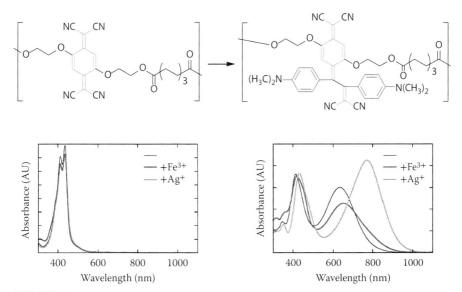

FIGURE 5.12
(See color insert.) Click postfunctionalization of the TCNQ polyester by dialkylaniline-appended alkynes and their spectral changes in response to metal ion recognition.

recovered to their original state by treatment with basic solutions. This feature highlights the advantages of colorimetric chemosensors without the use of expensive instrumentation.

5.6 Chemodocimetric Anion Sensors

The nonplanar donor–acceptor chromophores produced by the alkyne-acceptor click reactions were responsive to some anions. However, in contrast to the recognized metal ions, anions underwent nucleophilic addition to the donor–acceptor units, resulting in color changes. These detection modes are classified as chemodocimetric sensing.

The anion recognition ability of both TCNE-adducted and TCNQ-adducted polystyrenes was investigated in THF. Among the examined anions, it was found that both polystyrene derivatives showed colorimetric sensor performances when CN^-, F^-, and I^- ions were added. On the addition of these anions, the solutions of the polystyrene derivatives gradually became discolored. This was due to the nucleophilic addition of these anions to the donor–acceptor units, leading to the disruption of intramolecular CT interactions (Figure 5.13) [37]. The CT band decrease was most significant

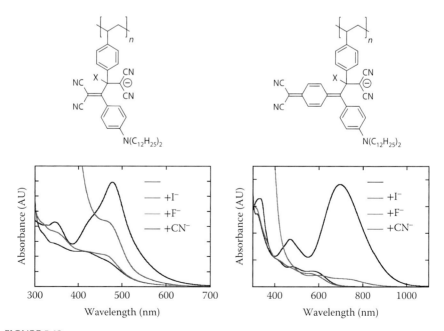

FIGURE 5.13
Spectral changes of clicked polystyrene derivatives on chemodocimetric anion recognition. (Adapted with permission from Michinobu and Hyakutake, 2623–2631, Copyright 2013 The Royal Chemical Society.)

when CN^- ion was added. It was postulated that the attack position of these anions was the dicyanovinyl moieties. The Job plots of the titration experiments clearly suggested 1:1 binding stoichiometry. However, as the small-molecular-weight counter units display a similar chemodocimetric sensing behavior of these anions, no noticeable advantages of using polymer ion sensors were proposed.

5.7 Conclusions and Future Perspectives

This chapter described the historical development of polymer ion sensors. Starting from conjugated polymer sensors, the recent trend toward the straightforward and atom-economic syntheses of polymer sensors was introduced. In this context, one of the most well-studied click reactions, the CuAAC, eventually found ion sensor applications through a molecular design suitable for intermolecular ion-dipole interactions. Finally, novel types of click reactions based on cycloaddition–electrocycloreversion between electron-rich alkynes and electron-deficient olefins were shown. These click reactions are effective tools in preparing highly colored polymers by post-functionalization, and they successfully afforded ion sensing abilities of the donor–acceptor products. In particular, visual detection of toxic metal ions or anions would be highly convenient and general goals.

The future of polymer ion sensors is bright due to the rapid growth of click chemistry research. Click reactions proceed very efficiently even at solid–liquid interfaces. This fact opens the door to the functionalization of polymer thin films and material surfaces. For example, covalently bonded layer-by-layer techniques using click reactions were applied to the preparation of thin films and capsules with controlled thicknesses. It is expected that these bottom-up nanofabrication techniques will contribute to the development of next-generation high-performance sensor devices.

References

1. Kim, H.M. and Cho, B.R. (2009) Two-Photon Probes for Intracellular Free Metal Ions, Acidic Vesicles, and Lipid Rafts in Live Tissues. *Acc. Chem. Res.*, 42, 863–872.
2. Que, E.L. and Chang, C.J. (2010) Responsive Magnetic Resonance Imaging Contrast Agents as Chemical Sensors for Metals in Biology and Medicine. *Chem. Soc. Rev.*, 39, 51–60.
3. Michinobu, T., Shinoda, S., Nakanishi, T., Hill, J.P., Fujii, K., Player, T.N., Tsukube, H., and Ariga, K. (2006) Mechanical Control of Enantioselectivity of Amino Acid Recognition by Cholesterol-Armed Cyclen Monolayer at the Air-Water Interface. *J. Am. Chem. Soc.*, 128, 14478–14479.

4. Mori, T., Okamoto, K., Endo, H., Hill, J.P., Shinoda, S., Matsukura, M., Tsukube, H., Suzuki, Y., Kanekiyo, Y., and Ariga, K. (2010) Mechanical Tuning of Molecular Recognition to Discriminate the Single-Methyl-Group Difference between Thymine and Uracil. *J. Am. Chem. Soc.*, 132, 12868–12870.

5. Michinobu, T., Shinoda, S., Nakanishi, T., Hill, J.P., Fujii, K., Player, T.N., Tsukube, H., and Ariga, K. (2011) Langmuir Monolayers of a Cholesterol-Armed Cyclen Complex That Can Control Enantioselectivity of Amino Acid Recognition by Surface Pressure. *Phys. Chem. Chem. Phys.*, 13, 4895–4900.

6. Ariga, K., Ito, H., Hill, J.P., and Tsukube, H. (2012) Molecular Recognition: From Solution Science to Nano/Materials Technology. *Chem. Soc. Rev.*, 41, 5800–5835.

7. Marsella, M.J. and Swager, T.M. (1993) Designing Conducting Polymer-Based Sensors: Selective Ionochromic Response in Crown Ether-Containing Polythiophenes. *J. Am. Chem. Soc.*, 115, 12214–12215.

8. Scheib, S. and Bäuerle, P. (1999) Synthesis and Characterization of Oligo- and Crown Ether-Substituted Polythiophenes-A Comparative Study. *J. Mater. Chem.*, 9, 2139–2150.

9. Rimmel, G. and Bäuerle, P. (1999) Molecular Recognition Properties of Crown Ether-Functionalized Oligothiophenes. *Synth. Met.*, 102, 1323–1324.

10. Wang, B. and Wasielewski, M.R. (1997) Design and Synthesis of Metal Ion-Recognition-Induced Conjugated Polymers: An Approach to Metal Ion Sensory Materials. *J. Am. Chem. Soc.*, 119, 12–21.

11. Thomas, S.W. III, Joly, G.D., and Swager, T.M. (2007) Chemical Sensors Based on Amplifying Fluorescent Conjugated Polymers. *Chem. Rev.*, 107, 1339–1386.

12. Kimura, M., Horai, T., Hanabusa, K., and Shirai, H. (1998) Fluorescence Chemosensor for Metal Ions Using Conjugated Polymers. *Adv. Mater.*, 10, 459–462.

13. Zhang, Y., Murphy, C.B., and Jones, W.E. Jr. (2002) Poly[*p*-(phenyleneethynylene)-*alt*-(thienyleneethynylene)] Polymers with Oligopyridine Pendant Groups: Highly Sensitive Chemosensors for Transition Metal Ions. *Macromolecules*, 35, 630–636.

14. Kolb, H.C., Finn, M.G., and Sharpless, K.B. (2001) Click Chemistry: Diverse Chemical Function from a Few Good Reactions. *Angew. Chem. Int. Ed.*, 40, 2004–2021.

15. Qin, A., Lam, J.W.Y., and Tang, B.Z. (2010) Click Polymerization: Progresses, Challenges, and Opportunities. *Macromolecules*, 43, 8693–8702.

16. Qin, A., Lam, J.W.Y., and Tang, B.Z. (2010) Click Polymerization. *Chem. Soc. Rev.*, 39, 2522–2544.

17. Qin, A., Tang, L., Lam, J.W.Y., Jim, C.K.W., Yu, Y., Zhao, H., Sun, J., and Tang, B.Z. (2009) Metal-Free Click Polymerization: Synthesis and Photonic Properties of Poly(aroyltriazole)s. *Adv. Funct. Mater.*, 19, 1891–1900.

18. Li, H., Wang, J., Sun, J.Z., Hu, R., Qin, A., and Tang, B.Z. (2012) Metal-Free Click Polymerization of Propiolates and Azides: Facile Synthesis of Functional Poly(aroxycarbonyltriazole)s. *Polym. Chem.*, 3, 1075–1083.

19. Schweinfurth, D., Hardcastle, K.I., and Bunz, U.H.F. (2008) 1,3-Dipolar Cyclo-addition of Alkynes to Azides. Construction of Operationally Functional Metal Responsive Fluorophores. *Chem. Commun.*, 2203–2205.

20. Brombosz, S.M., Appleton, A.L., Zappas, A.J. II, and Bunz, U.H.F. (2010) Water-Soluble Benzo- and Naphtho-Thiadiazole-Based Bistriazoles and Their Metal-Binding Properties. *Chem. Commun.*, 46, 1419–1421.

21. Bryant, J.J. and Bunz, U.H.F. (2013) Click to Bind: Metal Sensors. *Chem. Asian J.*, 8, 1354–1367.

22. Yuan, J., Fang, X., Zhang, L., Hong, G., Lin, Y., Zheng, Q., Xu, Y., Ruan, Y., Weng, W., Xia, H., and Chen, G. (2012) Multi-Responsive Self-Healing Metallo-Supramolecular Gels Based on "Click" Ligand. *J. Mater. Chem.*, 22, 11515–11522.

23. Huang, X., Meng, J., Dong, Y., Cheng, Y., and Zhu, C. (2010) Polymer-Based Fluorescence Sensor Incorporating Triazole Moieties for Hg^{2+} Detection via Click Reaction. *Polymer*, 51, 3064–3067.

24. Meudtner, R.M. and Hecht, S. (2008) Responsive Backbones Based on Alternating Triazole-Pyridine/Benzene Copolymers: From Helically Folding Polymers to Metallosupramolecularly Crosslinked Gels. *Macromol. Rapid Commun.*, 29, 347–351.

25. Happ, B., Pavlov, G.M., Perevyazko, I., Hager, M.D., Winter, A., and Schubert, U.S. (2012) Induced Charge Effect by Co(II) Complexation on the Conformation of a Copolymer Containing a Bidentate 2-(1,2,3-Triazol-4-yl)pyridine Chelating Unit. *Macromol. Chem. Phys.*, 213, 1339–1348.

26. Watanabe, W., Maekawa, T., Miyazaki, Y., Kida, T., Takeshita, K., and Mori, A. (2012) Quadruple Click: A Facile Pathway Leading to Tetrakis(4-(1,2,3-triazolyl) methyl)ethylenediamine Derivatives as a New Class of Extracting Agents for Soft Metal Ions. *Chem. Asian J.*, 7, 1679–1683.

27. Juwarker, H., Lenhardt, J.M., Castillo, J.C., Zhao, E., Krishnamurthy, S., Jamiolkowski, R. M., Kim, K.-H., and Craig, S.L. (2009) Anion Binding of Short, Flexible Aryl Triazole Oligomers. *J. Org. Chem.*, 74, 8924–8934.

28. Michinobu, T., Boudon, C., Gisselbrecht, J.-P., Seiler, P., Frank, B., Moonen, N.N.P., Gross, M., and Diederich, F. (2006) Donor-Substituted 1,1,4,4-Tetracyanobutadienes (TCBDs): New Chromophores with Efficient Intramolecular Charge-Transfer Interactions by Atom-Economic Synthesis. *Chem. Eur. J.*, 12, 1889–1905.

29. Kato, S.-i. and Diederich, F. (2010) Non-Planar Push–Pull Chromophores. *Chem. Commun.*, 46, 1994–2006.

30. Kivala, M., Boudon, C., Gisselbrecht, J.-P., Enko, B., Seiler, P., Müller, I.B., Langer, N., Jarowski, P.D., Gescheidt, G., and Diederich, F. (2009) Organic Super-Acceptors with Efficient Intramolecular Charge-Transfer Interactions by [2+2] Cycloadditions of TCNE, TCNQ, and F4-TCNQ to Donor-Substituted Cyanoalkynes. *Chem. Eur. J.*, 15, 4111–4123.

31. Michinobu, T. (2011) Adapting Semiconducting Polymer Doping Techniques to Create New Types of Click Postfunctionalization. *Chem. Soc. Rev.*, 40, 2306–2316.

32. Li, Y., Ashizawa, M., Uchida, S., and Michinobu, T. (2011) A Novel Polymeric Chemosensor: Dual Colorimetric Detection of Metal Ions Through Click Synthesis. *Macromol. Rapid Commun.*, 32, 1804–1808.

33. Li, Y., Ashizawa, M., Uchida, S., and Michinobu, T. (2012) Colorimetric Sensing of Cations and Anions by Clicked Polystyrenes Bearing Side Chain Donor–Acceptor. *Polym. Chem.*, 3, 1996–2005.

34. Washino, Y., Murata, K., Ashizawa, M., Kawauchi, S., and Michinobu, T. (2011) Creation of Persistent Charge-Transfer Interactions in TCNQ Polyester. *Polym. J.*, 43, 364–369.
35. Washino, Y. and Michinobu, T. (2012) Emergence of Colorimetric Chemosensor Ability of Metal Ions in TCNQ Polyester by Postfunctionalization. *J. Photopolym. Sci. Technol.*, 25, 267–270.
36. Shelton, R.D. and Chambers, J.Q. (1991) Preconcentration and Voltammetric Determination of Silver at TCNQ Polymer Film Electrodes. *J. Electroanal. Chem.*, 305, 217–228.
37. Michinobu, T., Li, Y., and Hyakutake, T. (2013) Polymeric Ion Sensors with Multiple Detection Modes Achieved by a New Type of Click Chemistry Reaction. *Phys. Chem. Chem. Phys.*, 15, 2623–2631.

6

Detection of Explosives and Metal Ions Using Fluorescent Polymers and Their Nanostructures

Akshay Kokil and Jayant Kumar

CONTENTS

6.1 Introduction

Sensors are devices that respond to external stimuli and convert them to a signal, which can be measured and recorded. In chemical sensors, the presence of specific chemical compounds or certain chemical reactions is detected through a variety of responses. Detection of environmentally toxic chemicals, especially metal ions, and chemicals that present a threat to safety and security, such as those used in explosive devices, has become increasingly important. A variety of nonoptical sensing platforms are being utilized. However, optical sensors, especially those that display a response in fluorescence, possess a matrix of beneficial attributes (see Section 6.1.1.3) and are at the forefront of technological investigations worldwide. This chapter focuses on the fluorometric detection of these analytes, that is, explosives and metal ions.

6.1.1 Conventional Techniques

6.1.1.1 Chemical Explosives

Different mixtures of chemical compounds that consist of nitrated organic molecules or peroxides are typically used in explosive devices [1]. The most commonly used explosive compounds are trinitrotoluene (TNT) and dinitrotoluene (DNT), and thus the detection of these compounds has been the focus of numerous investigations. Pentaerythritoltetranitrate and cyclotrimethylenetrinitramine (RDX) are also commonly used explosive compounds. However, the relatively low volatility of these chemicals makes their detection a challenge [1]. Hence, more volatile explosive compounds, such as DNT and TNT, with equilibrium vapor pressures around 100 and 5 ppb, respectively, have attracted significant attention [1]. DNT is typically present in TNT as an impurity and is often a target analyte due to its high vapor pressure [2]. The peroxide-based explosive compounds are strong oxidizers, which can result in a challenge as many of the common chemicals used in cleaning products can act as interferents and result in false positives [3].

Conventional nonoptical screening techniques used in the detection of these explosives include ionization mass spectrometry (IMS), x-ray dispersion, canine olfaction, and vibrational spectroscopy [4–6]. IMS can be highly sensitive as the samples collected are directly analyzed for the molecular mass. Typically, the measured signal is compared against a library of standards for positive identification. This technique is extremely sensitive and accurate when targeting clean samples of a few analytes. However, the instrumentation utilized in this technique is not portable and requires accurate calibration, which limits its scope and utility [7,8]. X-rays, which penetrate baggage and clothing, are used for obtaining high-resolution images [1]. These instruments are effective in detecting concealed devices; however, they are unable to detect any chemical residue.

Canine olfaction is a highly sensitive technique; research has shown that the canine olfactory system is four times larger than a human's and is highly responsive to trace vapor [9]. Well-trained canines can easily discriminate between target molecules [10]. However, this method is highly dependent on the training of canines and its extensive use can be a challenge. Vibrational spectroscopy is a well-developed method for explosives detection. However, limited specificity and false positives are its limitations. Raman spectroscopy and surface-enhanced Raman spectroscopy are promising techniques for sensitive detection; however, the instrumentation can be cumbersome and the detection requires a large sample size [4,11].

6.1.1.2 Metal Ions

The determination of metal ions is challenging with respect to the concentration ranges set by standards and guidelines for reasons of toxicity. In addition, the specificity of detection can also be a challenge. Conventional techniques used for the detection of toxic metals in aqueous samples include atomic absorption spectroscopy (AAS), inductively coupled plasma emission or mass spectrometry (ICP-ES or ICP-MS), total reflection x-ray fluorimetry (TXRF), and anodic stripping voltammetry [12,13]. AAS can be used for the detection of a single element. ICP-ES, ICP-MS, and TXRF can be readily utilized for multielement analysis. These techniques offer low detection limits; however, they can be prohibitive with regard to the instrumentation cost. The pretreatment of a sample also requires a considerable amount of time and can be a potential source of error. In the last decade portable and less expensive devices have been developed [14], but the cost of such portable devices limits the widespread use of these techniques for determination of toxic metal ions in potable groundwater.

6.1.1.3 Detection Using Optical Techniques

Sensors that display a change in color on detection of the analyte are termed colorimetric sensors. Such sensors can find widespread use due to their rapid detection, simple handling, and ease of interpretation [15]. However, such sensors can lack specificity and can be affected by many interferents, which result in false negatives or positives.

Fluorescence is the emission of photons following relaxation from an excited electronic state to the ground state. Sensors that display a change in fluorescence of the active material are the most promising and are being widely investigated [16]. A matrix of beneficial attributes is present in these sensors, such as the following:

- Sensitivity: Typically, the fluorescence signal is measured on a low background. The detection limit depends mostly on the properties of the analyte, and parts per billion or in some cases parts per trillion of the analytes can be readily detected. A small amount of untreated sample can be utilized for such sensitive detection.

- Specificity: Detection through fluorescence can be highly specific as many of the analytes do not absorb light in the same region as the sensing material. Specificity of the sensor can also be enhanced by readily tailoring the structure of sensing materials.

- Wide concentration range: Detection of a change in fluorescence can be measured over a wide concentration range of the analyte.

- Simplicity and speed: Fluorometry is a relatively simple analytical technique, and calibration is not required in most cases. The response of the sensor is also fast compared to many of the techniques discussed in Sections 6.1.1.1 and 6.1.1.2. Remote sensing is also possible with this technique.

- Low cost: Materials and instrumentation costs are relatively low compared to many other analytical techniques. Portable sensing modules can be fabricated at relatively low costs.

Some limitations of the fluorescence detection technique include photodegradation and photobleaching of the sensing materials, slow response times due to analyte diffusivity, as well nonspecific response due to interferents. Two different types of fluorescence detection can be utilized, one in which fluorescence intensity quenches (decreases) on detection (*turn-off*) and the other in which the fluorescence intensity increases on detection (*turn-on*).

6.2 Fluorescent Chemosensors

Fluorescent chemosensors are usually made up of three components: a receptor, a fluorophore, and in some cases a spacer linking them together (Figure 6.1). The receptor and the fluorophore in most of the fluorescent chemosensors are the same moiety in the sensing material. Typically, the spacer is not responsible for signal transduction. The readout of a fluorescent chemosensor is usually measured as a change in fluorescence intensity, a change in excited state lifetime, or a shift of the emission wavelength. In fluorescent chemosensors, the typical response times after the analyte binding event are short and the response occurs without any other intermediary process. This makes real-time and real-space detection of the analyte possible.

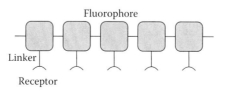

FIGURE 6.1
(See color insert.) Schematic of a typical fluorescent chemosensor.

6.2.1 Mechanism

There are several mechanisms of fluorescence sensing. Photoinduced electron transfer (PET) and electronic energy transfer (EET) are mechanisms that have been extensively studied and widely used in the design of chemosensors [16,17]. These mechanisms either result in a change in the fluorescence intensity or a change in the emission wavelength.

6.2.1.1 Photoinduced Electron Transfer

PET is the underlying mechanism in two categories of sensors: fluorescence turn-on or fluorescence turn-off. For fluorescence turn-on sensors, in the absence of analytes, the emission is quenched by rapid intramolecular electron transfer from the receptor to the excited fluorophore. On binding of the analyte, the highest occupied molecular orbital (HOMO) of the receptor is lowered. This decreases the driving force for the PET process, effectively turning on the fluorescence of the chromophore. In some cases, the receptor takes part only indirectly in the photophysical process. If the lowest unoccupied molecular orbital (LUMO) energy level of the analyte is between the energy levels of the fluorophore HOMO and LUMO, the analyte binding provides a nonradiative path to dissipate the excitation energy, resulting in a quenching of the fluorescence of the chemosensor (Figure 6.2). The difference between these two mechanisms is that the PET process takes place either before or after the analyte binding.

6.2.1.2 Electronic Energy Transfer

Energy transfer is another mechanism for fluorescence-based sensing. Two kinds of EET mechanisms result in the quenching of fluorescence or a change in fluorescence wavelength [17]. The double electron exchange (Dexter) energy transfer results in fluorescence quenching, or the dipole–dipole coupling (Förster) energy transfer results in a change in fluorescence wavelength. In the case of Dexter energy transfer, the fluorophore goes back to its ground state by nonradiative decay (Figure 6.3). The Dexter energy transfer requires a direct orbital overlap and close contact between the fluorophore and the analytes. This type of fluorescence quenching requires appropriate relative energy levels between fluorophore and analyte. The Förster resonance energy transfer

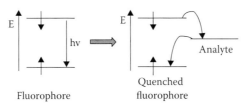

FIGURE 6.2
Mechanism for photoinduced electron transfer.

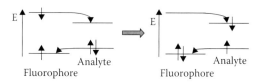

FIGURE 6.3
Mechanism for Dexter energy transfer.

mechanism involves long-range coupling of dipoles, allowing an exchange of excitation energy through space, that is, without direct orbital overlap. These two mechanisms of energy transfer differ primarily by their dependence on the distance between donor and acceptor.

6.2.1.3 Efficacy of Fluorescence Quenching

Generally, fluorescence quenching occurs through dynamic and/or static quenching processes [17]. Dynamic quenching, which is also called collisional quenching, depends on diffusion of the quencher. Dynamic quenching occurs when the excited state is deactivated on a diffusive encounter with the analyte (A) and relaxes to ground state without emission of a photon. Static quenching occurs through the formation of a stable nonfluorescent complex in the ground state. When this complex absorbs light, it immediately returns to the ground state without fluorescence emission. Because complex formation occurs in the ground state, the efficiency of static quenching is related to the association constant for the formation of the nonfluorescent complex.

Dynamic or static can be differentiated using several routes. In the first route, the dependence of fluorescence lifetime on the analyte concentration is determined. In dynamic quenching, the presence of an additional nonradiative pathway shortens the fluorescence lifetime [17]. In contrast, in static quenching the fluorescence lifetime is independent of the quencher concentration because the nonradiative complex formation takes place in the ground state. Uncomplexed fluorophore exhibits regular fluorescence behavior with the same lifetime. The effect of temperature on quenching efficiency can distinguish between the two mechanisms. Due to the increased diffusion of the analyte at a higher temperature, dynamic quenching is expected to increase with increasing temperature. In contrast, an increase in temperature would likely result in a decrease in the nonradiative complex stability, resulting in reduced static quenching. In static quenching, the nonradiative fluorophore–analyte complexes typically exhibit different absorption due to intermolecular interaction in the ground state [17], but in dynamic quenching such a change in absorption characteristics of the fluorophore is not observed. Hence, a change in the absorption of the fluorophore in the presence of the analyte can be indicative of a static quenching mechanism. However, both dynamic and static quenching processes can occur simultaneously in many systems.

6.2.1.4 Stern–Volmer Equation

Fluorescence quenching can be described by the Stern–Volmer (SV) equation, which is given by [17]:

$$I_0/I = 1 + K_{SV}[A] \tag{6.1}$$

where I_0 and I are the fluorescence intensities observed in the absence and presence of the quencher, respectively; [A] is the analyte concentration; and K_{SV} is the SV quenching constant. Hence, K_{SV} directly indicates the efficacy of analyte detection by the fluorophore. In dynamic quenching, the fluorescence intensity is quenched to same extent as the fluorescence lifetime. As a result, the SV equation can be written as follows:

$$\tau_0/\tau = I_0/I = 1 + k_q\tau_0[A] = 1 + K_D[A] \tag{6.2}$$

where τ_0 and τ are the fluorescence lifetimes measured without and with the analyte, respectively; k_q is the bimolecular quenching constant; and K_D is used to represent K_{SV} in the dynamic quenching process. The lifetime of the fluorophore in static quenching is not affected by the presence of the analyte, and the ratio of τ_0 to τ is 1. A plot of I_0/I versus [A] is known as an SV plot; it is typically a straight line with an intercept of 1 and a slope equal to K_{SV} for both pure dynamic and static quenching. However, a nonlinear SV plot indicates that in the presence of the same analyte the fluorophore is quenched by both dynamic and static quenching. In such circumstances, the original SV equation is modified to represent both the mechanisms:

$$I_0/I = (1 + K_D[A])(1 + K_S[A]) \tag{6.3}$$

This SV equation is second order in [A], and it combines the effects of both dynamic and static quenching. In such a case, the dynamic part of the observed quenching can be determined by a change in the excited state lifetime using Equation 6.2.

6.3 Conjugated Polymers

Conjugated polymers (CPs) have become an important class of materials in a wide variety of applications, including light-emitting diodes [18], solar cells [19], and field effect transistors [20]. Because of their ready tailorability and interesting photophysical properties, this class of materials has attracted considerable interest for applications in chemical or biological sensing [16,21–23]. The chemical structures of CPs can be tuned to obtain increased sensitivity. Depending on their structure, CPs exhibit strong absorption and fluorescence characteristics [24]. CP-based sensors have been fabricated in a variety of forms depending on the types of signals, such as electrical conductivity [25,26], chemical potential [27], absorption [28], and fluorescence [3,16,21–23].

As shown in Figure 6.4, various families of CPs have been synthesized, such as poly(*p*-phenylene) (PPP) [29], poly(*p*-phenylenevinylene) (PPV) [30], poly(*p*-phenyleneethynylene) (PPE) [31], polythiophene (PT) [32], polypyrrole (PPy) [33], polyaniline (PANI) [34], and polyfluorene (PF) [35].

6.3.1 Conjugated Polymers versus Small Molecules

A variety of small molecules (fluorescent dyes) have been extensively investigated as active materials in fluorescence sensors [36–41]. In CPs, fast transport of the photoexcited bound electron–hole pairs (excitons) along the conjugated backbone contributes to the increased sensitivity [42–44]. As opposed to the small molecules, fast exciton transport along the CP backbone results in *amplified quenching*, which aids in improving the detection limit as well as the response time [45,46].

The concept of amplified quenching in CPs was first described by Swager and coworkers in 1995 [45,46]. They reported that the fluorescence of a series of PPEs was efficiently quenched by methyl viologen salt (MV^{2+} or *N,N*-dimethyl-4,4-bipyridinium) (Figure 6.5). They investigated a fluorescent PPE model compound with a single cyclophane receptor and a PPE with the cyclophane moiety in each repeat unit (Figure 6.5). On interaction of MV^{2+} with the cyclophane moiety, the fluorescence of both the model compound and the polymer quenched through a static quenching process. However, compared

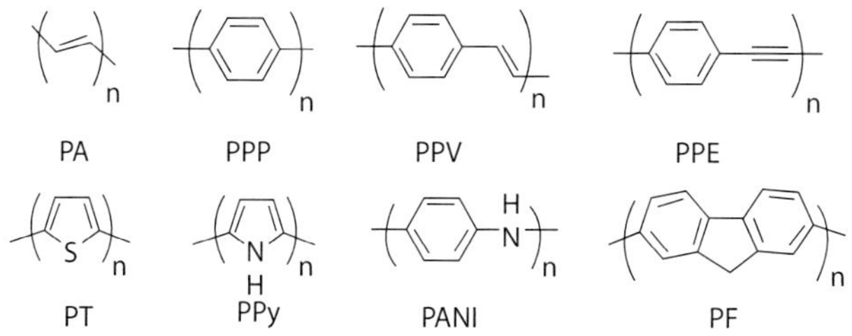

FIGURE 6.4
Chemical structures of different families of conjugated polymers.

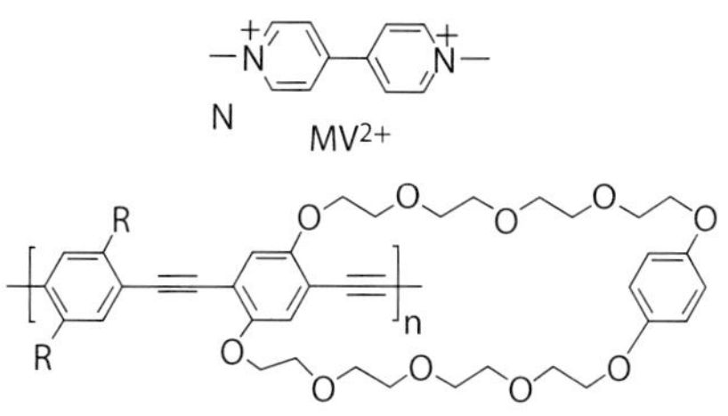

FIGURE 6.5
Chemical structure of the poly(phenylene ethynylene) derivative utlilized for detection of methyl viologen salt.

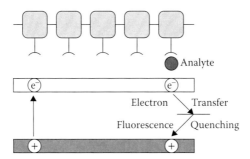

FIGURE 6.6
(See color insert.) Schematic for fluorescence quenching through the molecular wire effect.

to the model compound the PPE polymer exhibited a 67-fold enhancement in quenching efficiency with a K_{SV} approaching 10^5 M^{-1}. Swager attributed the enhanced fluorescence quenching of PPE to extended electronic communication and exciton transport on the CP chain, which was termed the *molecular wire effect* [47]. As shown in Figure 6.6, excitons can be generated randomly and migrate along the polymer chain. Each exciton passes multiple cyclophane receptors until it encounters an MV^{2+} bound receptor, where the excitation is quenched. Vicinity of multiple polymer chains also results in the transfer of this exciton from one chain to another. As a result, a single MV^{2+} binding event is able to quench multiple repeat units. Therefore, the fluorescence quenching response of the PPE polymer to MV^{2+} is greatly amplified.

Zhou and Swager reported that the amplified quenching effect is dependent on multiple factors such as polymer structure, delocalization length, and efficiency of fluorescence [46]. In addition, the quenching efficiency is also governed by the polymer molecular weight but plateaus after a certain polymer molecular weight. Because of the limited excited state lifetime, the excitons are not able to migrate through the entire length of higher-molecular-weight polymer chains. To maximize the number of receptors that an exciton can reach throughout its lifetime and increase the quenching efficiency, Swager and coworkers further designed a two-dimensional film that incorporates a rigid pentiptycene-derived PPE [48,49], and a three-dimensional film composed of multiple layers of different PPEs [50,51]. In the multidimensional PPE films, intermolecular energy transfer is greatly enhanced due to rapid diffusion of the exciton in three dimensions.

6.4 Conjugated Polymer Nanostructures

The sensitivity of CPs can be further enhanced by the formation of their nanostructures [52]. This has been achieved by functionalization of electrospun fibers or inorganic nanoparticles with CPs. In another approach, CP nanoparticles are directly fabricated.

6.4.1 Electrospinning

In a typical electrospinning experiment in a laboratory, a polymer solution or melt is pumped at low rates through a conducting thin nozzle, which can also serve as an electrode (Figure 6.7) [53,54]. A high electric field of 100–500 kV·m⁻¹ is applied to the nozzle, and the distance to the counter electrode, which is grounded, is typically maintained between 10 and 25 cm [55,56]. The currents that flow during electrospinning range between a few hundred nanoamperes and a few microamperes [57]. The substrate on which the electrospun fibers are collected is typically brought into contact with the grounded counter electrode. The alignment of the electrodes can be either horizontal or vertical. At higher voltages, a jet forms from the drop of polymer solution or melt that is at the tip of the conducting nozzle [58]. This jet moves toward the counter electrode and becomes narrower in the process [59]. On the way to the counter electrode, the solvent evaporates (or the melt solidifies), and a nonwoven web of solid fibers with diameters ranging from micrometers to nanometers is obtained on the counter electrode.

Although it appears straightforward, the electrospinning process is rather complex [61]. The jet only follows a direct path toward the counter electrode for a short distance, after which due to the electric field a cone-shaped envelope of the jet is formed [61,62]. The dimensions of the fibers formed depend on multiple parameters, such as polymer molecular weight, its solubility in the used solvent, solution viscosity, melt viscoelasticity, surface tension, and electrical conductivity [63,64]. The vapor pressure of the solvent and the relative humidity of the surroundings also have a significant impact otn the obtained fibers. The processing parameters, such as substrate properties, feed rate, strength of the applied electric field, and geometry of the electrodes, also play a major role [60].

6.4.2 Conjugated Polymer Nanoparticles

Conjugated polymer nanoparticles (CPNs) are prepared mainly by miniemulsion or reprecipitation methods [65,66].

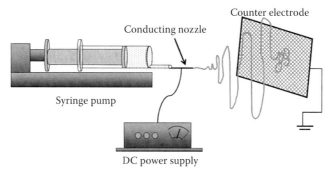

FIGURE 6.7
(See color insert.) Schematic of a typical electrospinning setup.

6.4.2.1 Miniemulsion

In this technique, CPNs can be obtained either by using presynthesized CPs or through polymerization of monomers in miniemulsion (Figure 6.8). Landfester and coworkers reported the fabrication of nanoparticles from various presynthesized CPs using this technique [67–71]. In a typical CPN fabrication methodology, the CP is first dissolved in a water-immiscible organic solvent. The solution is then introduced into an aqueous surfactant solution. Typically, the surfactant concentration is above the critical micelle concentration. Ultrasonication of the mixture effectively distributes the organic phase (i.e., polymer solution) into multiple droplets. However, the droplets can destabilize through Ostwald ripening and flocculation. To prevent destabilization of the distributed droplets, appropriate surfactants are used. The organic solvent is then slowly evaporated to obtain a stable dispersion of polymer nanoparticles in water. The size of obtained CPNs depends on a number of processing conditions, such as solution concentration, surfactant concentration, viscosity of the major phase, intensity of applied agitation, and the temperature used.

In another approach, Weder and coworkers reported the synthesis of millimeter-, micrometer-, and nanometer-sized cross-linked PPE particles (Figure 6.9) [72]. The particles were obtained by the polymerization of bifunctional monomers with small quantities of a trifunctional cross-linker in an oil-in-water emulsion containing the organic reaction mixture stabilized by sodium dodecyl sulfate in water. The polymerization was performed using Pd-catalyzed condensation reactions. The variation in obtained particle size was obtained by changing the reaction conditions, such as surfactant concentration and intensity of agitation.

Müllen and coworkers reported the synthesis of nanoparticles in nonaqueous oil-in-oil emulsions [73]. They reported the use of cyclohexane as the major phase and acetonitrile as the minor phase. Polyisoprene-block-poly

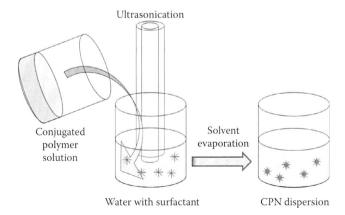

FIGURE 6.8
(See color insert.) Schematic of a typical miniemulsion protocol used for the synthesis of conjugated polymer nanoparticles.

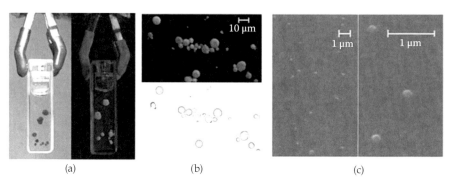

(a)   (b)   (c)

FIGURE 6.9
(See color insert.) (a) Photographs, (b) optical micrographs, and (c) scanning electron micrographs of cross-linked conjugated (a) milli-, (b) micro-, and (c) nanoparticles, respectively, prepared by method B (milli), method C (micro), and method D (nano). (Hittinger, E., Kokil, A., and Weder, C.: *Synthesis and Characterization of Cross-Linked Conjugated Polymer Milli-, Micro-, and Nanoparticles.* Angew. Chem. Int. Ed. 2004, 43, 1808–1811. Copyright Wiley-VCH Verlag GmbH & Co. KGaA. Reprinted with permission.)

(methyl methacrylate) was utilized as the emulsifying agent. Poly(3,4-ethylenedioxythiophene), polyacetylene, and poly (thiophene-3-yl-acetic acid) were synthesized to obtain CPNs by catalytic and oxidative polymerization. The emulsifying agent was removed from the surface of the nanoparticles by extracting it in tetrahydrofuran (THF).

Mecking and coworkers reported the synthesis of poly(arylene diethynylene) derivatives by Glaser coupling in miniemulsion to obtain CPNs [74]. They also covalently linked various dyes to the poly(arylene diethynylene) to tailor the fluorescence wavelength through energy transfer.

6.4.2.2 Reprecipitation

In this technique, a hydrophobic CP is dissolved in a good solvent (e.g., THF) and introduced into a miscible poor solvent (e.g., water) [75–78]. The resulting mixture is vigorously agitated, typically by ultrasonication, for fabrication of nanoparticles. After nanoparticle formation, the organic solvent is evaporated to obtain a water dispersion [79]. The driving force for the formation of nanoparticles is the hydrophobic effect. When the polymer solution is added to water, polymer chains aggregate to minimize their interface with water. Consequently, to obtain a minimum interface they form spherical shapes. The preparation does not involve the use of any additives such as surfactants and hydrophobes and can be applied to a wide variety of CPs that are soluble in organic solvents. In this method, it is possible to tune the size of nanoparticles by adjusting the polymer's molecular weight as well as the solution concentration [80–81].

Moon et al. reported the synthesis of nanoparticles from PPE derivatives. The synthesis involved dissolution of the polymer in DMSO and subsequent addition of the solution into an aqueous buffer [80]. The chemical structure of the polymer, especially the nature and density of the hydrophilic side groups, affected the particle formation. Side chains with hydrophilic protonated amine moieties and diethylene oxide side chains decorated the surface of the particles to hinder aggregation and precipitation. Smaller nanoparticles were reported, from similar hydrophilic CPs by altering the process parameters, such as pH, salt concentration, and mixing parameters [80,81].

6.5 Conjugated Polymers for Chemical Explosives Detection

Nitroaromatic compounds are electron deficient, and this property is utilized in most detection schemes. Substitution of the electron-withdrawing nitro groups on the aromatic ring lowers the LUMO energy level, which makes these compounds good electron acceptors. CPs are promising candidates for sensing because they are electron rich and good donors [16]. A wide variety of fluorescent organic and inorganic CPs have been developed for the detection of nitroaromatic explosives in solution and in the vapor phase [3]. Detection limits in the parts-per-billion, and even parts-per-trillion, range have been observed [82]. Fluorescence quenching is often achieved through a photoinduced electron transfer mechanism, as discussed previously. Thin films (a few nanometers) of CPs have been utilized in many of the reported investigations [42]. These results improve the diffusion of the analytes across the thickness of the sensing material.

PPEs display a significant fluorescence quenching response to explosive nitroaromatic compounds. Swager and coworkers have reported that thin films of PPEs (Figure 6.10a) display rapid quenching of their luminescence when exposed to TNT vapor at sub-part-per-trillion levels [83]. This polymer contained bulky pentiptycene moieties on each alternating phenyl unit of the backbone, which improved the rigidity of the polymer backbone to provide high exciton delocalization along the chain, resulting in their excellent sensitivity. The bulky pentiptycene groups also hinder interchain π stacking, which prevents the self-quenching of the luminescence. These bulky

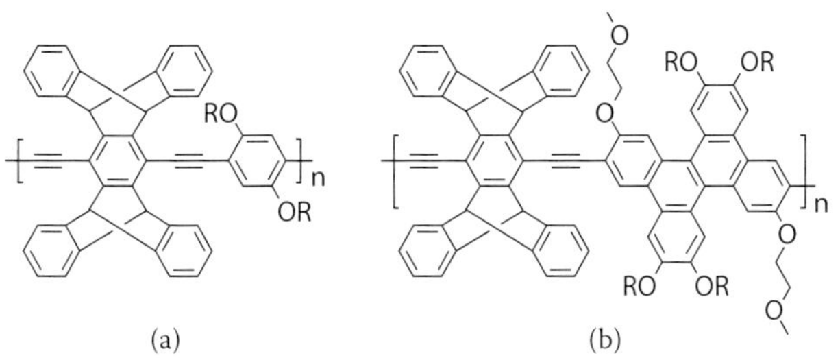

FIGURE 6.10
Chemical structures of poly(phenylene ethynylene) derivatives containing pentiptycene moieties.

groups also provide a porous packing arrangement in the solid state, which increases the free volume in the solid state and consequently improves the diffusion of the analytes. A 2.5-nm film displayed 50% quenching after 30 seconds of exposure to TNT vapor, and 75% at 60 seconds, and displayed a high K_{SV} (1.7×10^3 M^{-1}) for a solution of TNT. Due to the increased vapor pressure, DNT displayed a faster response with 91% and 95% quenching after 30 and 60 seconds of analyte exposure, respectively. Sensitivity to other interferents depends on the structure of the CP used for sensing. Thin films of this polymer displayed limited sensitivity to benzoquinone, even though the interferent had a higher vapor pressure. Compared to benzoquinone, the higher polarizability of the nitroaromatic compounds is one factor that improved the sensitivity of the PPE [84,85].

Short (200-ms) sampling times are needed to detect DNT, which is used in landmines and other explosive devices. Swager and coworkers have also copolymerized the pentiptycene monomer with a dibenzochrysene monomer (Figure 6.10b) to afford greater sensitivity to TNT than the parent pentiptycene-substituted PPEs [86]. These copolymers absorb in the visible (444 nm) spectral region and fluoresce near 474 nm with relatively long excited state lifetimes (~2 ns). Swager and coworkers have extended the application of CPs for the detection of less powerful electron-accepting analytes, such as 2,3-dimethyl-2,3-dinitrobutane (DMNB), which is a volatile taggant added to manufactured explosives to facilitate vapor detection of less volatile explosives [87]. For example, RDX has such a low volatility that its direct detection by ambient vapor sampling is not practical. Detection of DMNB was observed using polyphenylenes. This was possible because of

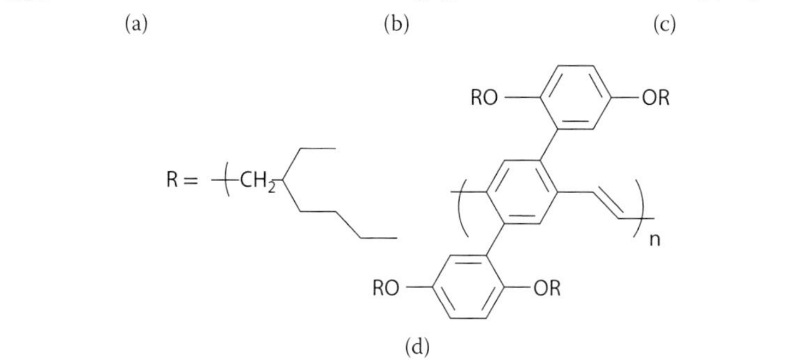

FIGURE 6.11
(a) Chemical structures for MEH-PPV, (b) DP10-PPV, (c) MPS-PPV, and (d) DPTEH-PPV.

the high-energy LUMOs and the large bandgap energies of polyphenylenes. Improved binding interactions between the nonplanar DMNB and these polymers should improve quenching efficiencies and detection limits.

Similar to PPEs, PPVs are strongly luminescent CPs with structures that result in high permeabilities to nitroaromatic small molecule analytes. As discussed previously, bulky phenyl substituents on PPVs (Figure 6.11) hinder self-quenching, resulting in a high sensitivity toward the detection of nitroaromatics [88]. Because it is more electron rich, Poly[2-methoxy-5-(2-ethylhexyloxy)-1,4-phenylenevinylene] (MEH-PPV) (Figure 6.11a) displayed greater fluorescence quenching in the presence of nitroaromatic compounds than DP10-PPV (Figure 6.11b). MEH-PPV also displays a more planar backbone compared to DP10-PPV, which results in improved exciton migration and improved amplification in fluorescence quenching. Similar to PPEs, both these PPVs displayed higher fluorescence quenching in the presence of DNT than TNT and relatively low sensitivity toward benzoquinone. Thin films of an anionic Poly[5-methoxy-2-(3-sulfopropoxy)-1,4-phenylenevinylene] (MPS-PPV) (Figure 6.11c) show similarly efficient luminescence quenching on exposure to DNT [89]. The adsorption of DNT is irreversible, due to a strong dipolar or charge transfer interaction between the polymer and the electron-deficient analyte. However, in the presence of a surfactant the films displayed reversible DNT binding on heating. The surfactant layer provided a permeable barrier that allowed the diffusion of the analyte and reduced the electrostatic interaction between the polymer and the analyte. PPVs (Figure 6.11d) with improved fluorescence quantum yield were utilized for highly sensitive detection of nitroaromatic compounds by inducing gain in the polymer films for lasing [90]. The lasing in the polymer films was achieved by pumping with a nitrogen laser. In the presence of TNT, the lasing signal was highly attenuated, resulting in enhanced sensitivities.

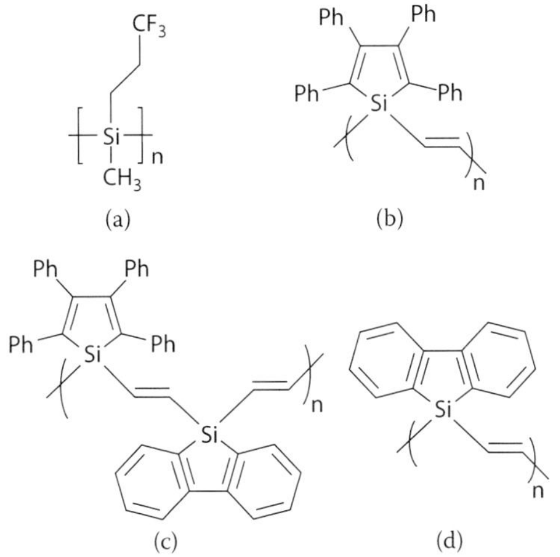

FIGURE 6.12
Chemical structures of polysilanes and polymetalloles used in fluorescent chemosensors.

Polysilanes are air-stable luminescent polymers with s-conjugation along the Si–Si chain. This imparts to polysilanes high exciton migration and efficient emission in the ultraviolet spectral region, which make them good candidates for explosives detection [91]. Poly(3,3,3-trifluoropropylmethyl silane) (PTFPMS) (Figure 6.12a) exhibits fluorescence quenching in the presence of nitroaromatics [92,93]. Interestingly, a thin film of the polymer was reported to detect nitroaromatic compounds in water and in vapor phase in the parts-per-million range. The rigidity of the polymer is imposed by the interaction between the Si atom of the backbone and an F atom of an adjacent pendant group. This interaction imparts long-range ordering of the polymer into a rod-like chain. PTFPMS solutions displayed enhanced photoluminescence quenching in the presence of nitroaromatic compounds, with $K_{SV} = 0.84 - 4.15 \times 10^4$ M^{-1}. The high sensitivity was attributed to the electron-withdrawing CF$_3$ groups, which increase the positive charge on silicon, thereby increasing its ability to interact with the nitro groups of the explosives.

Polymetalloles, which are polymers of silicon or germanium metallacyclopentadienes, are highly fluorescent and generally have a low-lying LUMO. The lower LUMO energy for silole (silacyclopentadiene) was reported to enhance the sensitivity of siloles toward nitroaromatic compounds and nitrate explosives [94]. Polymetalloles are well suited for explosives detection due to their high fluorescence quantum yield, relatively long excited state lifetimes (~0.7 ns), and helical structures, which facilitate quencher access to the polymer backbone. The structural control over polymetallole structures leads to tunable fluorescence quenching by specific nitroaromatics [94]. Silole-based polymers with silicon–vinylene functionalities (Figure 6.12b, c, and d) formed a more highly conjugated polymer and retained many of the electronic properties found in polysilanes [95]. Higher LUMO improved the PET, and these molecules exhibited higher quenching efficiencies toward TNT ($K_{SV} = 10^4$ M^{-1}). These polymers were also reported to detect RDX, which is an important analyte for detection for security purposes.

PTs are promising candidates for sensing nitroaromatic explosive compounds, due to high exciton migration length, absorption in the visible, and high Stokes shift in the fluorescence [96]. However, a low sensitivity for solution-based detection with these polymers was reported [96]. One of the reasons is that the side group attached to the polymer backbone prevents the approach of the quencher molecule to the chromophores. For a sensor to be efficient, it is imperative that the sensing element is accessible to the analytes. To overcome the low sensory response of PT-based polymers toward detection of nitroaromatics, we hypothesized that a dipolar 1,2,3-triazole moiety could enhance the analyte interaction.

The permeability of an analyte depends on its intermolecular interaction with the polymer and diffusivity in polymer thin films. The interaction of nitroaromatic compounds can be increased in thin films by appending dipolar and polarizable side chains onto the polymer backbone. As discussed previously, decreased packing of polymer chains on the one hand decreases self-quenching and, importantly, on the other hand significantly improves the diffusion of the analyte within the polymer film. These interchain interactions can be easily modulated by varying the side chains on the polymer. We reported that a dipolar 1,2,3-triazole with various side chains as pendant groups on the PT backbone can be used to enhance the fluorescence quenching of PTs by nitroaromatic analytes. The triazole moiety, obtained through the functionalization chemistry, was designed to interact with the nitroaromatics. Thiophene monomers were functionalized with the triazole moiety using Huisgen 1,3-dipolar cycloaddition and were then polymerized using Stille coupling to obtain the functionalized PT derivatives (Figure 6.13). The polymers displayed a good fluorescence quantum yield (~0.21).

Interestingly, all the functionalized polymers had similar LUMO energy levels, resulting in similar exergonicity for photon-induced electron transfer. Compared to the polymers with linear side chains, the polymer with branched side chains (P3TzdHT) displayed a lower redshift in solid state absorption compared to absorption in solution. This indicated that the branched side chain was indeed hindering the interchain π stacking, which resulted in reduced polymer aggregate formation. The sensor response of polymer thin films toward DNT and TNT was measured by the quenching in fluorescence intensity of polymer thin films on exposure to a saturated vapor of the analytes in a sealed chamber (Figure 6.14a and b). The percentage fluorescence quenching of the three polymers toward DNT was in the following order: P3TzdHT > P3TzHT > P3BSiT. It was observed that the aggregation of the polymer chains for P3TzHT and P3BSiT was similar. The diffusion of the analyte in the film largely depends on the degree of aggregation. Hence, it was expected that a similar fluorescence quenching response would be observed for both P3TzHT and P3BSiT. However, higher fluorescence quenching was observed for P3HTzHT. This indicated that the triazole moiety indeed helped improve the interaction of the polymers with DNT and TNT. Intermolecular interactions with nitroaromatic compounds can be improved through dipole–dipole, dipole-induced dipole, and hydrogen bonding interactions. The triazole moiety was reported to display a large

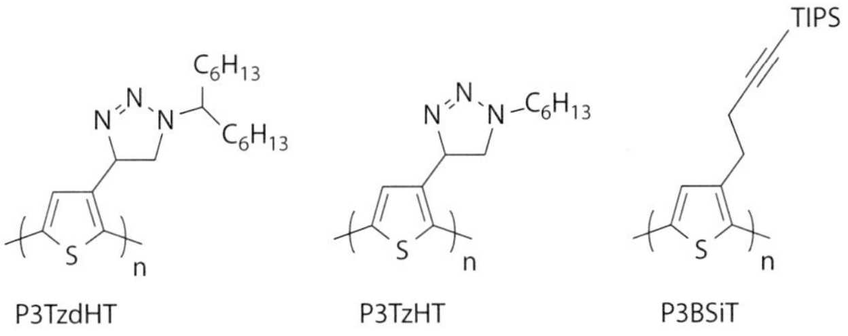

FIGURE 6.13
Chemical structure of polythiophenes containing triazole and silane moieties in the side chain.

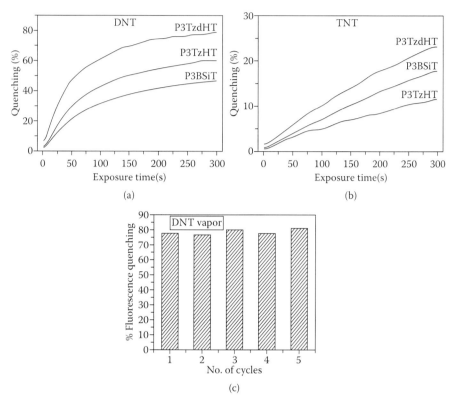

FIGURE 6.14

Thin film photoluminescence quenching of polythiophenes containing triazole and silane moieties in the side chain with (a) DNT, (b) TNT, and (c) the recycled sensitivity of P3TzdHT in the presence of DNT. (G. Nagarjuna, A. Kumar, A. Kokil, K. G. Jadhav, S. Yurt, J. Kumar, and D. Venkataraman, Enhancing sensing of nitroaromatic vapours by thiophene-based polymer films, *J. Mater. Chem.* 2011, *21*, 16597–16602. Reproduced by permission of The Royal Society of Chemistry.)

dipole moment (5 Debye) [97,98] and thus could interact with the analytes through either dipole–dipole interactions or hydrogen bonding interactions [99–101].

As discussed previously, PTs functionalized with branched side chains through the triazole moiety displayed reduced aggregation. This led to an increase in quenching efficiency by P3TzdHT. By increasing the molecular weight of P3TzdHT, the efficiency of fluorescence quenching was increased further by 10% and 24% for DNT and TNT, respectively. This can be attributed to the increased exciton diffusion length in the polymer with higher molecular weight. When the sensor was removed from the chamber containing the nitroaromatic vapors, the original fluorescence intensity was restored (Figure 6.14c). This indicated that the sensors were reversible and could be recycled for multiple sensing events.

6.6 Polymer Nanostructures for Chemical Explosives Detection

Utilization of ultra-high surface area CPNs could be an attractive alternative for improved sensitivities [52]. Yang and coworkers reported sensitive detection of TNT in solution using an anionic PPE derivative functionalized silica nanoparticles [102]. Using Pd-catalyzed cross coupling reactions, PPEs were grown from functionalized silica nanoparticles. A slight redshift in the fluorescence wavelength was reported on functionalization, which was attributed to aggregation of the polymer. A relatively low K_{SV} was reported for the detection of TNT by the free polymer in solution. However, on grafting onto the silica nanoparticles an increase in K_{SV} by around two orders of magnitude was reported. The authors also reported the effect of nanoparticle size on the determined K_{SV}. For the smallest nanoparticles (88 nm), the authors reported an almost three times higher K_{SV} compared to the largest particle (216 nm). This was attributed to the increase in surface area on reduction of particle size.

We reported the fabrication of CPNs using a quarterthiophene-based regioregular CP (Figure 6.15a) [103]. Compared to the polymer solution, a 10-nm redshift in the absorption maximum was observed for nanoparticles. This was attributed to the aggregation of the polymer in the nanoparticle. A similar redshift was also observed in the absorption spectrum of the solid state thin film. When excited at 491 nm, the nanoparticle dispersion fluoresced with two peaks around 613 and 655 nm (Figure 6.15b). A shift around 30 nm in the fluorescence maximum toward the longer wavelength was observed. An additional peak in the fluorescence spectrum was also observed compared to that of the polymer solution, which was attributed to the appearance of a new charge transfer band. The fabricated nanoparticles had a relatively broad particle size distribution, the average size of the nanoparticles being around 89 nm. The fluorescence intensity of the nanoparticles remarkably decreased in the presence of DNT. The fluorescence of nanoparticle dispersion in water also quenched on the addition of an aqueous DNT solution (Figure 6.15c). The first peak in the emission spectrum was utilized for obtaining the SV plot, and a K_{SV} value of around 2.7×10^3 M^{-1} was obtained (Figure 6.15d). Interestingly, the CPN dispersion also displayed strong two-photon fluorescence characteristics.

Using Poly[2-(3-thienyl)ethanol n-butoxycarbonylmethylurethane) (PURET) (Figure 6.16a), a substituted PT with urethane moieties in side chains, we reported the fabrication of CPNs [104]. PURET is an interesting candidate for the detection of nitroaromatic compounds as it displays low aggregation in solid films and has a high fluorescence quantum yield. The CPNs with PURET were fabricated using the miniemulsion protocols described previously. These nanoparticles display an absorption maximum around 440 nm, which is similar to thin film solid state absorption, and display a fluorescence maximum around 600 nm. The average size of the fabricated nanoparticles determined using dynamic light scattering was around 90 nm, and the obtained nanoparticles exhibited a fairly narrow size distribution (Figure 6.16b). Significant

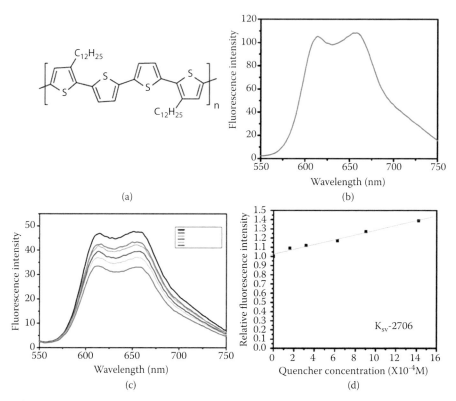

(a) (b)

(c) (d)

FIGURE 6.15

(a) Chemical structure of the quarterthiophene-based conjugated polymer. (b) Photoluminescence spectrum of the polymer nanoparticle dispersion. (c) Fluorescence quenching of the
nanoparticles in the presence of DNT. (d) The Stern–Volmer plot for the fluorescence quenching. (S. Satapathi, L. Li, R. Anandakathir, L. A. Samuelson, and J. Kumar, Sensory Response
and Two-Photon-Fluorescence Study of Regioregular Polythiophene Nanoparticles, *Journal of
Macromolecular Science, Part A; Pure and Applied Chemistry*, 2011, *48*, 1049–1054. Reprinted by
permission of the publisher: Taylor & Francis, Ltd, http://www.tandf.co.uk/journals.)

fluorescence quenching of the nanoparticle dispersion in water was observed
on the addition of aqueous DNT and TNT solutions.

As a control, the addition of equal amounts of pure water without DNT or TNT
displayed only a slight reduction in fluorescence intensity. This indicated that
the observed fluorescence quenching in the presence of nitroaromatic analytes
was not due to dilution of the nanoparticle dispersion. In comparison, the solution of PURET in THF displayed only minimal fluorescence quenching on addition of DNT and TNT. For detection of DNT with the nanoparticle dispersion
(Figure 6.16c), the K_{SV} value (4.5×10^3 M^{-1}) was two orders of magnitude higher
than that for PURET solution in THF (34 M^{-1}). For TNT, a high K_{SV} of around
8.6×10^3 M^{-1} was observed for detection using the nanoparticle dispersion. The
higher sensitivity displayed by the nanoparticles for detection of the nitroaromatic analytes can be attributed to improved polymer–analyte interaction.

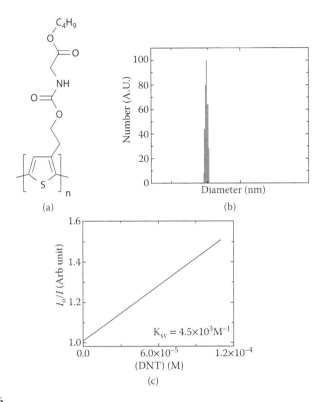

(a)   (b)

(c)

FIGURE 6.16
(a) Chemical structure of PURET, (b) DLS particle size distribution of PURET nanoparticles, and (c) Stern–Volmer plot for fluorescence quenching of PURET nanoparticles in the presence of DNT. (Reprinted, with permission from S. Satapathi, A. Kokil, B. H. Venkatraman, L. Li, D. Venkataraman, and J. Kumar, Sensitive Detection of Nitroaromatics with Colloidal Conjugated Polymer Nanoparticles, *IEEE Sensors Journal*, 2013, *13*, 2329–2333, © 2013 IEEE.)

The sensitivity of CP in solution toward DNT and TNT is limited by the poor polymer–analyte interactions. Typically, in solution the pendant side chains on the CPs form a corona around the polymer backbone, which helps in solvating these otherwise intractable polymers. However, the presence of the pendant side-chain corona limits the proximity of the analytes from the CP backbone. Hence, efficient fluorescence quenching cannot be obtained with nitroaromatic analytes in solution. In addition, the hydrophobic nature of most CPs poses a challenge for detection of the analytes in water. For nanoparticles dispersed in water, the pendant side chains do not form a corona around the polymer. Hence, the analytes are less hindered from coming close to the polymer backbone. Moreover, the charged surfactant head groups at the nanoparticle surface can interact with the analytes. Because ion–dipole interaction is strong, the density of analytes on the nanoparticle surface will be higher, which will enhance the PL quenching efficacy. In addition, the exciton diffusion occurring in the polymer nanoparticles may lead to better PL quenching.

We have also reported that PURET exhibits strong two-photon fluorescence as well as a large two-photon absorption cross section. Hence, we utilized the PURET nanoparticles for detecting DNT and TNT by two-photon fluorescence. Two-photon fluorescence quenching could provide additional selectivity and a sensing scheme to move out of the visible excitation region to the near infrared region. To determine the effect of the nitroaromatic analyte–fluorophore interaction on the two-photon-induced fluorescence, the nanoparticle fluorescence was investigated with an excitation at 800 nm. The 800-nm laser pulse was passed through the sample cuvette and the fluorescence signal collected at right angles. Interestingly, the two-photon fluorescence of the nanoparticle dispersion also quenched on introducing DNT and TNT (Figure 6.17). The obtained K_{SV} values for DNT and TNT were 3.5×10^3 and 9.5×10^3 M^{-1}, respectively. These values are very similar to the sensitivities observed for the one-photon process described previously. This suggests that the photoexcited electron transfer taking place from PURET to the analyte is not affected by the selection rule for two-photon induced fluorescence.

Swager and coworkers reported the use of electrostatic layer-by-layer (LbL) deposition for surface functionalization of silica nanoparticles with PPE derivative [105]. LbL deposition of polymeric thin films is a technique that uses the electrostatic forces between molecules of opposite charges to form the films [106]. In this technique, a charged substrate, for example, a substrate with anions at the surface, is taken. Polycations are then deposited on the surface and lead to complete charge neutralization, resaturation, and charge reversal, which results in the presence of cations at the surface. This is then dipped in a solution of polyanions, which after sufficient deposition again results in anions at the surface. Alternate deposition of polyelectrolytes with positive and negative charges increases the thickness of the film and also

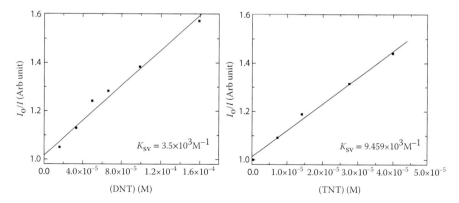

FIGURE 6.17
Stern–Volmer plots for two-photon-induced fluorescence quenching of PURET nanoparticles in the presence of DNT (left) and TNT (right). (Reprinted, with permission from S. Satapathi, A. Kokil, B. H. Venkatraman, L. Li, D. Venkataraman, and J. Kumar, Sensitive Detection of Nitroaromatics with Colloidal Conjugated Polymer Nanoparticles, *IEEE Sensors Journal*, 2013, *13*, 2329–2333, © 2013 IEEE.)

provides flexibility in choosing the number of bilayers while still maintaining nanoscale control of the thickness. One of the important advantages of the LbL film assembly is that it is very easy to use without the need of any kind of expensive or special apparatus. LbL deposition is advantageous not only because ultrathin films can be readily fabricated but also because it can be utilized for obtaining a conformal coating on a nonflat surface [107,108].

On deposition of the PPE on silica nanoparticles, Swager and coworkers reported a blueshift in the absorption maximum of the polymer, which was attributed to the restricted conformations of the PPE that resulted in a reduced conjugation. Also, the fluorescence of the PPE remained unaffected even after deposition on the silica, indicating that aggregate formation did not occur. The PPE-coated silica particles displayed fluorescence quenching in the presence of an aqueous DNT solution with a K_{SV} of around 70 M^{-1}.

Using LbL deposition, Tripathy and coworkers have reported efficient detection of DNT using the small molecular pyrene derivative 1-hydroxypyrene-3,6, 8-trisulfonate (HPTS) (Figure 6.18a) [109]. This dye was chosen due to its large Stokes shifts, high quantum yield, high absorbance, and excellent photostability. The dye was attached to poly(acryloyl chloride) to obtain an anionic polyelectrolyte. The LbL deposition was performed on anionically charged glass substrates, and the efficacy of DNT detection was reported. The LbL films displayed quenching of fluorescence upon introduction of DNT, and a high K_{SV} of 1.1×10^3 M^{-1} was reported.

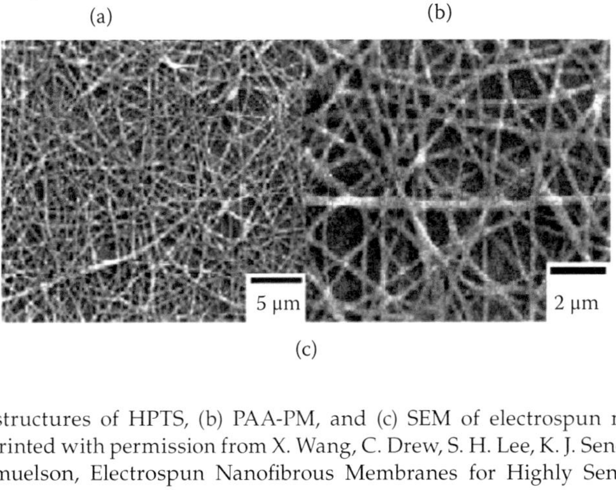

FIGURE 6.18
(a) Chemical structures of HPTS, (b) PAA-PM, and (c) SEM of electrospun membranes of PAA-PM. (Reprinted with permission from X. Wang, C. Drew, S. H. Lee, K. J. Senecal, J. Kumar, and L. A. Samuelson, Electrospun Nanofibrous Membranes for Highly Sensitive Optical Sensors, *Nano Lett.* 2002, *2*, 1273–1275. Copyright 2002 American Chemical Society.)

In another approach, we attached a pyrene moiety to polyacrylic acid using an ester linkage (Figure 6.18b) [110]. The polymer was then electrospun, using the technique described previously, to obtain nonwoven webs of fluorescent nanofibers (Figure 6.18c). The applied voltages during electrospinning ranged between 15 and 20 kV, the working distance between the nozzle and the grounded electrode was typically 15–20 cm, and the collection time was limited to 30–45 seconds. The diameters of the fibers were approximately 100–400 nm. The obtained porous structure of the electrospun membrane provides a surface area to volume ratio that is roughly one to two orders of magnitude higher than that known for continuous thin films, which can result in a considerable gain in sensitivity. Significant quenching of fluorescence was reported in the presence of a DNT solution. A K_{SV} for detection of DNT around 9.8×10^5 M^{-1} was obtained. Compared to the previous investigation on thin film sensors, the use of nanofibers in this investigation resulted in an almost three orders of magnitude higher sensitivity.

We also reported the LbL deposition of an anionic PURET on electrospun fibers for the efficient detection of MV^{2+} [111]. Electrospun cellulose acetate (CA) fibers were used as the porous scaffold for LbL deposition. The electrospun CA membrane is insoluble in water but carries a negative charge from the partial hydrolysis of surface ester groups. The deposition process was performed in two steps for each dipping cycle. In the first step, the electrospun CA membrane was immersed in an aqueous polycation solution for 5 minutes at room temperature and subsequently washed with distilled water for 3 minutes. After the deposition and washing step, the substrate was dried with nitrogen gas. In the second step, the substrate with a single layer of the polycation was immersed into a solution of the polyanion, anionic PURET, for five minutes, followed by the washing and drying procedures. This dipping sequence was repeated five times to build up the fluorescent sensing layers sequentially. On exposure to a solution containing MV^{2+}, the fluorescence intensity significantly decreased and a K_{SV} around 3.5×10^6 M^{-1} was calculated.

Although the fluorescence sensors possess the required sensitivity, their selectivity is still limited. The lack of selectivity can limit the application of these sensors in real-life environments, as many interferents could cause either fluorescence quenching or an increase in fluorescence. These interfering results could cause either false positives or false negatives. To utilize the sensitivity offered by optical sensors and overcome the limitations posed by nonspecificity, an alternative approach is to develop sensor arrays combined with appropriate pattern recognition methods.

We recently reported sensitive, fast, and selective recognition of explosives using fluorescent polymer sensors and pattern recognition methods [112]. The fluorescence-sensing array consisted of four elements based on PT-based CPs, such as ADS518PT, ADS2008P, Poly(thiophene-3-methyl acetate) (PTMA), and PURET (Figure 6.19). These polymers have the same PT backbone but different side groups. The fluorescence intensity of the

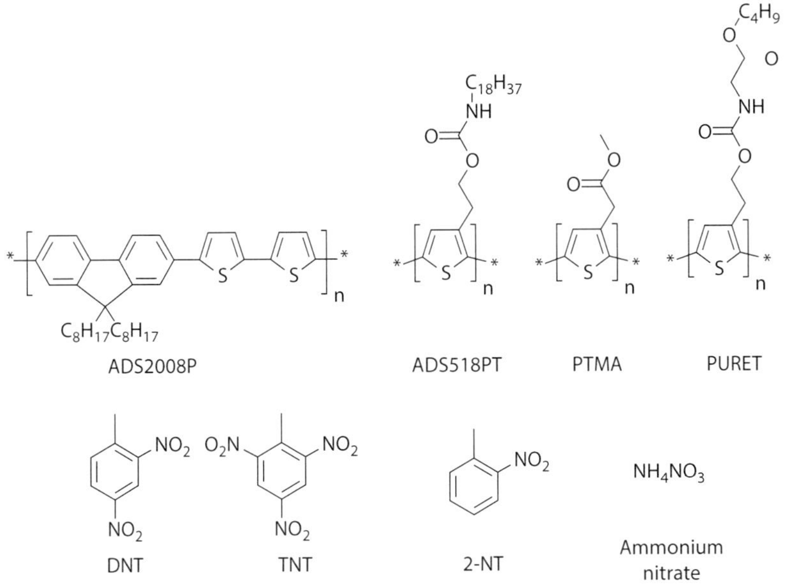

FIGURE 6.19
Chemical structures of thiophene-based conjugated polymers used in the sensor array and the target analytes.

thin films of these polymers was measured in the presence of various analyte vapors up to 300 seconds. From time response studies, percentage fluorescence quenching or the changes in normalized fluorescence intensity curves were extracted. The features obtained were used as input data for two popular pattern recognition methods, fuzzy Predictive Adaptive Resonance Theory (ARTMAP) and support vector machine (SVM) neural network, which are known for their excellent learning and prediction capabilities [113,114]. Fuzzy ARTMAP learns via a special leaning mechanism known as the stability–plasticity dilemma, which closely emulates human learning. SVM is also an excellent pattern recognition tool for handling electronic nose data because it has good generalization ability by maximizing the margin of hyperplane, which is the decision boundary, separating different classes.

The recognition rates of two pattern recognition methods were evaluated, and finally, by adding additional Gaussian noise to the original fluorescent data, robustness for the recognition of target explosives was reported. Four analytes were used for these investigations, namely, DNT, TNT, nitrotoluene, and ammonium nitrate. As expected, the fluorescence quenching response for each of the utilized CPs was different in the presence of different analytes (Figure 6.20). Two features were extracted from the time-dependent fluorescence response curves, the percentage fluorescence quenching, and three coefficients by fitting the experimental curve by a nonlinear curve. An

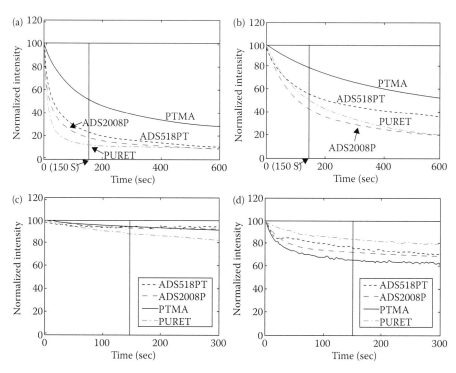

FIGURE 6.20

Time dependent fluorescence intensity of polymers (ADS518PT, ADS2008P, PTMA, and PURET) upon exposure of (a) DNT, (b) TNT, (c) ammonium nitrate, and (d) nitrotoluene vapors. (Reprinted from *Sensors and Actuators B*, 160, J. Cho, R. Anandakathir, A. Kumar, J. Kumar, and P. Kurup, Sensitive and fast recognition of explosives using fluorescent polymer sensors and pattern recognition analysis, 1237–1243, Copyright 2011, with permission from Elsevier.)

exponential function with three coefficients was used for fitting the time-based fluorescence quenching curves.

Principal components analysis (PCA) is a well-known method for reducing the dimensionality of a data set while preserving most of the variance present in the data. A data set of 20 measurements (five measurements per analyte for four analytes) was prepared, and PCA was performed separately using the four- and twelve-dimensional feature patterns consisting of percentage fluorescence quenching at different response times and the three coefficients of the estimated exponential function, respectively. Because the polymers displayed similar response, the analytes were not separable by PCA for fluorescence quenching after two seconds. However, five seconds onward the four different analytes could be separated (Figure 6.21). A 90% recognition rate for identifying the analyte was obtained using SVM neural network even after two seconds of analyte exposure. The recognition rate of SVM neural network was almost 100% after five seconds. Fuzzy ARTMAP network displayed perfect classification rates after six seconds.

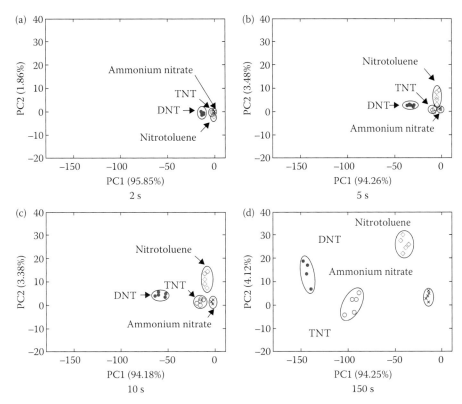

FIGURE 6.21
PCA results using percent fluorescence quenching at different response times: (a) 2 s, (b) 5 s, (c) 10 s and (d) 150 s. (Reprinted from *Sensors and Actuators B*, 160, J. Cho, R. Anandakathir, A. Kumar, J. Kumar, and P. Kurup, Sensitive and fast recognition of explosives using fluorescent polymer sensors and pattern recognition analysis, 1237–1243, Copyright 2011, with permission from Elsevier.)

6.7 Conjugated Polymers for Detection of Metal Ions

Reduced aggregation of CPs is desired for obtaining high sensitivity in fluorescence-based sensors, as discussed previously for the detection of nitroaromatic compounds. However, analyte-induced aggregation that leads to a response in fluorescence can be used as a sensitive transduction mechanism for analyte detection.

Swager and coworkers reported a PPE-based sensor for the detection of potassium ions via ion-induced aggregation (Figure 6.22a) [115]. The PPEs were functionalized with crown ethers (15-crown-5), which form a 2:1 complex with K+ ions. Fluorescence quenching of the functionalized polymer was reported on exposure to K+ ions, which generated potassium-bridged polymer aggregates. As seen in Section 6.6, a redshift in the absorption maximum

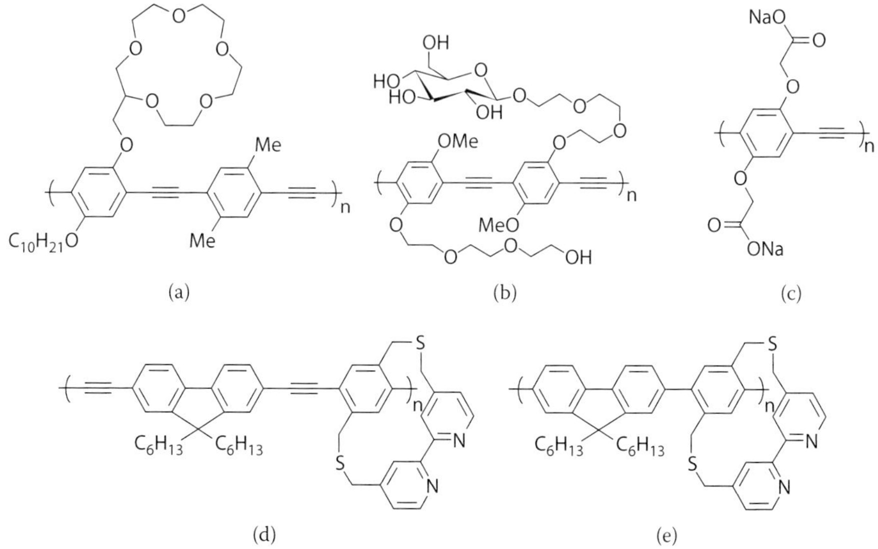

FIGURE 6.22
Chemical structures of (a) crown ether substituted PPE, (b) PPEs functionalized with sugar residues, (c) carboxylated PPEs, (d) cyclophane containing poly(arylene ethynylene)s, and (e) cyclophane containing polyarylenes, for detection of metal ions by fluorescence quenching.

was also reported on aggregation of the polymer. The K^+ ion–induced aggregation resulted in a decrease of 82% in the fluorescence intensity at 2:1 crown ether to K^+ mole ratio. Interestingly, absorption and emission spectra of the K^+-aggregated PPE were similar to those obtained from a Langmuir–Blodgett film, where the polymer chains were cofacially π stacked. These crown ether–containing PPEs were also selective, as a 1500-fold excess of either Li^+ or Na^+ did not significantly change polymer absorption or emission. This was attributed to the fact that Li^+ or Na^+ forms 1:1 complexes with 15-crown-5. Liu and coworkers reported triethylene glycol monomethyl ether functionalized PPE with groups that responded to Li^+ and Na^+ but not K^+ [116]. The fluorescence of PPE significantly quenched upon the addition of the ions; however, a change in the absorption spectrum was not observed. By increasing the steric bulk of the comonomer, the authors could obtain a selective response to one of the Li cations.

Bunz and coworkers utilized PPEs functionalized with sugar residues for fluorescent detection of Hg^{2+} (Figure 6.22b) [117]. The sugar-functionalized PPE displayed efficient fluorescence quenching in the presence of Hg^{2+}, with $K_{SV} = 1.1 \times 10^4$ M^{-1} (HgCl$_2$); $K_{SV} = 1.6 \times 10^4$ M^{-1} (Hg(OAc)$_2$); $K_{SV} = 3.8 \times 10^4$ M^{-1} (Hg(NO$_3$)$_2$); and $K_{SV} = 1.1 \times 10^4$ M^{-1} (Hg(tfa)$_2$). The sugar residues in the PPE side chains significantly improved the sensitivity of the polymer; a control polymer without the sugar moieties did not display any response. Using carboxylated PPEs (Figure 6.22c), which were soluble in water, the authors reported the detection of a variety of metal cations through fluorescence quenching [118]. The polymer, however, was most sensitive for Pb^{2+} cations over the other divalent cations, with $K_{SV} = 8.8 \times 10^5$ M^{-1} ($Pb(NO_3)_2$) and $K_{SV} = 6.9 \times 10^5$ M^{-1} ($Pb(OAc)_2$). The carboxylated polymer was more than 1500-fold more sensitive to Pb^{2+} than a low-molecular-weight model compound. The higher sensitivity of the PPE was attributed to a combination of multivalency, where Pb^{2+} could coordinate to multiple ligands in the polymer, and better exciton migration in the CP. The authors employed the same polymer and utilized an interesting route to report a highly selective sensor for Hg^{2+} [119]. The strategy was based on the complex formation between carboxylated PPE and papain, a cationic cysteine protease. The PPE–papain complex displayed a selective fluorescence, quenching only to Hg^{2+} over nine other control metal ions (Zn^{2+}, Ni^{2+}, Co^{2+}, Cu^{2+}, Ca^{2+} Cd^{2+}, Pb^{2+}, Fe^{2+}, and Mg^{2+}). Papain contains free thiol groups that bind strongly to Hg^{2+}. The polyplex that is formed by mixing anionic PPE and papain is more sensitive to the mercury cations compared to just the individual components. On addition of the mercury cations, cross-linking of the papain through Hg^{2+} and precipitation of the polyplex occurs, leading to quenched fluorescence.

Nitrogen-containing ligands such as 2,2′-bipyridine (bipy) are highly versatile and efficiently coordinate to a number of metal ions. These moieties are typically stronger ligands for transition metal centers compared to alkali and alkaline earth metals [120]. Incorporation of bipy in the PPE backbone has even been reported to cross-link in the presence of transition metal ions [121]. Wang and coworkers reported metal ion sensing using cyclophane-containing poly(arylene ethynylenes) with a fluorene comonomer (Figure 6.22d) [122]. An analogue of the copolymer without the ethynyl moieties in the repeat unit was also reported (Figure 6.22e). The fluorescence and absorbance of polymers were redshifted for these polymers compared to polymers without bipy-containing cyclophane. The fluorescence quenching of polymer without the ethynyl moieties was reported by Cu^{2+}, Ni^{2+}, Zn^{2+}, Mn^{2+}, Co^{2+}, and Ag^+. The bipy-containing poly(arylene ethynylene) displayed fluorescence quenching by Cu^{2+}, Co^{2+}, and Ni^{2+}. The quenching was also not as efficient. This difference in sensing behavior was attributed to the difference in flexibility of the polymer backbones. A shift in polymer absorbance was not observed on complexation of the metal ions as the bipy moieties were not in conjugation with the polymer backbone.

PPEs for the selective detection of Pd^{2+} ions were reported. The PPEs with metalinked pyridyl groups, which can selectively bind Pd^{2+} by self-assembly, were utilized [123]. A K_{SV} of 4.3×10^5 M^{-1} was observed, which was 56 times greater than the small molecular model compound. This indicated that the CP chain significantly amplified the quenching. The model polymers without the pyridyl moiety were 144 times less sensitive compared to the metalinked polymer. The authors reported that aggregation due to Pd^{2+} complexation resulted in the high selectivity and sensitivity of the polymer. Huang and coworkers reported the metal ion–sensing properties of three structurally related CPs that incorporated bipy moieties in the polymer backbone [124]. The three structures were polyarylene, poly(arylene vinyl-ene), and poly(arylene ethynylene) (Figure 6.23a, b, and c, respectively). The authors reported the highest sensitivity for polyarylene compared to the two other backbone structures. The authors attributed the higher sensitivity to weaker resistance for obtaining coplanarity of the bipy moiety when joined by a single bond, compared to a double or triple bond.

Ma and coworkers compared the efficacy of the sensing by polyarylene with bipy moieties to those that contained coplanar phenanthroline moieties [125]. The authors reported that the binding constants of the metal ions did not correlate with the redshifts in the absorption maxima on complexation. The fluorescence quenching by the phenanthroline moiety by all the transition metal ions used was higher than the polymers with bipy, which was attributed to the planarity and rigidity of the phenanthroline unit.

Zhu and coworkers reported an interesting strategy for the synthesis of polyarylenes containing a triazole moiety in the polymer backbone using a Huisgen dipolar cycloaddition reaction between two bifunctional monomers [126]. This polymer displayed selective fluorescence quenching by Hg^{2+} ions compared to Co^{2+}, Ni^{2+}, Ag^+, Cd^{2+}, Cu^{2+}, and Zn^{2+}. Addition of Hg^{2+} to the

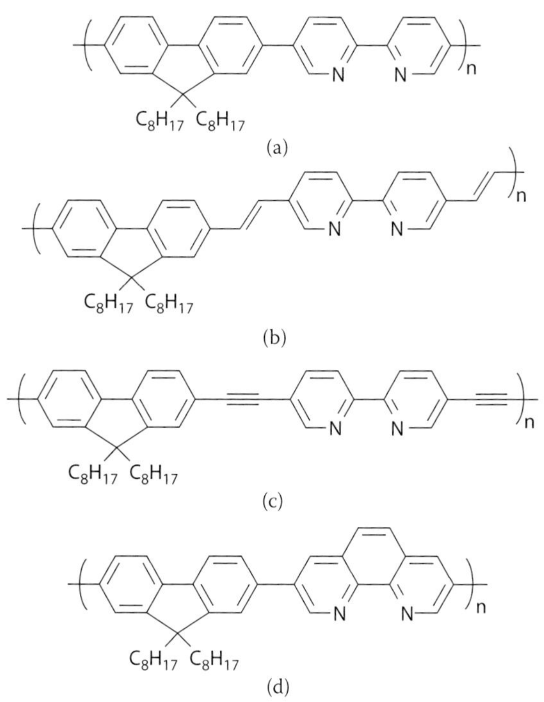

FIGURE 6.23
Chemical structures of (a) bipy containing polyarylenes, (b) bipy containing poly(arylene vinylene)s, (c) bipy containing poly(arylene ethynylene)s, and (d) phenanthroline containing polyarylenes, for detection of metal ions by fluorescence quenching.

polymer solution quenched the fluorescence by about 74.3% polymer at a concentration of 4:1 molar ratio. The authors attributed the quenching efficacy to the enhanced rigidity of the polymer backbone with the triazole moiety. On the other hand, Hg^{2+} is a soft acid and the use of soft donor atoms, including nitrogen, sulfur, or phosphorus, in a chelating moiety results in an increase in its affinity and selectivity for Hg^{2+} cation.

Leclerc and coworkers reported a 15-crown-5-functionalized poly(thiophene) as a fluorescent probe for K^+ [127]. Similar to the PPEs, this material exhibited selectivity for K^+ over Li^+ and Na^+, although all three cations did quench polymer emission at millimolar concentrations. The authors attributed the quenching to potassium ion–induced aggregation. Jones and coworkers reported poly[*p*-(phenyleneethynylene)-*alt*-(thienyleneethynylene)] with different oligopyridine pendant groups as receptors for transition metals (Figure 6.24) [128]. These polymers were reported to take advantage of the strong conjugation and luminescence properties of the polyarylene ethynylene backbone together with the multidentate Lewis base coordination ability of oligopyridines. The polymer with the tolylterpyridine moiety displayed the highest sensitivity. The different chelating abilities of terpyridine with different transition metals resulted in this polymer displaying selectivity toward different transition metals. The authors reported a 79.6% fluorescence quenching with Cd^{2+} and a 29% fluorescence quenching with Ni^{2+}. However, sensory response to the presence of

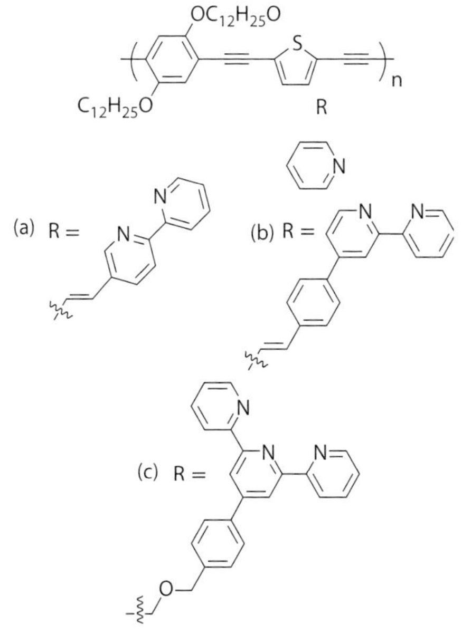

FIGURE 6.24
Chemical structures of conjugated polymers with bipyridine and terpyridine moieties in the side chain for detection of metal ions by fluorescence quenching.

common cations such as Na^+ or Ca^{2+} was not observed. The polymer was also not responsive to competing contaminants such as Pb^{2+} or Hg^{2+}.

Xi and coworkers reported an assay for K^+ using crown ether-functionalized PPVs [129]. The crown ether utilized was a larger benzo-18-crown-6. The authors also observed redshifted and diminished emission in the presence of the potassium ion. However, the polymer did not display selectivity for K^+ over Na^+ or Li^+. Although the emission was responsive to metal ions, the absorption spectra remained unchanged. Smith and coworkers synthesized benzo-15-crown-5-containing PPVs that responded to Eu^{3+} [130]. The polymer fluorescence intensity increased more than eightfold on exposure to Eu^{3+}, and the absorbance remained unchanged. The authors proposed that the increase in emission was due to the inhibition of an internal change transfer process on cation binding.

6.8 Polymer Nanostructures for Detection of Metal Ions

Similar to nitroaromatic compound detection, use of nanostructured polymers for detection of metal ions is expected to improve the sensitivity of metal ion detection. Tian and coworkers fabricated CPNs with copolymers of fluorene and benzothiadiazole [131]. The CPNs displayed fluorescence quenching in the presence of Ag^+ ions. Amphiphilic sulfonated polystyrene copolymers were utilized for the stabilization of CPNs. The sulfonated styrene units in the amphiphilic polymer located at the surface of the CPNs were involved in the binding of the Ag^+ ions. The CPNs fabricated from fluorene homopolymers displayed only limited fluorescence quenching. The slight decrease in fluorescence was attributed to the cation-induced aggregation of the CPNs. Increasing the concentration of benzothiadiazole in copolymers significantly increased fluorescence quenching. This indicated that increased sulfur and nitrogen in the copolymers improved the Ag^+ binding. The observed Ag^+-induced fluorescence quenching was attributed to the combination of aggregation-induced fluorescence static quenching and Ag^+-triggered fluorescence dynamic quenching.

Using LbL deposition, Tripathy and coworkers have reported efficient detection of Fe^{3+} and Hg^{2+} using the small molecular pyrene derivative HPTS [109]. Thin films of HPTS-containing polymers displayed fluorescence quenching of the LbL films on a glass substrate. High K_{SV} values around 3.6×10^3 and 3.2×10^3 M^{-1} were reported for Fe^{3+} and Hg^{2+} ions, respectively. We attached a pyrene moiety to polyacrylic acid using an ester linkage [110]. The polymer was then electrospun, using the technique described in Section 6.4.1, to obtain nonwoven webs of fluorescent nanofibers. These nanofibers displayed efficient fluorescence quenching in the presence of Fe^{3+} and Hg^{2+} ions. Significantly higher K_{SV} values, almost three orders of magnitude higher values, around 1.1×10^6 and 8.9×10^5 M^{-1} were observed for Fe^{3+} and Hg^{2+} ions, respectively.

6.9 Summary

A number of CPs have been developed for sensing chemical analytes such as nitroaromatic explosive compounds and toxic metal ions. Fluorescence quenching has been used for detection of analytes in solution as well as in vapor form. The detection of analytes in the solid state is enhanced by the use of ultra-thin films. Fabrication of polymer nanostructures significantly improves the sensor response to the analytes. Although selectivity toward certain analytes has been reported for metal ions, it is still desired in the detection of chemical explosive compounds. Increased selectivity by CPs is expected to reduce false negatives and false positives. Another promising approach toward increasing the selectivity and identification of analytes is the development of sensor arrays and utilization of neural networks for analysis of the obtained data from the sensor array.

Acknowledgment

We thank the Polymer-Based Materials for Harvesting Solar Energy (PHaSE), an energy frontier research center funded by the U.S. Department of Energy, Office of Science, Basic Energy Sciences under Award DE-SC0001087 for their support.

References

1. S. Singh, Sensors—An effective approach for the detection of explosives, *J. Haz. Mater.* 2007, *144*, 15–28.
2. T. F. Jenkins, D. C. Leggett, P. H. Miyares, M. E. Walsh, T. A. Ranney, J. H. Cragin, and V. George, Chemical signatures of TNT-filled landmines, *Talanta* 2001, *54*, 501–513.
3. M. E. Germain and M. J. Knapp, Optical explosives detection: From color changes to fluorescence turn-on, *Chem. Soc. Rev.* 2009, *38*, 2543–2555.
4. D. S. Moore, Instrumentation for trace detection of high explosives, *Rev. Sci. Instrum.* 2004, *75*, 2499–2512.

5. A. W. Czarnik, A sense for landmines, *Nature* 1998, *394*, 417–418.

6. T. L. Pittman, B. Thomson, and W. J. Miao, Ultrasensitive detection of TNT in soil, water, using enhanced electrogenerated chemiluminescence, *Anal. Chim. Acta* 2009, *632*, 197–202.

7. E. Wallis, T. M. Griffin, N. Popkie, Jr., M. A. Eagan, R. F. McAtee, D. Vrazel, and J. McKinly, Instrument response measurements of ion mobility spectrometers in situ: Maintaining optimal system performance of fielded systems, *Proc. SPIE–Int. Soc. Opt. Eng.* 2005, *5795*, 54–64.

8. G. A. Eiceman and H. Schmidt, Advances in ion mobility spectrometry of explosives, in: *Aspects of Explosives Detection*, Elsevier Science, Amsterdam, the Netherlands, 2009, pp. 171–202.

9. J. Yinon, Detection of explosives by electronic noses. *Anal. Chem.* 2003, *75*, 98a–105a.

10. K. G. Furton and L. J. Myers, The scientific foundation and efficacy of the use of canines as chemical detectors for explosives, *Talanta* 2001, *54*, 487–500.

11. J. I. Steinfeld and J. Wormhoudt, Explosives detection: A challenge for physical chemistry, *Ann. Rev. Phys. Chem.* 1998, *49*, 203–232.

12. W. Fresenius, K. E. Quentin, and W. Schneider, eds., *Water Analysis*, Springer Verlag, Berlin, Germany, 1988.

13. R. Klockenkämpfer, *Total-Reflection X-ray Fluorescence Analysis*, Wiley, Hoboken, NJ, 1997.

14. R. J. C. Brown and M. J. T. Milton, Analytical techniques for trace element analysis: An overview, *TRAC Trend. Anal. Chem.* 2005, *24*, 266–274.

15. C. McDonagh, C. S. Burke, and B. D. MacCraith, Optical chemical sensors, *Chem. Rev.* 2008, *108*, 400–422.

16. D. T. McQuade, A. E. Pullen, and T. M. Swager, Conjugated polymer-based chemical sensors, *Chem. Rev.* 2000, *100*, 2537–2574.

17. J. R. Lakowicz, *Principles of Fluorescence Spectroscopy*, Kluwer Academic/Plenum, Dordrecht, the Netherlands, 1999.

18. M. S. Alsalhi, J. Alam, L. A. Dass, and M. Raja, Recent advances in conjugated polymers for light-emitting devices. *Int. J. Mol. Sci.* 2011, *12*, 2036–2054.

19. B. C. Thompson and J. M. J. Fréchet, Polymer-fullerene composite solar cells, *Angew. Chem. Int. Ed.* 2008, *47*, 58–77.

20. S. Holliday, J. E. Donaghey, and I. McCulloch, Advances in charge carrier mobilities of semiconducting polymers used in organic transistors, *Chem. Mater.* 2014, *26*, 647–663.

21. Y. Liu, K. Ogawa, and K. S. Schanze, Conjugated polyelectrolytes as fluorescent sensors, *J. Photochem. Photobio. C Photochem. Rev.* 2009, *10*, 173–190.

22. A. Alvarez, A. Salinas-Castillo, J. M. Costa-Fernández, R. Pereiro, and A. Sanz-Medel, Fluorescent conjugated polymers for chemical and biochemical sensing, *TRAC Trend. Anal. Chem.* 2011, *30*, 1513–1525.

23. S. Rochat and T. M. Swager, Conjugated amplifying polymers for optical sensing applications, *ACS Appl. Mater. Interfaces* 2013, *5*, 4488–4502.

24. T. A. Skotheim and J. R. Reynolds, Eds., *Handbook of Conducting Polymers*, CRC Press, Boca Raton, FL, 2007.

25. E. W. Paul, A. J. Ricco, and M. S. Wrighton, Resistance of polyaniline films as a function of electrochemical potential and the fabrication of polyaniline-based microelectronic devices, *J. Phys. Chem.* 1985, *89*, 1441–1447.

26. V. Dua, S. P. Surwade, S. Ammu, X. Zhang, S. Jain, and S. K Manohar, Chemical vapor detection using parent polythiophene nanofibers, *Macromolecules*, 2009, *42*, 5414–5415.

27. J. Janata and M. Josowicz, Chemical modulation of work function as a transduction mechanism for chemical sensors, *Acc. Chem. Res.* 1998, *31*, 241–248.

28. Y. L. Tang, F. Teng, M. H. Yu, L. L. An, F. He, S. Wang, Y. L. Li, D. B. Zhu, and G. C. Bazan, Direct visualization of glucose phosphorylation with a cationic polythiophene, *Adv. Mater.* 2008, *20*, 703–705.

29. J. K. Stille and Y. Gilliams, Poly(p-phenylene), *Macromolecules*, 1971, *4*, 515–517.

30. S. Shi and F. Wudl, Synthesis and characterization of water-soluble poly(*p*-phenylenevinylene) derivative, *Macromolecules* 1990, *23*, 2119–2124.

31. U. H. F. Bunz, Poly(aryleneethynylene) s: syntheses, properties, structures, and applications, *Chem. Rev.* 2000, *100*, 1605–1644.

32. I. F. Perepichka, D. F. Perepichka, H. Meng, and F. Wudl, Light-emitting polythiophenes, *Adv. Mater.* 2005, *17*, 2281–2305.

33. R. Bouldin, S. Ravichandran, A. Kokil, R. Garhwal, S. Nagarajan, J. Kumar, F. F. Bruno, L. A. Samuelson, and R. Nagarajan, Synthesis of polypyrrole with fewer structural defects using enzyme catalysis, *Synth. Met.* 2011, *161*, 1611.

34. A. G. MacDiarmid, "Synthetic metals": A novel role for organic polymers (Nobel Lecture), *Angew. Chem. Int. Ed.* 2001, *40*, 2581–2590.

35. U. Scherf and E. J. W. List, Semiconducting polyfluorenes—towards reliable structure–property relationships, *Adv. Mater.* 2002, *14*, 477–487.

36. B. Valeur and I. Leray, Design principles of fluorescent molecular sensors for cation recognition, *Coord. Chem. Rev.* 2000, *205*, 3–40.

37. L. Basabe-Desmonts, D. N. Reinhoudt, and M. Crego-Calama, Design of fluorescent materials for chemical sensing, *Chem. Soc. Rev.* 2007, *36*, 993–1017.

38. A. P. de Silva, H. Q. N. Gunaratne, T. Gunnlaugsson, A. J. M. Huxley, C. P. McCoy, J. T. Rademacher, and T. E. Rice, Signaling recognition events with fluorescent sensors and switches, *Chem. Rev.* 1997, *97*, 1515–1566.

39. R. Martinez-Manez and F. Sancenon, Fluorogenic and chromogenic chemosensors and reagents for anions, *Chem. Rev.* 2003, *103*, 4419–4476.

40. T. Gunnlaugsson, M. Glynn, G. M. Tocci, P. E. Kruger, and F. M. Pfeffer, Anion recognition and sensing in organic and aqueous media using luminescent and colorimetric sensors, *Coord. Chem. Rev.* 2006, *250*, 3094–3117.

41. P. Jiang and Z. Guo, Fluorecent detection of zinc in biological systems: Recent development on the design of chemosensors and biosensors, *Coord. Chem. Rev.* 2004, *248*, 205–229.

42. S. W. Thomas, G. D. Joly, and T. M. Swager, Chemical sensors based on amplifying fluorescent conjugated polymers, *Chem. Rev.* 2007, *107*, 1339–1386.

43. T. Q. Nguyen, J. J. Wu, V. Doan, B. J. Schwartz, and S. H. Tolbert, Control of energy transfer in oriented conjugated polymer–mesoporous silica composites, *Science* 2000, *288*, 652–656.

44. J. Gierschner, J. Cornil, H. J. Egelhaaf, Optical bandgaps of pi-conjugated organic materials at the polymer limit: Experiment and theory, *Adv. Mater.* 2007, *19*, 173–191.

45. Q. Zhou and T. M. Swager, Methodology for enhancing the sensitivity of fluorescent chemosensors—energy migration in conjugated polymers, *J. Am. Chem. Soc.* 1995, *117*, 7017–7018.

46. Q. Zhou and T. M. Swager, Fluorescent chemosensors based on energy migration in conjugated polymers: The molecular wire approach to increased sensitivity, *J. Am. Chem. Soc.* 1995, *117*, 12593–12602.
47. T. M. Swager, The molecular wire approach to sensory signal amplification, *Acc. Chem. Res.* 1998, *31*, 201–207.
48. J. S. Yang and T. M. Swager, Fluorescent porous polymer films as TNT chemosensors: Electronic and structural effects, *J. Am. Chem. Soc.* 1998, *120*, 11864–11873.
49. J. S. Yang and T. M. Swager, Porous shape persistent fluorescent polymer films: An approach to TNT sensory materials, *J. Am. Chem. Soc.* 1998, *120*, 5321–5322.
50. J. S. Kim, D. T. McQuade, A. Rose, Z. G. Zhu, and T. M. Swager, Directing energy transfer within conjugated polymer thin films, *J. Am. Chem. Soc.* 2001, *123*, 11488–11489.
51. I. A. Levitsky, J. S. Kim, and T. M. Swager, Energy migration in a poly (phenyleneethynylene): Determination of interpolymer transport in anisotropic Langmuir–Blodgett films, *J. Am. Chem. Soc.* 1999, *121*, 1466–1472.
52. Z. Tian, J. Yu, C. Wu, C. Szymanski, and J. McNeill, Amplified energy transfer in conjugated polymer nanoparticle tags and sensors, *Nanoscale*, 2010, *2*, 1999–2011.
53. A. E. Zacharides, R. S. Porter, J. Doshi, G. Srinivasan, and D. H. Reneker, High modulus polymers. A novel electrospinning process, *Polymer News*, 1995, *20*, 206–207.
54. D. H. Reneker and I. Chun, Nanometre diameter fibres of polymer, produced by electrospinning, *Nanotechnology*, 1996, *7*, 216–223.
55. R. Dersch, A. Greiner, and J. H. Wendorff, Polymer nanofibers prepared by electrospinning, in: *Dekker Encyclopedia of Nanoscience and Nanotechnology*, edited by J. A. Schwartz, C. J. Contesen, and K. Putgern, Marcel Dekker, New York, 2004, pp. 2931–2938.
56. D. Li and Y. Xia, Electrospinning of nanofibers: Reinventing the wheel? *Adv. Mater.* 2004, *16*, 1151–1170.
57. I. S. Chronakis, Novel nanocomposites and nanoceramics based on polymer nanofibers using electrospinning process—a review, *J. Mater. Process. Technol.* 2005, *167*, 283–293.
58. A. L. Yarin, S. Koombhongse, and D. H. Reneker, Taylor cone and jetting from liquid droplets in electrospinning of nanofibers, *J. Appl. Phys.* 2001, *90*, 4836–4846.
59. T. Subbiah, G. S. Bhat, R. W. Tock, S. Parameswaran, and S. S. Ramkumar, Electrospinning of nanofibers, *J. Appl. Polym. Sci.* 2005, *96*, 557–569.
60. A. Greiner and J. H. Wendorff, Electrospinning: A fascinating method for the preparation of ultrathin fibers, *Angew. Chem. Int. Ed.* 2007, *46*, 5670–5703.
61. D. H. Reneker, A. L. Yarin, H. Fong, and S. Koombhongse, Bending instability of electrically charged liquid jets of polymer solutions in electrospinning, *J. Appl. Phys.* 2000, *87*, 4531–4547.
62. A. L. Yarin, S. Koombhongse, and D. H. Reneker, Bending instability in electrospinning of nanofibers, *J. Appl. Phys.* 2001, *89*, 3018–3026.
63. M. M. Hohman, M. Shin, G. Rutledge, and M. P. Brenner, Electrospinning and electrically forced jets. I. Stability theory, *Phys. Fluids* 2001, *13*, 2201–2220.
64. M. M. Hohman, M. Shin, G. Rutledge, and M. P. Brenner, Electrospinning and electrically forced jets. I. *Applications Phys. Fluids* 2001, *13*, 2201–2220.
65. D. Tuncel and H. V. Demir, Conjugated polymer nanoparticles, *Nanoscale*, 2010, *22*, 484–494.

66. J. Pecher and S. Mecking, Nanoparticles of conjugated polymers, *Chem. Rev.* 2010, *110*, 6260–6279.
67. K. Landfester, Polyreactions in miniemulsions, *Macromol. Rapid Commun.* 2001, *22*, 896–936.
68. K. Landfester, R. Montenegro, U. Scherf, R. Güntner, U. Asawapirom, S. Patil, D. Neher, and T. Kietzke, Semiconducting polymer nanospheres in aqueous dispersion prepared by a miniemulsion process, *Adv. Mater.* 2002, *14*, 651–655.
69. K. Landfester, The generation of nanoparticles in miniemulsions, *Adv. Mater.* 2001, *13*, 765–768.
70. K. Landfester, Synthesis of colloidal particles in miniemulsions, *Annu. Rev. Mater. Res.* 2006, *36*, 231–279.
71. K. Landfester, Miniemulsion polymerization and the structure of polymer and hybrid nanoparticles, *Angew. Chem. Int. Ed.* 2009, *48*, 4488–4507.
72. E. Hittinger, A. Kokil, and C. Weder, Synthesis and characterization of crosslinked conjugated polymer milli-, micro-, and nanoparticles, *Angew. Chem. Int. Ed.* 2004, *43*, 1808–1811.
73. K. Müller, M. Klapper, and K. Müllen, Synthesis of conjugated polymer nanoparticles in non-aqueous emulsions, *Macromol. Rapid Commun.* 2006, *27*, 586–593.
74. M. C. Baier, J. Huber, and S. Mecking, Fluorescent conjugated polymer nanoparticles by polymerization in miniemulsion, *J. Am. Chem. Soc.* 2009, *131*, 14267–14273.
75. C. Wu, C. Szymanski, and J. McNeill, Preparation and encapsulation of highly fluorescent conjugated polymer nanoparticles, *Langmuir*, 2006, *22*, 2956–2960.
76. C. Wu, C. Szymanski, Z. Cain, and J. McNeill, Conjugated polymer dots for multiphoton fluorescence imaging, *J. Am. Chem. Soc.* 2007, *129*, 12904–12905.
77. C. Wu, B. Bull, C. Szymanski, K. Christensen, and J. McNeill, Multicolor conjugated polymer dots for biological fluorescence imaging, *ACS Nano*, 2008, *2*, 2415–2423.
78. C. Wu, H. Peng, Y. Jiang, and J. McNeill, Energy transfer mediated fluorescence from blended conjugated polymer nanoparticles, *J. Phys. Chem. B*, 2006, *110*, 14148–14154.
79. N. Kurokawa, H. Yoshikawa, N. Hirota, K. Hyodo, and H. Masuhara, Size-dependent spectroscopic properties and thermochromic behavior in poly (substituted thiophene) nanoparticles, *ChemPhysChem*, 2004, *5*, 1609–1615.
80. J. H. Moon, P. MacLean, W. McDaniel, and L. F. Hancock, Conjugated polymer nanoparticles for biochemical protein kinase assay, *Chem. Commun.* 2007, *46*, 4910–4912.
81. N. A. A. Rahim, W. McDaniel, K. Bardon, S. Srinivasan, V. Vickerman, P. T. C. So, and J. H. Moon, Conjugated polymer nanoparticles for two-photon imaging of endothelial cells in a tissue model, *Adv. Mater.*, 2009, *21*, 3492–3496.
82. S. J. Toal and W. C. Trogler, Polymer sensors for nitroaromatic explosives detection, *J. Mater. Chem.* 2006, *16*, 2871–2883.
83. J. S. Yang and T. M. Swager, Porous shape persistent fluorescent polymer films: An approach to TNT sensory materials, *J. Am. Chem. Soc.* 1998, *120*, 5321–5322.
84. J. W. Grate and M. H. Abraham, Solubility interactions and the design of chemically selective sorbent coatings for chemical sensors and arrays, *Sensor Actuat. B* 1991, *3*, 85–111.
85. W. M. Haynes, *CRC Handbook of Chemistry and Physics*, CRC Press, Boca Raton, FL, 2013.

86. S. Yamaguchi and T. M. Swager, Oxidative cyclization of bis(biaryl) acetylenes: Synthesis and photophysics of dibenzo[*g,p*]chrysene-based fluorescent polymers, *J. Am. Chem. Soc.* 2001, *123*, 12087–12088.

87. S. W. Thomas, III, J. P. Amara, R. E. Bjork, T. M. Swager, Amplifying fluorescent polymer sensors for the explosives taggant 2,3-dimethyl-2,3-dinitrobutane (DMNB), *Chem. Commun.* 2005, *36*, 4572–4574.

88. C. Chang, C. Chao, J. H. Huang, A. Li, C. Hsu, M. Lin, B. Hsieh, and A. Su, Fluorescent conjugated polymer films as TNT chemosensors, *Synth. Met.* 2004, *144*, 297–301.

89. L. Chen, D. McBranch, R. Wang, and D. Whitten, Surfactant-induced modification of quenching of conjugated polymer fluorescence by electron acceptors: Applications for chemical sensing, *Chem. Phys. Lett.* 2000, *330(1–2)*, 27–33.

90. A. Rose, Z. G. Zhu, C. F. Madigan, T. M. Swager, and V. Bulovic, Sensitivity gains in chemosensing by lasing action in organic polymers, *Nature*, 2005, *434*, 876–879.

91. T. Ichikawa, Y. Yamada, J. Kumagai, and M. Fujiki, Suppression of the Anderson localization of charge carriers on polysilane quantum wire, *Chem. Phys. Lett.* 1999, *306*, 275–279.

92. A. Saxena, M. Fujiki, R. Rai, and G. Kwak, Fluoroalkylated polysilane film as a chemosensor for explosive nitroaromatic compounds, *Chem. Mater.* 2005, *17*, 2181–2185.

93. H. Sohn, M. J. Sailor, D. Magde, and W. C. Trogler, Detection of nitroaromatic explosives based on photoluminescent polymers containing metalloles, *J. Am. Chem. Soc.* 2003, *125*, 3821–3830.

94. S. J. Toal, J. C. Sanchez, R. E. Dugan, and W. C. Trogler, Visual detection of trace nitroaromatic explosive residue using photoluminescent metallole-containing polymers, *J. Forensic Sci.* 2007, *52*, 79–83.

95. J. C. Sanchez, A. G. DiPasquale, A. L. Rheingold, and W. C. Trogler, Synthesis, luminescence properties, and explosives sensing with 1,1-tetraphenylsilole- and 1,1-silafluorene-vinylene polymers, *Chem. Mater.* 2007, *19*, 6459–6470.

96. G. Nagarjuna, A. Kumar, A. Kokil, K, Jadhav, S. Yurt, J. Kumar, and D. Venkataraman, Enhancing sensing of nitroaromatic vapours by thiophene-based polymer films, *J. Mater. Chem.* 2011, *21*, 16597–16602.

97. Y. R. Hua and A. H. Flood, Click chemistry generates privileged CH hydrogen-bonding triazoles: The latest addition to anion supramolecular chemistry, *Chem. Soc. Rev.* 2010, *39*, 1262–1271.

98. K. P. McDonald, Y. Hua, and A. H. Flood, 1,2,3-Triazoles and the expanding utility of charge neutral CH⋯ anion interactions, *Top. Heterocycl. Chem.* 2010, *24*, 341–366.

99. J. W. Grate, Hydrogen-bond acidic polymers for chemical vapor sensing, *Chem. Rev.* 2008, *108*, 726–745.

100. E. J. Houser, T. E. Mlsna, V. K. Nguyen, R. Chung, R. L. Mowery, and R. A. McGill, Rational materials design of sorbent coatings for explosives: Applications with chemical sensors, *Talanta* 2001, *54*, 469–485.

101. R. A. McGill, T. E. Mlsna, R. Chung, V. K. Nguyen, and J. Stepnowski, The design of functionalized silicone polymers for chemical sensor detection of nitroaromatic compounds, *Sensor Actuat. B* 2000, *65*, 5–9.

102. J. Feng, Y. Li, and M. Yang, Conjugated polymer-grafted silica nanoparticles for the sensitive detection of TNT, *Sensors Actuat. B* 2010, *145*, 438–443.

103. S. Satapathi, L. Li, R. Anandakathir, L. A. Samuelson, and J. Kumar, Sensory response and two-photon-fluorescence study of regioregular polythiophene nanoparticles, *J. Macromol. Sci. A Pure Appl. Chem.* 2011, *48*, 1049–1054.

104. S. Satapathi, A. Kokil, B. H. Venkatraman, D. Venkataraman, and J. Kumar, Sensitive detection of nitroaromatics with colloidal conjugated polymer nanoparticles, *IEEE Sensors J.* 2013, *13*, 2329.

105. J. H. Wosnick, J. H. Liao, and T. M. Swager, Layer-by-layer poly (phenylene ethynylene) films on silica microspheres for enhanced sensory amplification, *Macromolecules*, 2005, *38*, 9287–9290.

106. G. Decher, Fuzzy nanoassemblies: Toward layered polymeric multicomposites, *Science* 1997, *277*, 1232–1237.

107. Y. Lvov, G. Decher, and G. Sukhorukov, Assembly of thin films by means of successive deposition of alternate layers of DNA and poly (allylamine), *Macromolecules* 1993, *26*, 5396–5399.

108. G. Decher, Y. Lvov, and J. Schmitt, Proof of multilayer structural organization in self-assembled polycation-polyanion molecular films, *Thin Solid Films* 1994, *244*, 772–777.

109. S. H. Lee, J. Kumar, and S. K. Tripathy, Thin film optical sensors employing polyelectrolyte assembly, *Langmuir* 2000, *16*, 10482–10489.

110. X. Wang, C. Drew, S. H. Lee, K. J. Senecal, J. Kumar, and L. A. Samuelson, Electrospun nanofibrous membranes for highly sensitive optical sensors, *Nano Lett.* 2002, *2*, 1273–1275.

111. X. Wang, Y. G. Kim, C. Drew, B. C. Ku, J. Kumar, and L. A. Samuelson, Electrostatic assembly of conjugated polymer thin layers on electrospun nanofibrous membranes for biosensors, *Nano Lett.* 2004, *4*, 331–334.

112. J. Choa, R. Anandakathir, A. Kumarb, J. Kumarb, and P. U. Kurup, Sensitive and fast recognition of explosives using fluorescent polymer sensors and pattern recognition analysis, *Sensor Acutat. B Chem.* 2011, *160*, 1237–1243.

113. Z. Xu, X. Shi, L. Wang, J. Luo, C. J. Zhong, and S. Lu, Pattern recognition for sensor array signals using fuzzy ARTMAP, *Sensor Actuat. B Chem.* 2009, *141*, 458–464.

114. M. Pardo and G. Sberveglieri, Classification of electronic nose data with support vector machines, *Sensor Actuat. B Chem.* 2005, *107*, 730–737.

115. J. Kim, D. T. McQuade, S. K. McHugh, and T. M. Swager, Ion-specific aggregation in conjugated polymers: Highly sensitive and selective fluorescent ion chemosensors *Angew. Chem. Int. Ed.* 2000, *39*, 3868–3872.

116. Z. Chen, C. Xue, W. Shi, F. T. Luo, S. Green, J. Chen, and H. Liu, Selective and sensitive fluorescent sensors for metal ions based on manipulation of side-chain compositions of poly (p-phenyleneethynylene)s, *Anal. Chem.* 2004, *76*, 6513–6518.

117. L. B. Kim, B. Erdogan, J. N. Wilson, and U. H. F. Bunz, Sugar–poly (*para*-phenylene ethynylene) conjugates as sensory materials: Efficient quenching by Hg^{2+} and Pb^{2+} ions, *Chem. Eur. J.* 2004, *10*, 6247–6254.

118. L. B. Kim, A. Dunkhorst, J. Gilbert, and U. H. F. Bunz, Sensing of lead ions by a carboxylate-substituted PPE: Multivalency effects, *Macromolecules* 2005, *38*, 4560–4562.

119. L. B. Kim and U. H. F. Bunz, Modulating the sensory response of a conjugated polymer by proteins: An agglutination assay for mercury ions in water, *J. Am. Chem. Soc.* 2006, *128*, 2818–2819.

120. L. Prodi, F. Bolletta, M. Montalti, and N. Zaccheroni, Luminescent chemosensors for transition metal ions, *Coord. Chem. Rev.* 2000, *205*, 59–83.
121. A. Kokil, P. Yao, and C. Weder, Organometallic networks based on 2,2′-bipyridine-containing poly (*p*-phenylene ethynylene)s, *Macromolecules* 2005, *38*, 3800–3807.
122. W. L. Wang, J. W. Xu, and Y. H. Lai, Bipyridinophane-containing conjugated polymers modulated with an intramolecular aromatic C—H/π interaction, *J. Polym. Sci., Part A: Polym. Chem.* 2006, *44*, 4154–4164.
123. H. Huang, K. Wang, W. Tan, D. An, X. Yang, S. Huang, Q. Zhai, L. Zhou, and Y. Jin, Design of a modular-based fluorescent conjugated polymer for selective sensing, *Angew. Chem. Int. Ed.* 2004, *43*, 5635–5638.
124. B. Liu, W. L. Yu, J. Pei, S. Y. Liu, Y. H. Lai, and W. Huang, Design and synthesis of bipyridyl-containing conjugated polymers: Effects of polymer rigidity on metal ion sensing, *Macromolecules* 2001, *34*, 7932–7940.
125. M. Zhang, P. Lu, Y. Ma, and J. Shen, Metal ionochromic effects of conjugated polymers: Effects of the rigidity of molecular recognition sites on metal ion sensing, *J. Phys. Chem. B* 2003, *107*, 6535.
126. X. Huang, J. Meng, Y. Dong, Y. Cheng, and C. Zhu, Polymer-based fluorescence sensor incorporating triazole moieties for Hg^{2+} detection via click reaction, *Polymer* 2010, *51*, 3064–3067.
127. A. Boldea, I. Lévesque, and M. Leclerc, Controlled ionochromism with polythiophenes bearing crown ether side chains, *J. Mater. Chem.* 1999, *9*, 2133–2138.
128. Y. Zhang, C. B. Murphy, and W. E. Jones, Poly[*p*-(phenyleneethynylene)-*alt*-(thienyleneethynylene)] polymers with oligopyridine pendant groups: Highly sensitive chemosensors for transition metal ions, *Macromolecules* 2002, *35*, 630–636.
129. H. Liu, S. Wang, Y. Luo, W. Tang, G. Yu, L. Li, C. Chen, Y. Liu, and F. Xi, Synthesis and properties of crown ether containing poly (*p*-phenylenevinylene), *J. Mater. Chem.* 2001, *11*, 3063–3067.
130. G. Ramachandran, G. Simon, Y. Cheng, T. A. Smith, and L. Dai, The dependence of benzo-15-crown-5 ether-containing oligo paraphenylene vinylene (CE-OPV) emission upon complexation with metal ions in solution, *J. Fluoresc.* 2003, *13*, 427–436.
131. H. Yang, C. Duan, Y. Wu, Y. Lv, H. Liu, Y. Lv, D. Xiao, F. Huang, H. Fu, and Z. Tian, Conjugated polymer nanoparticles with Ag^+-sensitive fluorescence emission: A new insight into the cooperative recognition mechanism, *Part. Part. Syst. Charact.* 2013, *30*, 972–980.

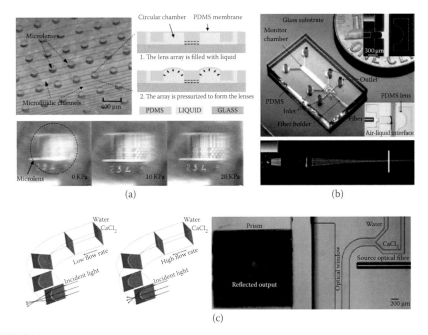

FIGURE 2.1
(a) Microscopic image of the microlens array with underlying microfluidic network for pressure regulation (left) and schematic of the microlens actuation via pressure control. The focusing results can be varied due to change of the pressure. (b) A microfluidic lens that is tunable through active pressure control on an air–liquid interface. The experimental results of the device demonstrate good focusing and allow focal length adjustment through pressure control. (c) Mechanism of in-plane tunable fluid–fluid lens for focusing light in the z-direction with respect to device plane (where the plane of the device is the x–y plane). Change of the lens curvature and focal distance can be achieved by changing the flow rate. (Images are recompiled with permission from *Optics Express*, The Optical Society of America; *Microfluidics and Nanofluidics*, Springer; and *Lab on a Chip*, Royal Society of Chemistry.)

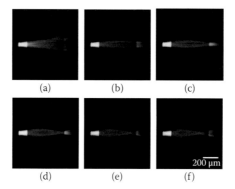

(a)　　　　(b)　　　　(c)

(d)　　　　(e)　　　　(f)

200 μm

FIGURE 2.5
Ray-tracing experiments for the translation mode to characterize the variable focal length at different flow conditions. (a) Ray tracing for stagnant flow (homogenous refractive index). (b–f) Ray tracing for dynamic flows. The CaCl$_2$ flow rates were fixed at 3 μL/min and the H$_2$O flow rates were (b) 0.6, (c) 1.2, (d) 1.8, (e) 2.4, and (f) 3 μL/min. (Images are reproduced with permission from *Lab on a Chip,* Royal Society of Chemistry.)

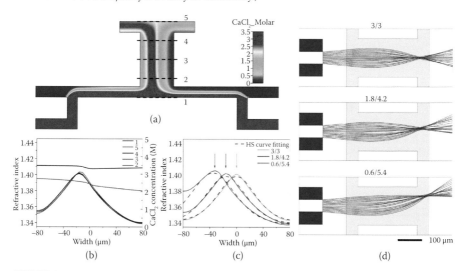

FIGURE 2.7
Computational fluid dynamics (CFD) and ray-tracing simulation of the swing mode. (a) Simulated CaCl$_2$ concentration distribution (CaCl$_2$ flow rates = 3 μL/min on each side, and the H$_2$O flow rate on the left = 1.8 μL/min, and the H$_2$O flow rate on the right = 4.2 μL/min, abbreviated as 1.8/4.2 μL/min) in a short-version liquid gradient refractive index (L-GRIN) lens. The color bar represents the molar concentration of CaCl$_2$. (b) Cross-sectional CaCl$_2$ concentration/refractive index profiles at different locations (top to bottom: cross-sections 1, 5, 4, 3, and 2 defined in [a]). (c) Refractive index profile at the middle of the L-GRIN lens (cross-section 3) for different flow conditions (left to right: CaCl$_2$ solution flow rates were fixed at 3 μL/min for both sides, and the H$_2$O flow rates were 0.6/5.4, 1.8/4.2, and 3/3 μL/min, respectively). Dotted lines are hyperbolic secant curve fitting. (d) Ray-tracing simulation in a short-version L-GRIN lens using the parameters obtained from the CFD simulation. The flow conditions are indicated in the graph (i.e., 1.8/4.2 represents H$_2$O flow rate 1.8 μL/min on the left and 4.2 μL/min on the right). (Images are reproduced with permission from *Lab on a Chip,* Royal Society of Chemistry.)

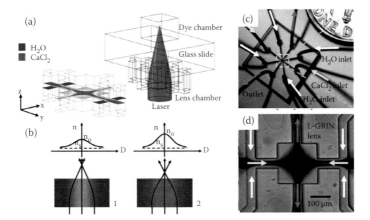

FIGURE 2.9
Principle of the out-of-plane liquid gradient refractive index (L-GRIN) lens. (a) The out-of-plane L-GRIN lens structure is composed of an L-GRIN lens chamber, two inlets for $CaCl_2$ solution, four inlets for DI water, and two outlets. The diffusion of $CaCl_2$ inside the L-GRIN lens chamber results in a two-dimensional (2D) axis-symmetric hyperbolic secant (HS) refractive index profile in the x–y plane, which can be used to focus the light beam along the z-direction. The lens is bonded to a glass substrate and a dye chamber is bonded on the opposite side of the glass substrate for visualization purposes. The input light is generated via a laser diode aligned vertically to the device plane. (b) The side view of the refractive index distribution in the L-GRIN lens chamber. The focal point of the out-of-plane L-GRIN lens can be shifted along the z-axis (e.g., from b1 to b2 by adjusting the refractive index gradient within the liquid medium). (c) The fluid injection setup of the out-of-plane L-GRIN lens (dye chamber is not included in this image). (d) A microscopic image of the 2D diffusion pattern in the lens chamber. (Images are reproduced with permission from *Lab on a Chip*, Royal Society of Chemistry.)

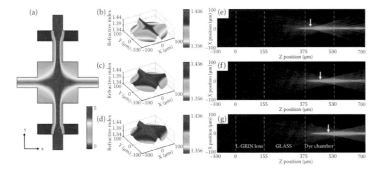

FIGURE 2.10
Computational fluid dynamics (CFD) and ray-tracing simulation of the variable light focusing process. (a) Top view of the CFD simulation from mixing of $CaCl_2$ and DI water. (b–d) The CFD simulated two-dimensional hyperbolic secant refractive index profile within the lens chamber for different flow conditions (DI water flow rate is 8 µL/min, 3.2 µL/min, and 2.4 µL/min, respectively, and $CaCl_2$ solution flow rate is 8 µL/min in all three cases). (e–g) The simulated trajectories of the light beam during the focusing process with three different fluidic flow conditions shown in (b–d), respectively. The structure in the simulation is composed of three parts: the liquid gradient refractive index lens, the glass substrate, and the dye chamber. (Images are reproduced with permission from *Lab on a Chip*, Royal Society of Chemistry.)

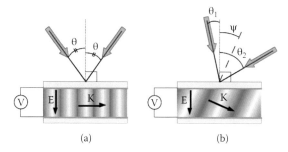

FIGURE 3.9
Holographic grating transmission geometries. (a) Symmetric. (b) Tilted.

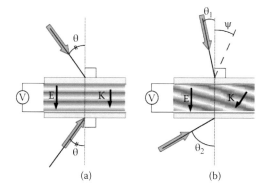

FIGURE 3.10
Holographic grating reflection geometries. (a) Symmetric. (b) Tilted.

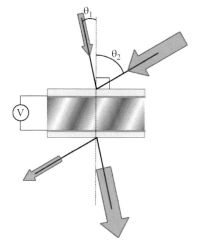

FIGURE 3.11
Two-beam coupling measurement configuration.

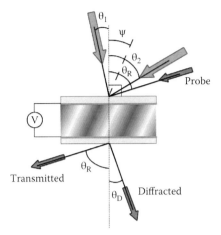

FIGURE 3.12
Four-wave mixing measurement configuration.

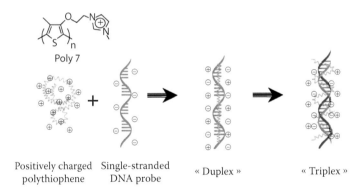

Poly 7

Positively charged polythiophene · Single-stranded DNA probe · « Duplex » · « Triplex »

FIGURE 4.6
Schematic description of the formation of polythiophene/single-stranded probe deoxyribonucleic acid (DNA) duplex and polythiophene/double-stranded probe DNA triplex complex. (Reprinted from *Tetrahedron*, 60, Bera-Aberem et al., Functional polythiophenes as optical chemo- and biosensors, 11169, Copyright 2004, with permission from Elsevier.)

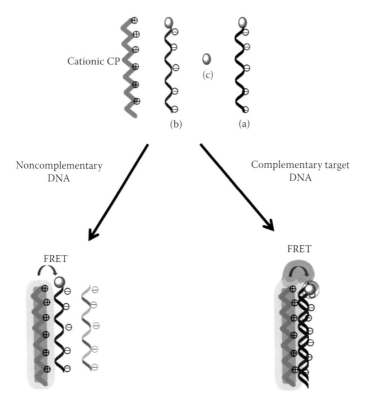

FIGURE 4.8
The components involved in cationic conjugated-polymer-based fluorescence resonance energy transfer a DNA sensor with (a) labeled target, (b) labeled probe, and (c) intercalating dye.

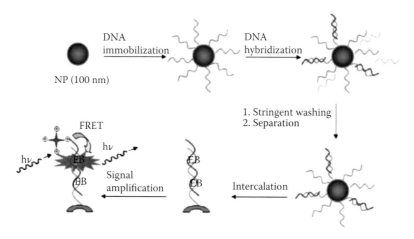

FIGURE 4.9
Schematic cationic conjugated-polymer-assisted nanoparticle-based DNA assay. (Reprinted with permission from Wang et al., 7214. Copyright 2007 American Chemical Society.)

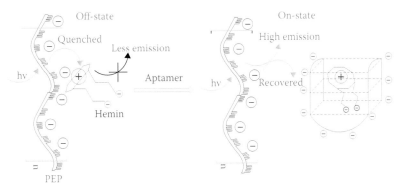

FIGURE 4.12

Schematics for the principle of fluorescent switch combining poly(fluorene-co-phenylene), hemin, and hemin aptamer. (Reprinted with permission from Li et al., 3544. Copyright 2009 American Chemical Society.)

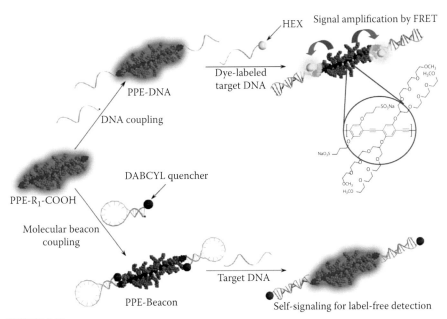

FIGURE 4.14

Homogenous molecular-beacon-based DNA sensing system with signal transduction afforded by the poly(*p*-phenylene ethynylene). (Lee et al.: *Advanced Functional Materials*. 2007. 17. 2580. Copyright Wiley-VCH Verlag GmbH & Co. KGaA. Reproduced with permission.)

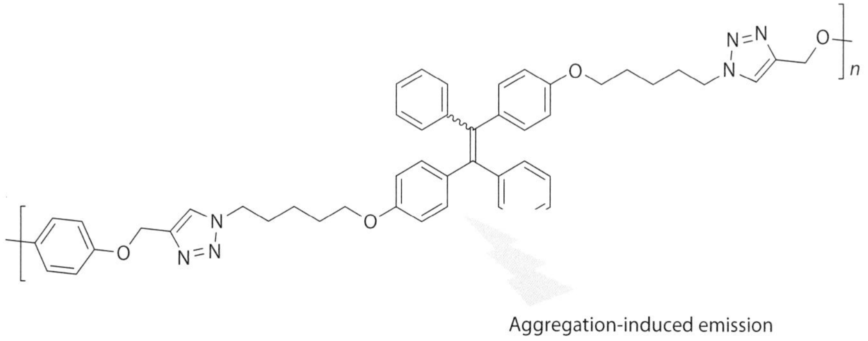

Aggregation-induced emission

FIGURE 5.5
Chemical structure of the poly(triazole) sensor containing aggregation-induced emission chromophores.

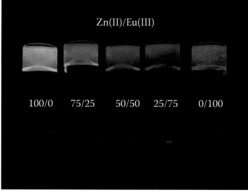

Zn(II), Eu(III)

Zn(II)/Eu(III)

100/0   75/25   50/50   25/75   0/100

FIGURE 5.6
Multicolored organogels of poly(triazole) with Zn^{2+} and Eu^{3+} ions. (Adapted with permission from Yuan et al., 11515–11522, Copyright 2012 The Royal Chemical Society.)

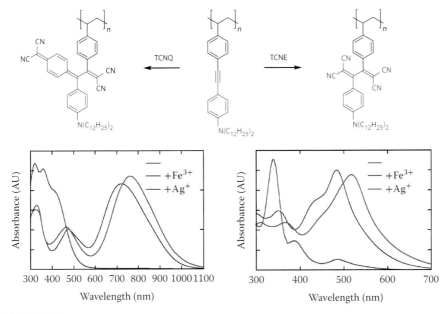

FIGURE 5.11
Click postfunctionalization of polystyrene derivative by TCNE and TCNQ and their spectral changes in response to metal ion recognition. (Adapted with permission from Li et al., 1996–2005, Copyright 2012 The Royal Chemical Society.)

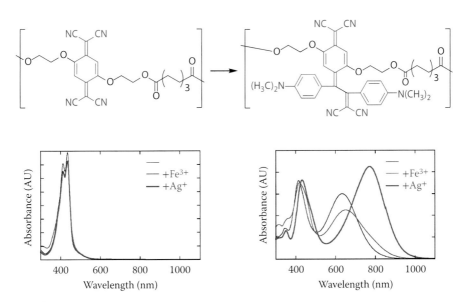

FIGURE 5.12
Click postfunctionalization of the TCNQ polyester by dialkylaniline-appended alkynes and their spectral changes in response to metal ion recognition.

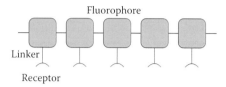

FIGURE 6.1
Schematic of a typical fluorescent chemosensor.

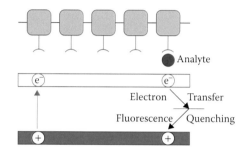

FIGURE 6.6
Schematic for fluorescence quenching through the molecular wire effect.

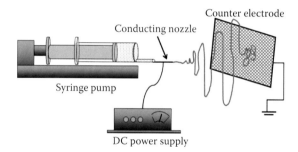

FIGURE 6.7
Schematic of a typical electrospinning setup.

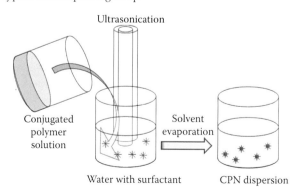

FIGURE 6.8
Schematic of a typical miniemulsion protocol used for the synthesis of conjugated polymer nanoparticles.

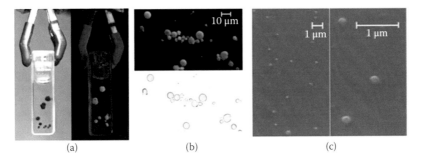

FIGURE 6.9
(a) Photographs, (b) optical micrographs, and (c) scanning electron micrographs of cross-linked conjugated (a) milli-, (b) micro-, and (c) nanoparticles, respectively, prepared by method B (milli), method C (micro), and method D (nano). (Hittinger, E., Kokil, A., and Weder, C.: *Synthesis and Characterization of Cross-Linked Conjugated Polymer Milli-, Micro-, and Nanoparticles.* Angew. Chem. Int. Ed. 2004, 43, 1808–1811. Copyright Wiley-VCH Verlag GmbH & Co. KGaA. Reprinted with permission.)

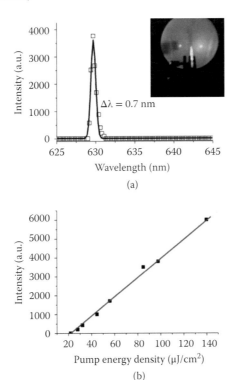

$\Delta\lambda = 0.7$ nm

(a)

(b)

FIGURE 7.22
(a) Emission spectra from the sample with conductive polymer layer and HPDLC layer (curve: Gaussian fit; inset: laser emission profile image) and (b) emission laser intensity versus pump energy density. (Reprinted from *Org. Electron.*, 13, W. Huang, Z. Diao, Y. Liu, Z. Peng, C. Yang, J. Ma, and L. Xuan, Distributed feedback polymer laser with an external feedback structure fabricated by holographic polymerization technique, 2307–2311, Copyright 2012, with permission from Elsevier.)

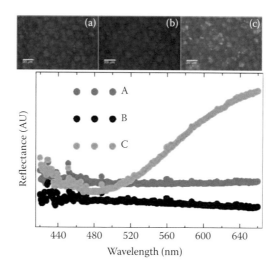

FIGURE 8.9
Optical micrographs of the same graphite particles at different stages of lithium content: (a) no lithium, (b) LiC_{12}, and (c) LiC_6. The scale bars are 20 µm in all pictures. Plot of the corresponding reflectance versus wavelength spectra. (From A. Timmons et al., Quantification of color changes in graphite upon the electrochemical intercalation of Li, *212th ECS Meeting*, Abstract 755.)

FIGURE 8.16
Electrochromic materials in architectural pilot projects: automatically dimming smart windows from a Berkeley lab (a) 10:30 a.m. and (b) 10:50 a.m. (Adapted with permission from Windows and Daylighting Group, Lawrence Berkeley National Laboratory, Berkeley, California.)

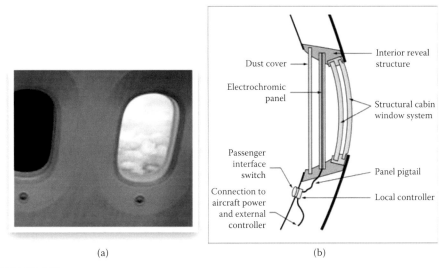

(a)   (b)

FIGURE 8.17
(a) Electrochromic dimmable aircraft window developed by Gentex Corporation; (b) schematic of the electrochromic window fitting inside the aircraft. (Adapted from Gentex Corporation, http://www.gentex.com.)

FIGURE 8.18
Electrochromic mirror for automobiles for glare reduction: automatically dimming rearview mirror from Gentex (http://www.gentex.com/automotive_products2.html).

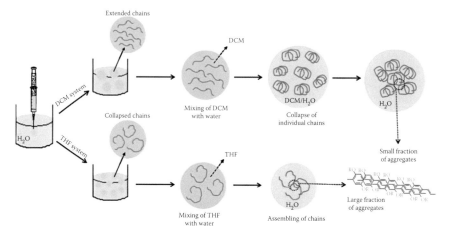

FIGURE 9.4
Proposed schematic for nanoparticle formation when changing initial solvents. (Reprinted from *Journal of Colloid and Interface Science*, 403, Potai and Traiphol, Controlling chain organization and photophysical properties of conjugated polymer nanoparticles prepared by reprecipitation method: The effect of initial solvent, 58–66, Copyright 2013, with permission from Elsevier.)

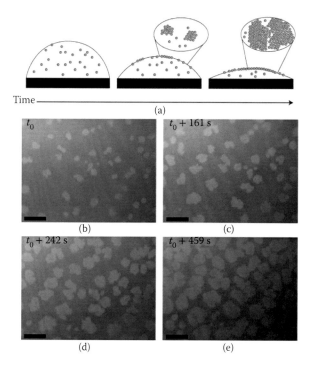

FIGURE 9.14
Schematic of migration of nanoparticles to the liquid–air interface during drying (a) and optical images of nanoparticle island growth (b–e). (Reprinted with permission from Macmillan Publishers Ltd. *Nature Materials*, Bigioni et al. 2006, Copyright 2006.)

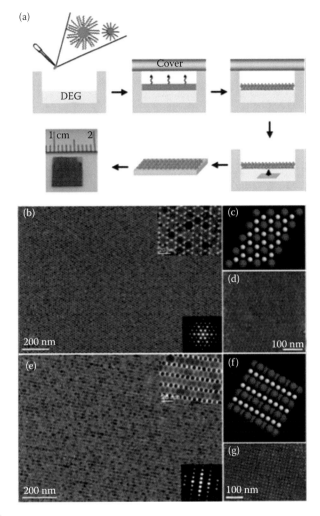

FIGURE 9.15
Schematic of nanoparticle assembly at the DEG–air interface (a), and AlB2-type BNSL membranes self-assembled from 15-nm Fe_3O_4 and 6-nm FePt nanocrystals (b–g). (Reprinted with permission from Macmillan Publishers Ltd. *Nature*, Shevchenko et al. 2006b, Copyright 2010.)

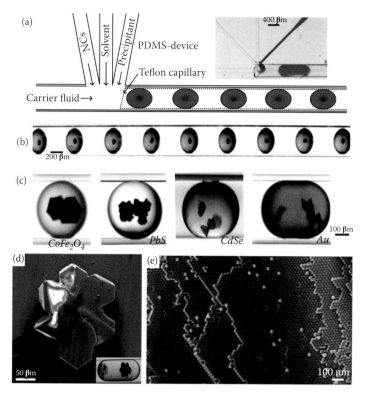

FIGURE 9.17

Apparatus for microfluidic self-assembly (a) in nanoliter plugs (b). Crystallization within the plugs (c) results in ordered assemblies (d and e). (Reprinted with permission from Bodnarchuk et al. 2011a, 8956–8960. Copyright 2011 American Chemical Society.)

7

Holographic Polymer-Dispersed Liquid Crystals: From Materials and Morphologies to Applications

Ji Ma, Wenbin Huang, Li Xuan, and Hiroshi Yokoyama

CONTENTS

7.1 Introduction

Polymers and liquid crystals (LCs) are intensively used for various optical devices. They are convenient to process and easy to integrate with other optical components [1–7]. LCs have liquid-like fluidity and crystal-like molecular order in which the optical and electrical properties along the direction parallel and perpendicular to the LC molecule are anisotropic. This unique property makes LC devices easily respond to external fields such as electric field, temperature, mechanical force, and magnetic or optical field [8–18]. The combination between LCs and functional polymers can bring more opportunities for novel practical applications [19–22]. The LC and polymer systems or composites are mainly divided into two classes: polymer-dispersed liquid crystals (PDLCs) [23,24] and polymer-stabilized liquid crystals (PSLCs) [25–28]. In PDLCs, the LC exists in the polymer matrix in the form of microsized droplet. The concentration of LC is from 20 to 80 wt %. In PSLCs, the LC forms a continuous medium, while the polymer is in the form of a fiber-like or sponge-like structure. The concentration of LC is usually less than 10 wt %.

Holographic polymer-dispersed liquid crystals (HPDLCs) are one of the well-known LC and polymer composites that have been widely investigated since 1993 [29–31]. HPDLCs or HPDLC gratings are formed by photo-induced phase separation (PIPS) and recorded in a holographic interference optical field created by two or multiple coherent laser beams. The monomers or prepolymers in the bright interference fringes undergo photopolymerization. The polymers formed in those regions eject the LCs to the dark fringes in the interference optical pattern. The achieved HPDLC gratings consist of alternated LC-rich and polymer-rich layers. Compared to other microstructure fabrication techniques, such as nanoimprint lithography [32], holographic lithography [33], electron beam lithography [34], or self-assembly of block copolymers [35], HPDLCs exhibit the merits of simple optical construction, one-step fabrication, large-area processing capability, and tunability by external fields [31]. The potential applications of HPDLCs include displays, optical switches, optical communications, organic lasers, and so on.

In this chapter, we first introduce HPDLC fabrication including optical setups, materials, and curing conditions, and then discuss three types of HPDLC morphologies. The optical applications of distributed feedback (DFB) lasers based on HPDLCs, electrically tunable HPDLC lasing behaviors, and anisotropic waveguide theory for HPDLC DFB lasers are presented in Section 7.4. The development and progress of HPDLCs and their optical applications on organic lasers are specified here.

7.2 Holographic Polymer-Dispersed Liquid Crystals

HPDLCs are formed in the interference pattern created by two or multiple laser beams by the PIPS method. The prepolymer mixture contains photocurable monomer(s)/oligomer(s) or prepolymer(s) and LC(s) (typical LC concentration: 20–40 wt %). Unlike PDLCs, in which the photopolymerization happens everywhere because of uniform illumination [29–31,36–40], the photopolymerization in HPDLCs starts preferentially in the bright fringes in the interference pattern. The monomer consumption in the bright fringes induces a chemical potential gradient, causing the diffusion of monomers from the dark fringes to the bright fringes and the diffusion of LCs from the bright fringes to the dark fringes [41–45]. Depending on different writing setups, materials, and curing conditions, HPDLC gratings with different morphologies can be achieved that would exhibit different optical properties and performances.

7.2.1 HPDLC Writing Setup

HPDLC transmission gratings and reflection gratings can be obtained by using different writing setups. For transmission gratings, the writing beams are incident onto the same side of the sample and the formed grating planes are perpendicular to the cell surface (Figure 7.1a). For reflection gratings, the writing beams are incident onto the both sides of the sample and the formed grating planes are parallel to the cell surface (Figure 7.1b). The designed grating periods (grating pitches) are easy to control by adjusting the intersection beam angle (θ_i), according to

$$\Lambda = \frac{\lambda_w}{2\bar{n}\sin(\theta_i/2)} \tag{7.1}$$

where Λ is the grating period, λ_w is the writing wavelength, and $\bar{n}$ is the average refractive index of the material mixture [31].

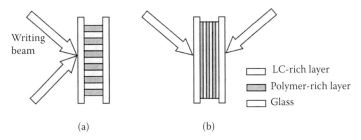

(a)   (b)

FIGURE 7.1
Typical optical writing setups for HPDLCs: (a) for transmission gratings and (b) for reflection gratings.

7.2.2 Raman–Nath- and Bragg-Type HPDLC

Depending on the grating period and thickness, HPDLC can be regarded as Raman–Nath- or Bragg-type grating on the basis of Cook–Klein parameter Q [46]:

$$Q = 2\pi \frac{\lambda_i d}{\bar{n}\Lambda^2} \quad (7.2)$$

where d is the thickness of the grating and λ_i is the incident light wavelength. For Raman–Nath-type grating ($Q < 1$), that is, *thin* grating, multiple diffraction orders can be observed. For Bragg-type grating ($Q > 1$), that is, *thick* or volume grating, only zero order and first order of diffraction can be observed when the light is incident to the sample and satisfies the Bragg condition. Because optical losses of diffraction are low, Bragg-type HPDLCs are preferable for practice applications. In this chapter, only Bragg-type HPDLCs are discussed.

7.2.3 HPDLC Prepolymer Mixture

In general, HPDLC prepolymer mixture consists of monomer(s), LC(s), and photoinitiator(s). Two types of monomers, acrylate-based and thiol-ene-based monomers, are used in HPDLCs. Acrylate-based monomers are extensively used for transmission, reflection, and two-dimensional (2D) and three-dimensional (3D) HPDLC gratings [47–83]. Acrylate-based materials with different functionalities have been developed, including dipentaerythrol hydroxy pentaacrylate (DPHPA) [29,30,62,84], pentaerythritol tetraacrylate (PETA-3) [85,86], 2-ethylhexyl acrylate (EHA) [87,88], trimethylolpropane triacrylate (TMPTA) [88–91], aliphatic urethane acrylate oligomer/monomer blend (such as hexafunctional EB8301 [92–94] and trifunctional EB4866 [93,94]), dipentaerythritol pentaacrylate (SR399) [95], polyurethane acrylate oligomers (PUA) [84,96–98], phthalic diglycol diacrylate (PDDA) [99,100], and so forth. Thiol-ene-based monomers, as alternative prepolymers, have virtues of high monomer conversion, good stability, and less shrinkage, and have also been widely used in HPDLCs [101–105]. Thiol-ene monomers are composed of multifunctional thiol and vinyl monomers, such as Norland optical adhesives (NOA 61 [102], NOA 65 [106], and NOA 68). Initially thiol-ene-based monomers could only be cured by UV light. Later, more material systems were developed for visible-light curing [104,105]. Some examples of prepolymers/monomers are shown in Figure 7.2.

As for photoinitiators, Rose Bengal (RB) and coinitiator *N*-phenylglycine (NPG) are usually used in HPDLCs to generate free radicals and initiate photopolymerization for acrylate-based material systems. Because the absorption peak of RB is 490 nm, blue-green lasers (488 nm, 514 nm, and 532 nm) are adopted as writing light sources. Later, photoinitiators for UV laser (351 nm) [103], He–Ne laser (633 nm) [107,108], and infrared (IR) lasers (830

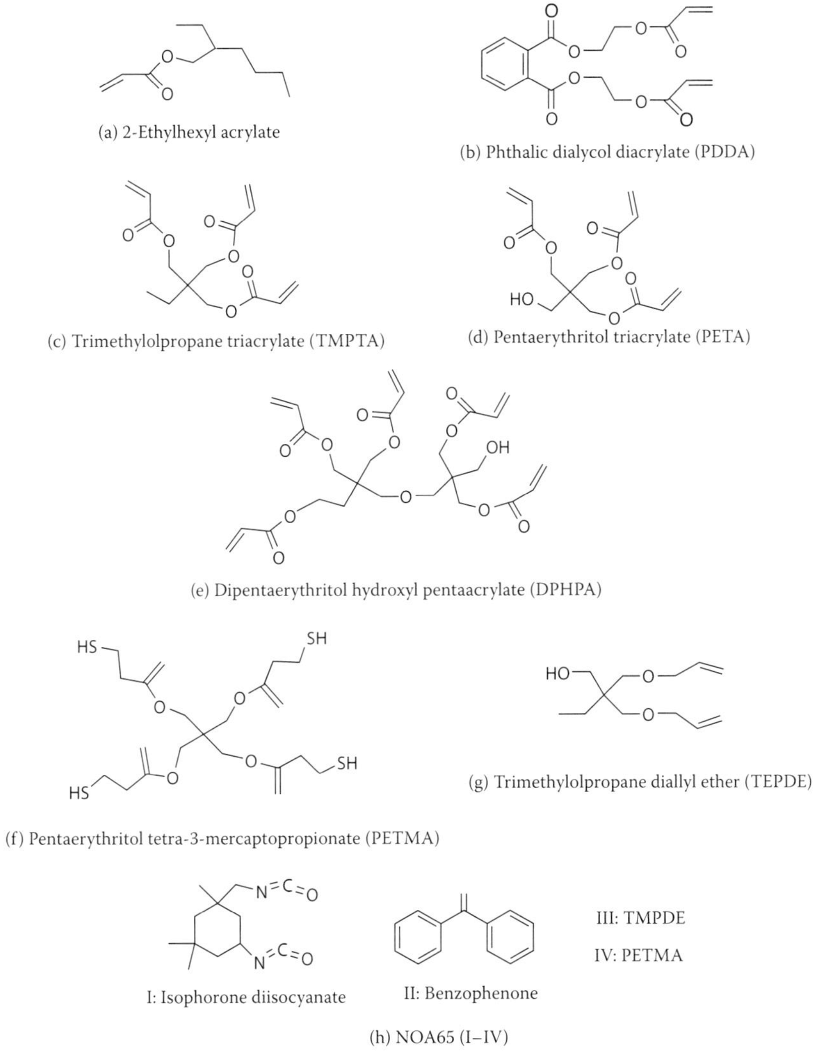

(a) 2-Ethylhexyl acrylate

(b) Phthalic dialycol diacrylate (PDDA)

(c) Trimethylolpropane triacrylate (TMPTA)

(d) Pentaerythritol triacrylate (PETA)

(e) Dipentaerythritol hydroxyl pentaacrylate (DPHPA)

(f) Pentaerythritol tetra-3-mercaptopropionate (PETMA)

(g) Trimethylolpropane diallyl ether (TEPDE)

III: TMPDE

IV: PETMA

I: Isophorone diisocyanate

II: Benzophenone

(h) NOA65 (I–IV)

FIGURE 7.2
Chemical structures of typical prepolymers/monomers used in HPDLCs: (a–e) for acrylate-based material systems and (f–h) for thiol-ene-based material systems.

nm, 834 nm, and 850 nm) [109,110] were developed. The design of material systems for different writing laser beams is meaningful. For reflection HPDLCs, longer writing laser wavelengths can allow reflection bands of the gratings to extend up to 2 μm, whereas for transmission HPDLCs, the development of shorter writing wavelengths can allow gratings with narrow periods below 200 nm. Some examples of photoinitiators are shown in Figure 7.3.

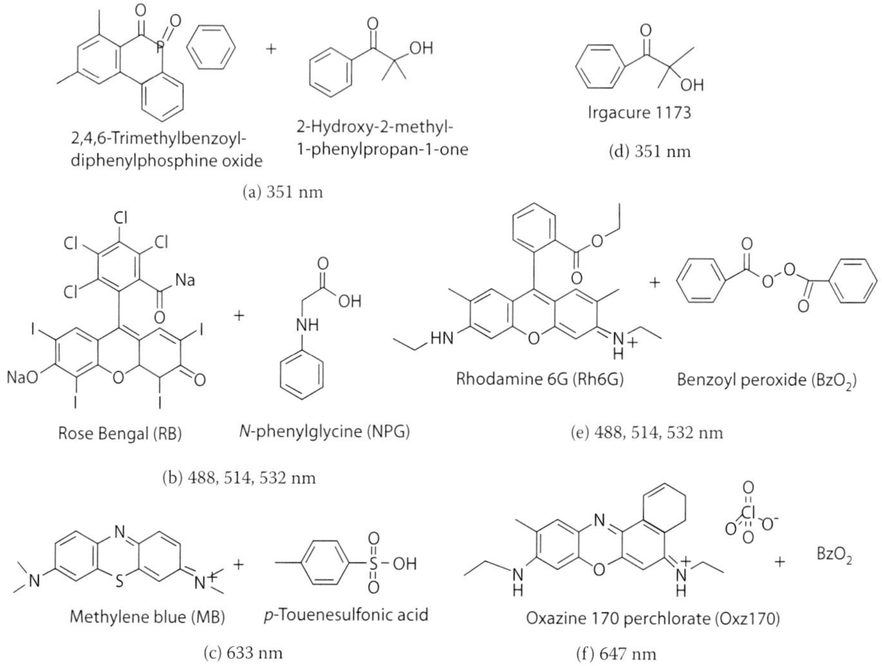

FIGURE 7.3
Chemical structures of typical photoinitiators used in HPDLCs: (a–c) for acrylate-based material systems and (d–f) for thiol-ene-based material systems.

7.3 HPDLC Morphology

In this section, we divide HPDLC grating morphologies into three categories and introduce each type in detail. The first category is LC droplet-like morphology, in which the transmission or reflection gratings are made from acrylate-based monomers or thiol-ene-based monomers with a fast curing process. The second is polymer scaffolding morphology, in which the transmission gratings are made from acrylate-based monomers with a slow curing process. The third is sliced polymer morphology, in which the transmission gratings are made from thiol-ene-based monomers with a slow curing process. It should be mentioned that all the gratings in the three categories discussed here are highly efficient, with better phase separations and high diffraction efficiencies.

7.3.1 LC Droplet-Like Morphologic HPDLC Transmission Gratings

In LC droplet-like morphologic transmission gratings, the phase-separated LC is in the form of distinct, elongated droplets embedded in the polymer matrix. Scanning electron microscopy (SEM) images of LC droplet-like morphologic HPDLC transmission gratings made from acrylate-based material systems [88] are shown in Figure 7.4. The transmission gratings made from thiol-ene-based material systems also show similar morphology [105]. Compared with acrylate-based HPDLCs, LC droplets in thiol-ene-based HPDLCs are more spherical and uniform. The diameter of LC domains in this type HPDLC is about half grating pitch ($\approx$500 nm), which is comparable to the visible-light wavelength. Therefore, when visible light is incident on the gratings, light scattering will be induced by the refractive index mismatch between LC droplets and surrounding polymer (up to 15%).

To better understand the formation mechanism of LC droplet-like morphologic transmission gratings, the curing conditions from some examples are listed in Table 7.1. These three examples are selected here because the achieved transmission gratings exhibit high diffraction efficiencies. High diffraction efficiency means better phase separation between LCs and polymers. We can see that the curing time is fast ($\approx$30 seconds) and the curing intensity per each beam is strong ($\approx$100 mW/cm^2). This indicates that the polymerization rates in these samples are high. High curing light intensity and high average functionality of monomers (>3.5) are the reasons for LC droplet-like morphology formation.

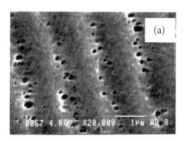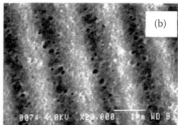

FIGURE 7.4

SEM images of typical LC droplet-like morphologic transmission gratings made from acrylate-based material systems: (a) hexafluoroisopropyl acrylate (HFIPA) and (b) trifluoroethyl acrylate (TFEA). (Reprinted from *Polymer*, 43, M. D. Sarkar, J. Qi, and G. P. Crawford, Influence of partial matrix fluorination on morphology and performance of HPDLC transmission gratings, 7335–7344, Copyright 2002, with permission from Elsevier.)

TABLE 7.1

Curing Conditions for LC Droplet-Like Morphologic HPDLC Transmission Gratings

Monomer	Effective Functionality	Writing Wavelength (nm)	Exposure Intensity (per beam, mW/cm^2)	Curing Time (seconds)
DPHPA [31]	5	488	95	30–120
EB4866 and EB8301 [112]	4.5	514	150	30
NOA65 [104]	—	647	100–500	45

In Sections 7.3.4 and 7.3.5 we list and compare the curing conditions for other morphologic gratings.

The LC alignment is very critical in LC-involved devices because specific LC director configuration or LC orientation will directly influence the capabilities to modulate the light of the devices. For HPDLC transmission gratings, it has been discovered that the diffraction efficiency for p polarized light is always different from that for s polarized light, indicating the LCs in the grating have a certain orientation. However, the LC director configuration details are not easy to identify because the sizes of LC droplets are not uniform, the LC director configuration in each droplet is puzzling, and the droplet sizes are relatively small (<1 μm). Direct observation using polarization optical microscopy (POM) to analyze the LC director configuration inside the droplets is also difficult, and even though the LC droplets can be regarded as uniaxial domains, the distribution of these domains in the grating is unrevealed. Previous studies on this topic used some fitting parameters to achieve agreement between experimental diffractive data and anisotropic coupled wave theory. The fitting parameters used in the models will more or less give some clues about the LC domain alignment inside the droplet-like domain.

The first analysis of LC droplet orientation was proposed through SEM image analysis [111]. It assumed that the LC droplets were uniaxial domains and were *squeezed* along the grating vector during the curing process, so most LC molecules in the droplets would align along the main axis of droplet domains. Another precise characterization was carried out by measuring the diffraction efficiency at a temperature above the LC clearing point to fit the filling fraction of LC droplets and confirmed that the LC molecules in the droplets are highly ordered along the grating vector [112]. A more direct way to determine the LC alignment was using linear optical birefringence measurement and rotational second harmonic generation technique, and it was found that the LCs are aligned approximately along the grating vector [113]. On the other hand, another theoretical model that was proposed found that the distribution of LC domains varies with different LCs [85,86]. For the transmission gratings made from LC E7 and BL045, the LC domains would tend to align along the grating vector. For fluorinated biphenyls LC TL216, TL213, and MLC2048, the LC domains would tend to align in the grating plane with a high amount of disorder. From the viewpoint of application, the diffraction efficiency in the latter is too low to exploit.

7.3.2   LC Droplet-Like Morphologic HPDLC Reflection Gratings

So far, HPDLC reflection gratings with efficient phase separation exhibit LC droplet-like morphology [114,115]. Typical images of LC droplet-like morphologic HPDLC reflection gratings [115,116] are shown in Figure 7.5. The diameter of the LC droplet is approximately 100 nm in these images, which is much smaller than that in the transmission gratings because of the narrower grating period. Similarly, compared with acrylate-based reflection

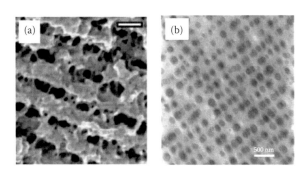

FIGURE 7.5
Images of LC droplet-like morphologic HPDLC reflection gratings: (a) made from multifunctional acrylate-based material system (scale bar: 150 nm) (R. Jakubiak, T. J. Bunning, R. A. Vaia, L. V. Natarajan, and V. P. Tondiglia: Electrically switchable, one-dimensional polymeric resonators from holographic photopolymerization: a new approach for active photonic bandgap materials. *Adv. Mater.* 2003. 15. 241–244. Copyright Wiley-VCH Verlag GmbH & Co. KGaA. Reproduced with permission.) and (b) made from thiol-ene-based material system (scale bar: 500 nm). (Reprinted from *Polymer*, 47, L. V. Natarajan, D. P. Brown, J. M. Wofford, V. P. Tondiglia, R. L. Sutherland, P. F. Loyd, and T. J. Bunning, Holographic polymer dispersed liquid crystal reflection gratings formed by visible light initiated thiol-ene polymerization, 4411–4420, Copyright 2006, with permission from Elsevier.)

HPDLCs, the LC droplet domains in thiol-ene-based reflection HPDLCs are more spherical and uniform.

Usually the diffraction efficiency is nonmonotonically dependent on the product of the film thickness and phase separation degree for transmission HPDLCs, whereas it will increase monotonically with this product for reflection HPDLCs [117]. Hence, when the cell gap is fixed, it is convenient to optimize fabrication conditions to achieve reflection gratings with higher diffraction efficiency. For example, the diffraction efficiency of reflection gratings as a function of exposure intensity [118] is shown in Figure 7.6. Higher exposure intensity gives higher diffraction efficiency. However, the polymer chain length obtained at a low photopolymerization rate will negatively influence the grating formation kinetics [119]. The effective diffusion constant of monomers could be decreased by two orders of magnitude blocked by the long polymer chains. This nonlocal effect is extremely influential for reflection gratings, as the polymer chain lengths are in 10^1 nm and the reflection grating periods are in 10^2 nm. Consequently, the slow curing process with a low photopolymerization rate is not suitable for reflection HPDLC gratings. To obtain good performance reflection gratings, the curing process has to use fast polymerization process to suppress the nonlocal effect.

Earlier, the distribution of LC droplets in reflection gratings was regarded as random and the achieved gratings were isotropic, because the diffraction efficiency for p and s polarized light was the same when the light was incident in the normal direction to the grating. Later, the diffraction efficiency for p and s polarized light was found to be different when the

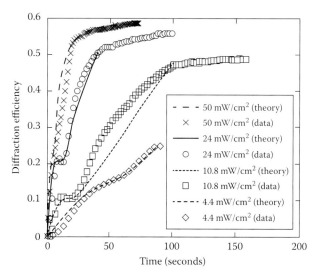

FIGURE 7.6

Diffraction efficiencies of HPDLC reflection gratings (normal incidence) versus exposure times with different exposure intensities. (Reprinted with permission from R. L. Sutherland, V. P. Tondiglia, L. V. Natarajan, and T. J. Bunning, Phenomenological model of anisotropic volume hologram formation in liquid-crystal-photopolymer mixtures, *J. Appl. Phys.* [2004], 96: 951–965. Copyright 2004, American Institute of Physics.)

light was in 45° to the grating, indicating that the LC droplets have a sort of order and the gratings are anisotropic [120]. Highly spherical and uniform-sized LC droplets in thiol-ene-based reflection gratings make it easier to build a theoretical model to analyze the distribution of symmetry axes of the LC droplets [121]. The parallel and perpendicular components of effective dielectric tensor of the LC droplet were obtained by assuming a bipolar LC director configuration in the spherical droplets. The grating was analyzed by the diffraction efficiency as a function of the incidence angle and anisotropic theory [121], as shown in Figure 7.7.

The result indicated that $\varepsilon_{xx}^{(1)} \approx \varepsilon_{yy}^{(1)} \gg \varepsilon_{zz}^{(1)}$, meaning the orientation of symmetry axes of the LC droplets are confined to the grating plane and randomly distributed. With this assumption, the grating, on average, will be isotropic for normal incidence, and the anisotropy of the grating for oblique incidence might be induced by polymer shrinkage [121]. However, this does not mean that the orientation of these LC droplets is not unchangeable. For example, it could be controlled by applying an in situ shear during the recording of thiol-ene-based reflection gratings [122,123]. The strain imposed in the fluidic state could be locked by polymer after gelation, leading to LC droplets orienting at ≈45° with respect to the grating vector. The diffraction efficiency could be 99% for the polarized light parallel to the shear, and almost zero for the polarized light perpendicular to the shear, producing a high degree of polarization dependence gratings.

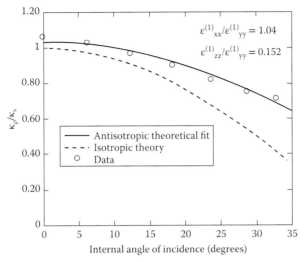

FIGURE 7.7
Coupling coefficient ratio κ_p/κ_s as a function of internal angle of incidence. κ_p and κ_s are the coupling coefficients for p and s polarized light in the anisotropic coupled wave theory. (Reproduced with permission from R. L. Sutherland, V. P. Tondiglia, L. V. Natarajan, P. F. Lloyd, and T. J. Bunning, Coherent diffraction and random scattering in thiol-ene-based holographic polymer-dispersed liquid crystal reflection gratings, *J. Appl. Phys.* [2006], 99: 123104. Copyright 2004, American Institute of Physics.)

7.3.3 Polymer Shrinkage

In this section, we will discuss polymer shrinkage for HPDLC transmission and reflection gratings. During the photopolymerization process, carbon–carbon double bonds of monomers are opened by free radicals and linked to each other. Because the intramolecular distance is smaller than the intermolecular distance, polymerization will inevitably lead to a reduction in volume, which is called *polymerization shrinkage* [85,86,118,124,125]. An in situ shrinkage measurement for HPDLC transmission gratings [126] is shown in Figure 7.8. It was found that the position of the diffractive beam is not changed during the transmission grating formation, which means the shrinkage does not influence the period of transmission gratings. The reason might be that the polymer shrinkage occurs at the direction of the cell normal. However, for reflection gratings, polymer shrinkage will reduce the grating period and result in a blue shift of the reflection wavelength. For example, the dependence of the reflection wavelength on the exposure time [118] is shown in Figure 7.9. After the recording beam was off, the blue shift of the Bragg wavelength continued with time by a certain amount. The polymer shrinkage ratio also followed a linear relationship with effective functionality of monomers for reflection gratings [94]. Furthermore, it was found that the acrylate-based reflection grating shrinkage (>5%) is larger than the thiol-ene-based reflection grating shrinkage (≈3%).

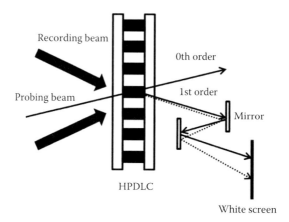

FIGURE 7.8
Experimental setup to detect polymer shrinkage during the HPDLC transmission grating formation.

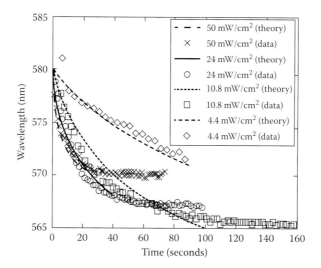

FIGURE 7.9
Blue shift of the Bragg wavelength in HPDLC reflection gratings as a function of time with different exposure intensities. Writing times are 20, 40, 100, and 200 seconds for 50, 24, 10.8, and 4.4 mW/cm² light intensities, respectively. (Reprinted with permission from R. L. Sutherland, V. P. Tondiglia, L. V. Natarajan, and T. J. Bunning, Phenomenological model of anisotropic volume hologram formation in liquid-crystal-photopolymer mixtures, *J. Appl. Phys.* [2004], 96: 951–965. Copyright 2004, American Institute of Physics.)

7.3.4 Polymer Scaffolding Morphologic HPDLC Transmission Gratings

The slow curing process would produce efficient transmission HPDLC gratings with polymer scaffolding morphology in the acrylate-based material systems. An SEM image of typical polymer scaffolding morphologic transmission HPDLC [89] is shown in Figure 7.10. The gratings with large

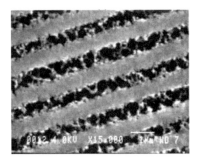

FIGURE 7.10
SEM image of polymer scaffolding morphologic transmission grating made from an acrylate-based material system. (Reprinted with permission from Sarkar et al., 630–638. Copyright 2003 American Chemical Society.)

LC domains or LC layers instead of small-sized LC droplets can be easily recognized. Sometimes the border between polymer scaffolding morphology and LC droplet-like morphology is ambiguous when the curing process is vague (not too fast but not too slow). The polymer scaffolding model was first proposed for nondroplet-like morphologic gratings fabricated by low functionality acrylate-based material systems [87], as shown in Figure 7.11, in which polymer filaments are transverse between the neighboring polymer-rich planes. Other studies have also been performed [89,127,128]. The curing conditions from two examples are listed in Table 7.2. Either relatively low-exposure light intensity or low effective functionality of monomers is necessary to realize a low polymerization rate. Polymer scaffolding morphologic HPDLCs have only been formed from acrylate-based material systems because polymer filaments are easy to form in this case.

To understand the alignment of LC molecules in polymer scaffolding morphologic transmission gratings, the real-time detection of diffraction efficiency for s and p polarized light [128] is shown in Figure 7.12. At the initial stage of reaction (<10 seconds), the amount of diffused LC to the dark fringes is small. The LC is still miscible with monomers, the dissolved LCs are randomly distributed, and the phase grating is isotropic in this stage. The diffraction efficiency for each polarized light is equal. As the polymerization and diffusion processes proceed, there is a moment (10 seconds) when the diffused LCs start phase separation from mixture. The diffractive efficiency for p polarized light is much higher than that for s polarized light in the following stage, indicating the final alignment of LC molecules along the grating vector. The scattering losses are low in this type of HPDLC as the LC domain size is not compared to the visible-light wavelengths.

When light is propagating through an anisotropic medium, its polarization state can be changed because of different phase retardation in different directions. The extent of the change relies on the birefringence of the medium and the distance of light travel. This principle can be exploited to determine the amount of phase-separated LCs [128,129]. The experimental setup to

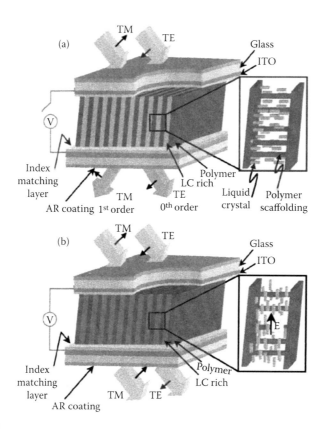

FIGURE 7.11

Model of polymer scaffolding morphologic transmission HPDLCs: (a) without electric field and (b) with electric field (TE: transverse electric wave; TM: transverse magnetic wave). (Reprinted with permission from K. K. Vardanyan, J. Qi, J. N. Eakin, M. D. Sarkar, and G. P. Crawford, Polymer scaffolding model for holographic polymer-dispersed liquid crystals, *Appl. Phys. Lett.* [2002], 81: 4736–4738. Copyright 2002, American Institute of Physics.)

TABLE 7.2

Curing Conditions for Polymer Scaffolding Morphologic HPDLC Transmission Gratings

Monomer	Effective Functionality	Writing Wavelength (nm)	Exposure Intensity (per beam, mW/cm²)	Curing Time (seconds)
EB8301, TMPTA, and EHA [89]	1.5	351	22	40
DPHPA [128]	5	532	4	180

measure the HPDLC grating birefringence is shown in Figure 7.13. Polarizer and analyzer are placed before and behind the sample. The transmitted intensity for vertically aligned polarizers ($I_\perp$) and parallel aligned polarizers ($I_\parallel$) are measured and the phase retardation of the grating (ϕ_{probe}) can be detected as $\phi_{\text{probe}} = 2\arcsin\sqrt{\dfrac{I_\perp}{I_\perp + I_\parallel}}$. The birefringence of the grating can be

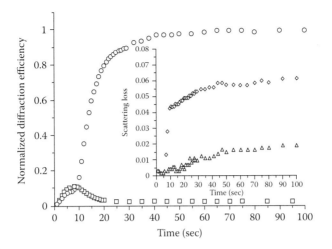

FIGURE 7.12

Evolution of normalized diffraction efficiency for *p* polarized light (open circle) and *s* polarized light (open square) with time. The inset shows the evolution of scattering losses for *p* polarized light (open diamond) and *s* polarized light (open triangle) with time. (W. Huang, Y. Liu, Z. Diao, C. Yang, L. Yao, J. Ma, and L. Xuan, Theory and characteristics of holographic polymer dispersed liquid crystal transmission grating with scaffolding morphology, *Appl. Opt.* [2012], 51: 4013–4020. With permission from Optical Society of America.)

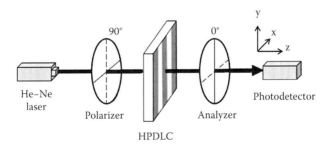

FIGURE 7.13

Experimental setup to measure the HPDLC grating birefringence.

regarded as $\Delta n = n_{\text{eff}} - n_{\text{o}}$, where n_{eff} is the effective refractive index, and the average birefringence of the grating can be obtained by $\Delta n = \dfrac{\lambda_{\text{probe}} \phi_{\text{probe}}}{2\pi d}$, where λ_{probe} is the wavelength of incident probe light. As the polymer in the grating is isotropic and does not induce phase retardation, the amount of phase separation LCs (α) can be calculated by using $\alpha = \dfrac{\Delta n}{n_{\text{e}} - n_{\text{o}}}$.

Once the alignment and amount of phase-separated LC are known, the anisotropic coupled wave theory can be used to explain the diffractive properties of the grating. The diffraction efficiency for *p* polarized light as a function of the deviation from the Bragg angle [128] is shown in Figure 7.14.

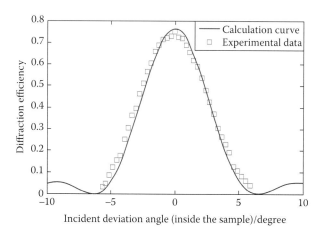

FIGURE 7.14

Diffraction efficiency of polymer scaffolding morphology transmission HPDLC grating for *p* polarized light as a function of the deviation incident angle from Bragg angle (curve: calculation fitting data; open square: experimental data). (W. Huang, Y. Liu, Z. Diao, C. Yang, L. Yao, J. Ma, and L. Xuan, Theory and characteristics of holographic polymer dispersed liquid crystal transmission grating with scaffolding morphology, *Appl. Opt.* [2012], 51: 4013–4020. With permission from Optical Society of America.)

The experimental data were a good fit with anisotropic coupled wave theory [128,130]. Compared with other modeling methods, this method does not use fitting parameters to determine the phase-separated LC amount using birefringence measurement, which is effective and straightforward. The theoretical analysis confirms that the LC molecules in the scaffolding morphologic transmission HPDLC grating are uniformly aligned along the grating vector.

Compared with LC droplet-like morphologic transmission gratings, polymer scaffolding morphologic transmission gratings made from acrylate-based monomers are formed with a slow curing process. About 50%–70% LCs can phase separate out to form a pure LC layer and the scattering losses are low because of the absence of LC droplets. The polarization anisotropy is high ($\eta_p/\eta_s > 20$) [87,128], meaning that the LC alignment is along the grating vector. Furthermore, a high electric field is required to align LCs when switching the grating (critical field: ≈ 3 V/μm; switching-off field: ≈ 12 V/μm [89]). In the meantime, the anchoring effect from the polymer filament will assist LCs to recover back to the initial state when the electric field is turned off, causing a faster fall response time (≈ 0.15 minutes [87]).

7.3.5 Sliced Polymer Morphologic Transmission HPDLC Gratings

When thiol-ene-based material systems are used with a slow curing process, sliced polymer morphologic transmission HPDLCs can be achieved (also so-called polymer-liquid-crystal-polymer slices, POLICRYPS) [102,129,131–136].

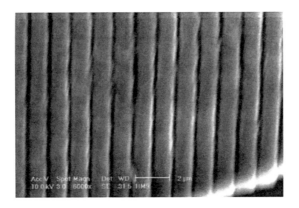

FIGURE 7.15
SEM image of sliced polymer morphologic HPDLC transmission grating made from thiol-ene-based material system. (R. Caputo, A. Veltri, C. Umeton, and A. V. Sukhov, Kogelnik-like model for the diffraction efficiency of POLICRYPS gratings, *J. Opt. Soc. Am. B: Opt. Phys.* [2005], 22: 735–742. With permission from Optical Society of America.)

This grating type is composed of polymer slices separated by uniformly aligned nematic LC layers. Figure 7.15 shows an SEM image of a typical sliced polymer morphologic HPDLC transmission grating [133]. The LC layers are narrower than the polymer layers, which can be expected due to the low LC concentration (15–30 wt % LC). Compared with other types of HPDLC gratings, the fabrication of sliced polymer morphologic transmission HPDLCs needs a special formation process: First, the whole sample is heated to a temperature above the LC clearing point, and then the sample is cured by two interfering UV beams (e.g., 351 nm; 5 mW/cm^2 per beam; 15 minutes). The high LC diffusion constant at high temperature and the low polymerization rate of thiol-ene monomers at low exposure intensity prevent the formation of LC droplets, bringing about sliced, phase-separated polymer layers and LC layers. When the reaction ends, the sample is slowly cooled to room temperature, and the isotropic LCs turn into the nematic state. Thus, the HPDLC grating with sliced polymer layers and nematic LC layers is formed. The smooth and clean microstructure of polymer sliced morphologic gratings will certainly bring advantages for practical applications because the scattering losses are low and the phase separation degree is high.

There are several evidences showing that the LC director in sliced polymer morphologic transmission grating is along the grating vector. No change was observed for the diffraction efficiency for *s* polarized light when applying the electric field along the cell normal, which indicates that the LC director is distributed in the grating plane [134]. The POM and birefringence measurements further showed that the LCs are aligned along the grating vector [137]. Meanwhile, the high diffraction efficiency for *p* polarized light means that the *p* wave experiences a larger refractive index modulation than the *s* wave [134]. Considering that the refractive mismatch between the refractive index of NOA polymer (1.559) and n_e (1.706) is larger than that between the refractive index of NOA polymer and n_o (1.528), it can be inferred that the LCs are homogenously aligned along the grating vector. Although the alignment of LC in sliced polymer morphologic transmission gratings is the same as that in the polymer scaffolding morphologic transmission gratings, the physical reasons are different. It was suspected that there is a hybrid layer with some polymer networks at the LC/polymer interface, the LC molecules are anchored by the hybrid layer, and then LCs are aligned along the grating vector in a sliced polymer morphologic transmission grating [102].

7.3.6 Degree of Phase Separation

The degree of phase separation in sliced polymer morphologic transmission gratings is higher than that in LC droplet-like and polymer scaffolding morphologic transmission gratings. Experiments regarding temperature dependence on the diffraction efficiency of HPDLC transmission gratings with different morphologies showed that the phase-separated LCs in the LC droplet-like or polymer scaffolding morphologic gratings normally have a much decreased N-I point compared with pure LCs [111,128,138,139]. This is attributed to the unreacted monomers, which are dissolved in the LC regions. For LC droplet-like morphologic gratings [111] and polymer scaffolding morphologic gratings [128], about 5%–10% unreacted monomer is dissolved in the phase-separated LCs. For sliced polymer morphologic gratings, no LC N-I transition point change was found [137], which means the degree of phase separation is very high in this grating type.

7.4 HPDLC Applications for Organic Lasers

During the first decade of HPDLC developments, the applications have primarily been focused on the diffractive properties of gratings. For transmission HPDLCs, the high diffraction efficiency and high angular selectivity of Bragg-type gratings have been exploited for beam steering devices [140]. The ability to switch the diffraction efficiency by electric field has proved potentials in optical communications [141–144], beam splitters [145], optical switches, photonic crystals, and so forth. For reflection HDPLCs, the high reflection efficiency, narrow bandwidth, and electric switchability make it promising in reflective color displays [146–152]. As the Bragg reflection wavelength depends sensitively on the grating period and average HPDLC refractive index, it is also sensitive to various external stimuli for gratings. Reflection gratings have demonstrated potentials as pressure sensors [153–155] and chemical sensors [156]. Nowadays, more novel devices based on HPDLC systems have been developed. In this section, we mainly focus our discussion on the applications for organic DFB lasers.

7.4.1 Lasers

A laser (light amplification by stimulated emission of radiation) device emits light through the process of optical amplification based on the stimulated emission of radiation. The device has three main building blocks: stimulated emission source (energy source), gain medium, and laser resonator (or cavity) [157]. The energy level configuration of gain media can be a three-level or four-level laser system. The energy level configuration will enable the population of atoms to be transformed into excited states when pumped by the energy source. By using a laser resonator or cavity, the oscillation of a few resonated light modes can be built up, and the modes fallen in the gain spectrum of the gain medium will be amplified through the stimulated emission process. When the gain (amplification) exceeds the resonator losses (mainly caused by self-absorption and light scattering), the device emits lasing output. The photons generated by the stimulated emission have the same frequency and phase. Therefore, the lasing output is formed with narrow wavelength width, spatial coherence, and small divergence.

The most widely used Fabry–Perot laser cavity is constructed by two end mirrors [157]. There are a lot of light modes in the macroscopic Fabry–Perot cavity that could satisfy the phase matching condition to form a standing wave, but the gain spectrum of the inorganic solid laser medium is inherently narrow, making the single-mode laser operation possible. However, for organic gain media such as laser dyes, the gain spectrum is broadened by various intermolecular or intramolecular vibration and rotational energy levels. Therefore, an additional wavelength selection mechanism has to be introduced into organic laser devices to produce lasing actions.

7.4.2 HPDLCs for External Feedback Elements

The Bragg stack can be used to replace one mirror in organic laser devices. HPDLCs are one alternative to external feedback Bragg stacks [158]. The optical setup of a two-photon-pumped cavity lasing output using an HPDLC grating as an angle-tunable feedback element is shown in Figure 7.16. The partial visible reflector served as the front and output mirror, whereas the HPDLC grating served as the rear mirror and angle-tunable feedback element. This system demonstrated that a laser around 561 nm was achieved and the lasing wavelength could be tunable from 561.5 to 548.5 nm by rotating the HPDLC grating, showing the angle tunability of HPDLCs used in lasers. An electrical tuning of the lasing wavelength without mechanical movement was also demonstrated [159] by using a similar concept in Figure 7.16.

7.4.3 HPDLCs for DFB Laser Cavities

There are several advantages of using a DFB structure as a laser cavity. The light feedback is distributed through the gain medium, and the inter-action length between the feedback light and the gain medium is long, so

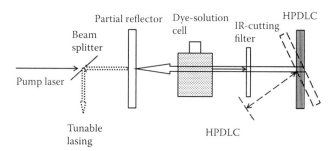

FIGURE 7.16
Optical setup of the two-photon pumped cavity lasing output using an HPDLC grating as an angle-tunable feedback element.

the thresholds of laser devices are low. The filter-like wavelength selection mechanism in the DFB structure would narrow the laser bandwidth and give a pure lasing output. Furthermore, this structure is wavelength-stable during the operation and is convenient to be integrated with other optical devices. In the DFB laser cavity, the feedback for lasing modes is not provided by conventional mirror reflections, but relies on the Bragg scattering by periodic microstructures [160]. Each period of microstructure will scatter light and the light wavelength Λ_B satisfies the Bragg laser equation:

$$m_B\lambda_B = 2N\Lambda \tag{7.3}$$

where m_B is the Bragg order and N is the effective refractive index of the lasing mode. The scattered lights by each period would recombine coherently and form a standing wave inside the periodic structure. In general, DFB lasers from a lower Bragg order will give better laser performance, such as narrower full width at half maximum (FWHM) and lower threshold. For this reason, the first- and the second-order DFB lasers are most widely investigated.

HPDLC reflection gratings were first demonstrated to be used in surface-emitting DFB lasers [115,161–163]. The laser dye coumarin 485 (C485) was combined into the periodic structure through the permeation of C485 butyl acetate solution into the grating [115]. Other laser dyes such as rhodamine 6G (R6G) and 4-(dicyanomethylene)-2-methyl-6-(*p*-dimethylaminostyryl)-4H-pyran (DCM) were then directly doped into LCs or HPDLCs [161,164–167]. Several examples of laser dyes used in HPDLC systems are shown in Figure 7.17. It should be mentioned that the absorption of photoinitiators in the material system at the writing light wavelength needs to be much higher than that of laser dyes. This not only ensures the formation of a better polymer structure, but also prevents laser dyes from bleaching during the writing process.

HPDLC transmission gratings were then used for DFB lasing action [164,168]. The laser dye (e.g., DCM) was mixed in the prepolymer syrup. The formed HPDLC was pumped by an external pumping source (e.g., Q-switched Nd:YAG pulsed laser). A pumping beam with a 10-mm strip could cover thousands of grating periods and such a long gain length would

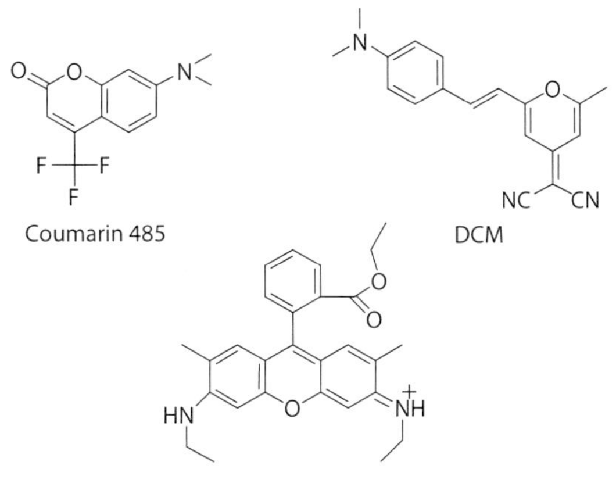

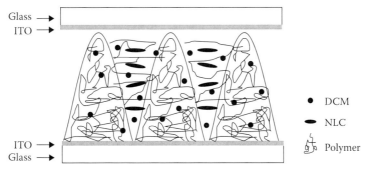

FIGURE 7.17
Chemical structures of typical laser dyes used in HPDLCs.

FIGURE 7.18
Schematic configuration of HPDLC transmission grating for DFB lasers.

produce efficient mode coupling in the HPDLC period structure. The schematic configuration of the transmission grating (e.g., with polymer scaffolding morphology) for a DFB laser is shown in Figure 7.18. The HPDLC film is usually sandwiched by two glass substrates (sometimes with an indium tin oxide [ITO] layer). As the average refractive index of the HPDLC (≈1.55) is larger than that of the glass substrate (≈1.50), this simple three-layer structure can be regarded as a waveguide. The modes propagating along the grating vector will be effectively confined between the two glass substrates. Efficient confinement in the cavity is essential for efficient lasing output [169–171].

The representative lasing spectrum and the dependence of lasing intensity on the pumping energy from an HPDLC transmission grating [168] are shown in Figure 7.19. The linewidth of this lasing peak was ≈1.8 nm and the threshold was ≈120 µJ at the pumping wavelength of 532 nm. The decreased threshold can be achieved by lowering the LC droplet size [171] or using scaffolding morphologic transmission HPDLC gratings. Because of the better

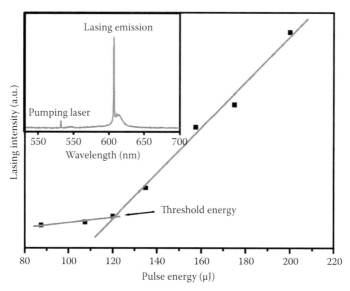

FIGURE 7.19

Output lasing intensity from a DCM-doped HPDLC transmission grating as a function of the laser pumping energy. The inset shows the lasing and the pump laser spectra. (Reprinted with permission from Y. J. Liu, X. W. Sun, P. Shum, H. P. Li, J. Mi, and W. Ji, Low-threshold and narrow-linewidth lasing from dye-doped holographic polymer-dispersed liquid crystal transmission gratings, *Appl. Phys. Lett.* [2006], 88: 061107. Copyright 2006, American Institute of Physics.)

confined mode in the waveguide structure and the better mode coupling induced by the long gain length, the transmission HPDLC gratings are more suitable for DFB laser applications.

7.4.4 2D HPDLCs for DFB Lasers

Besides 1D HPDLC gratings, 2D HPDLC gratings have also been exploited for laser applications [161,172–177]. The initial motivation for using 2D HPDLC grating as a DFB cavity was not only for scientific interests, but also due to the theoretical predictions that multidirection distributed feedback within a 2D structure would show an enhanced coupling coefficient [178,179]. The laser beam from a 2D DFB HPDLC laser shows small divergence and is closed to the diffraction limit [180]. The threshold of the 2D DFB HPDLC laser is currently in the range of 0.2–2 mJ/cm² [172–176]. An example of a square lattice HPDLC grating and its SEM images [172] is shown in Figure 7.20. Although several 2D HPDLC structures fabricated by different writing configurations exist [93,181–184], the physical insights, such as LC existing forms, LC configurations, and degrees of phase separation, still need further investigation [185]. Generally speaking, using 2D HPDLC gratings could be an effective way to improve laser performance with optimal parameters such as grating pitch, film thickness, lower scattering loss, and so forth.

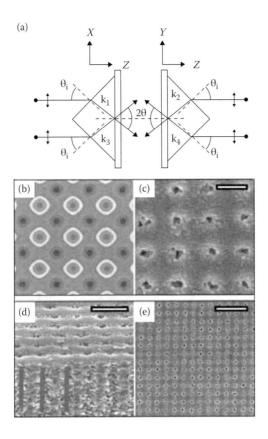

FIGURE 7.20
(a) Diagram of 2D square lattice HPDLC recording, (b) light-intensity pattern calculation, (c) SEM image of 2D columnar grating (scale bar: 375 nm), (d) SEM image of 2D grating in transverse view (scale bar: 500 nm), and (e) SEM image of the grating (scale bar: 2 µm). (R. Jakubiak, V. P. Tondiglia, L. V. Natarajan, R. L. Sutherland, P. Lloyd, T. J. Bunning, and R. A. Vaia: Dynamic lasing from all-organic two-dimensional photonic crystals. *Adv. Mater.* 2005. 17. 2807–2811. Copyright Wiley-VCH Verlag GmbH & Co. KGaA. Reproduced with permission.)

7.4.5 HPDLCs as External Feedbacks

In most cases, laser dyes are doped in HPDLC material systems as gain media for laser applications. Semiconducting materials, as gain media, can be also used to combine with HPDLC gratings. Poly[2-methoxy-5-(2-ethylhexyloxy)-1,4-phenylenevinylene] (MEH-PPV) as a gain medium has been demonstrated in HPDLCs [186,187], whose device configuration is schematically shown in Figure 7.21. In this structure, the MEH-PPV was placed adjacent to the HPDLC layer through the spin coating on the glass inner surface. The HPDLC grating is used as external feedback. The MEH-PPV layer also serves as the waveguide core layer.

In this device, the feedback comes from the evanescent wave spread from the MEH-PPV layer (into the HPDLC layer); thus, optimized laser performance

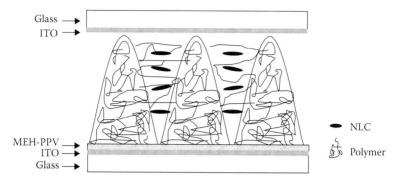

Glass →
ITO →

MEH-PPV →
ITO →
Glass →

● NLC

Polymer

FIGURE 7.21
Schematic configuration of the device containing conductive polymer as a gain medium and HPDLC transmission grating as an external feedback.

can be achieved by adjusting the MEH-PPV thickness. Figure 7.22 shows laser performance from the device with a 75-nm thick MEH-PPV layer [186]. The working threshold was 0.02 mJ/cm², which was one-tenth of that for dye-doped DFB HPDLC laser. Such a combination between semiconducting materials and HPDLCs can be an alternative for incomplex, low-cost organic lasers.

7.4.6 HPDLCs for Multiwavelength DFB Lasers

HPDLCs can be used for multiple or dual-wavelength DFB lasers [188,189]. Because of their unique lasing properties, dual-wavelength lasers show potential applications in fiber sensor systems, atmospheric monitoring, differential absorption lidar (DIAL), and optical communications [190–192]. In general, two or more dispersion elements such as gratings or mirrors are needed to establish a discrete feedback path for each lasing wavelength [193]. The laser configurations are very complex and the controls or demands of the optical apparatus are rigorous. A dual-wavelength DFB laser constructed by dye-doped HPDLC grating was achieved [188]. According to the Bragg equation (Equation 7.3), the output lasing wavelength is also related to the Bragg order and the workable DCM-doped HPDLC grating period ranges can be calculated for different Bragg orders, as shown in Figure 7.23. In this figure, there is an overlap for possible lasing output ranges. For the seventh and eighth Bragg orders, the grating period ranges are in 1301–1553 nm and 1487–1775 nm, respectively. As a result, a grating with a period between 1487 nm and 1553 nm can be selected to generate dual-wavelength lasing action via these two Bragg orders simultaneously.

A dual-wavelength laser can be experimentally obtained in which the lasing positions are at two lasing wavelengths (586.6 nm and 670.2 nm) via the seventh and eighth Bragg order [188], as shown in Figure 7.24. Because the threshold and effective energy transfer are directly related to the Bragg order, the threshold is high in this device. The higher the Bragg order, the higher the threshold. Above the threshold, the lasing intensities of both

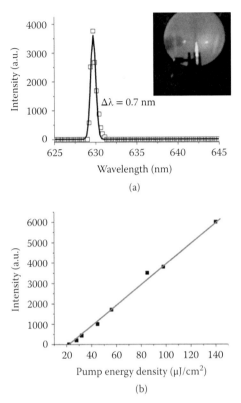

FIGURE 7.22
(See color insert.) (a) Emission spectra from the sample with conductive polymer layer and HPDLC layer (curve: Gaussian fit; inset: laser emission profile image) and (b) emission laser intensity versus pump energy density. (Reprinted from *Org. Electron.*, 13, W. Huang, Z. Diao, Y. Liu, Z. Peng, C. Yang, J. Ma, and L. Xuan, Distributed feedback polymer laser with an external feedback structure fabricated by holographic polymerization technique, 2307–2311, Copyright 2012, with permission from Elsevier.)

components will increase nearly linearly with pump energy (Figure 7.24b). To further lower the threshold and enhance the effective transfer of the pump energy, high-efficiency dyes or material systems, new device configurations, and different optical pumping manners can be developed. The combination of semiconducting material and doped-dye in HPDLC has also shown dual-wavelength DFB lasing behavior, in which DCM-doped HPDLC grating was formed on the top of the MEH-PPV layer [189]. The output wavelength positions of the two lasers can be modulated simultaneously or individually.

7.4.7 Electrically Tunable HPDLC Lasers

Lasing actions and electrically tunable lasers have been achieved in dye-doped LCs with 1D periodic structures, including HPDLCs [194–208]. The lasing wavelengths can be tuned by applying an external electric field across

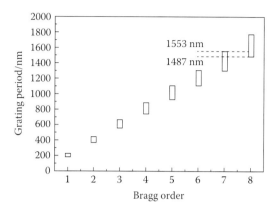

FIGURE 7.23
Prediction of grating period range for different Bragg-order lasing. (Z. Diao, S. Deng, W. Huang, L. Xuan, L. Hu, Y. Liu, and J. Ma, Organic dual-wavelength distributed feedback laser empowered by dye-doped holography, *J. Mater. Chem.* [2012], 22: 23331–23334. Reproduced by permission of The Royal Society of Chemistry.)

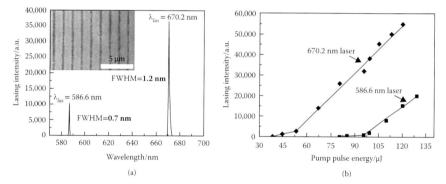

FIGURE 7.24
(a) Spectra of the dual wavelength laser based on DCM-doped HPDLC (grating period: 1520 nm; inset: SEM of the grating); and (b) output lasing intensity as a function of the pump energy. (Z. Diao, S. Deng, W. Huang, L. Xuan, L. Hu, Y. Liu, and J. Ma, Organic dual-wavelength distributed feedback laser empowered by dye-doped holography, *J. Mater. Chem.* [2012], 22: 23331–23334. Reproduced by permission of The Royal Society of Chemistry.)

the cell to realign the LCs or change the effective refractive index of the laser cavity in the transmission, reflection, and 2D HPDLCs [115,161,172,209,210]. Figure 7.25 shows the emission spectra from a DCM-doped polymer scaffolding morphologic HPDLC transmission grating under different electric fields [210]. On the electric field over the threshold (2 V/μm), all transverse-magnetic (TM) modes started to shift to longer wavelengths. At 14 V/μm, the maximum redshift of wavelengths was achieved and the tuning ranges of TM0, TM1, and TM2 modes were 12, 10.6, and 5.9 nm, respectively.

An anisotropic waveguide theory has been developed for HPDLC lasers to explain the shifting of lasing wavelength [210], as different TM modes will

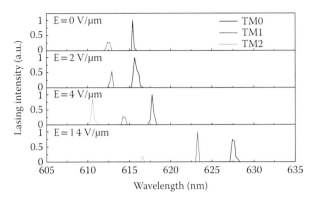

FIGURE 7.25
Emission spectra from a dye-doped polymer scaffolding morphologic HPDLC laser under different electric fields.

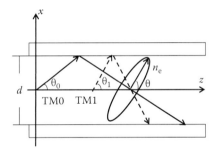

FIGURE 7.26
Schematic representation of light propagation path for TM0 and TM1 modes.

experience different LC refractive indices in their respective propagation path, as shown in Figure 7.26. With the LC tilt angle θ, the relative permittivity of pure LC layer $\overline{\varepsilon}_{LC}$ is expressed [211] as

$$\overline{\varepsilon}_{LC} = \begin{bmatrix} \varepsilon_e \sin^2 \theta + \varepsilon_o \cos^2 \theta & 0 & \varepsilon_a \sin \theta \cos \theta \\ 0 & \varepsilon_o & 0 \\ \varepsilon_a \sin \theta \cos \theta & 0 & \varepsilon_e \cos^2 \theta + \varepsilon_o \sin^2 \theta \end{bmatrix} \quad (7.4)$$

where $\varepsilon_a = \varepsilon_e - \varepsilon_o$, $\varepsilon_o = n_o^2$, and $\varepsilon_e = n_e^2$. The average relative permittivity of the HPDLC core layer $\overline{\varepsilon}_{core}$ can be expressed as

$$\overline{\varepsilon}_{core\_layer} = \begin{bmatrix} \varepsilon_{xx} & 0 & \varepsilon_{xz} \\ 0 & \varepsilon_{yy} & 0 \\ \varepsilon_{zx} & 0 & \varepsilon_{zz} \end{bmatrix} = (1-\alpha)\varepsilon_p + \alpha\overline{\varepsilon}_{LC} \quad (7.5)$$

where α is the amount of phase-separated LCs and ε_p the relative permittivity of polymer matrix, which is given by $\varepsilon_p = (2n_o^2 + n_e^2)(1 - \phi_p - \alpha) + n_{pp}^2 \phi_p / (1-\alpha)$.

Here, n_{pp} is the refractive index of the pure polymer and φ_p is the volume proportion of the pure polymer. Because the off-diagonal elements (ε_{xz} and ε_{zx}) are so small compared with the diagonal ones and only contribute to TM–TM coupling, these elements can be ignored.

A TM-polarized light wave propagating in the HPDLC core layer can be expressed by $\vec{H} = H_y(x)\exp\left[j(k_0 Nz - \omega t)\right]$, where k_0 and ω are the wave vector and the angular frequency of light, respectively, and N is the effective refractive index of the TM mode. We can use Maxwell's equation $\nabla\left(\nabla \bullet \vec{H}\right) - \nabla^2\vec{H} = -\mu_0\varepsilon_0\bar{\varepsilon}_{core\_layer}\dfrac{\partial^2\vec{H}}{\partial t^2}$ to find the solution for $H_y(x)$ as

$$H_y(x) = \cos\left[k_0\sqrt{\frac{\varepsilon_{zz}}{\varepsilon_{xx}}}\,(\varepsilon_{xx} - N^2)x\right] \tag{7.6}$$

Further, in the symmetric waveguide configuration composed by the HPDLC core layer and two glass substrates with refractive index n_g, the magnetic fields in these regions can be obtained [212] by

$$H_{yI}(x) = \cos\left[k_0\sqrt{\frac{\varepsilon_{zz}}{\varepsilon_{xx}}}\,(\varepsilon_{xx} - N^2)x + \phi\right]\left(-\frac{d}{2} \leq x \leq \frac{d}{2}\right)$$

$$H_{yII}(x) = \cos\left[-k_0\frac{d}{2}\sqrt{\frac{\varepsilon_{zz}}{\varepsilon_{xx}}}\,(\varepsilon_{xx} - N^2) + \phi\right]\exp\left[k_0\sqrt{N^2 - n_g^2}\left(x + \frac{d}{2}\right)\right]\left(x < -\frac{d}{2}\right)$$

$$H_{yIII}(x) = \cos\left[k_0\frac{d}{2}\sqrt{\frac{\varepsilon_{zz}}{\varepsilon_{xx}}}\,(\varepsilon_{xx} - N^2) + \phi\right]\exp\left[k_0\sqrt{N^2 - n_g^2}\left(\frac{d}{2} - x\right)\right]\left(x > \frac{d}{2}\right) \tag{7.7}$$

where H_{yI}, H_{yII}, and H_{yIII} are the magnetic fields in the core layer ($-d/2 \leq x \leq d/2$), bottom glass substrate ($x < -d/2$), and upper glass substrate ($x > d/2$), respectively, and φ is a constant about phase. By putting Equation 7.7 into the boundary conditions for transverse field components [213]:

$$\frac{1}{\varepsilon_{zz}}\frac{\partial H_{yI}}{\partial x}\left(-\frac{d}{2}\right) = \frac{1}{n_g^2}\frac{\partial H_{yII}}{\partial x}\left(-\frac{d}{2}\right)$$

$$\frac{1}{\varepsilon_{zz}}\frac{\partial H_{yI}}{\partial x}\left(\frac{d}{2}\right) = \frac{1}{n_g^2}\frac{\partial H_{yIII}}{\partial x}\left(\frac{d}{2}\right) \tag{7.8}$$

Then the anisotropic waveguide equation for TM mode can be solved:

$$2\tan^{-1}\left[\frac{\varepsilon_{zz}}{n_g^2}\sqrt{\frac{N^2 - n_g^2}{\frac{\varepsilon_{zz}}{\varepsilon_{xx}}(\varepsilon_{xx} - N^2)}}\right] + mx = k_0 d\sqrt{\frac{\varepsilon_{zz}}{\varepsilon_{xx}}(\varepsilon_{xx} - N^2)} \tag{7.9}$$

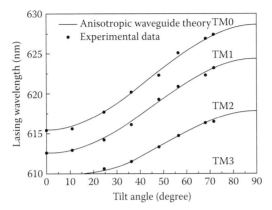

FIGURE 7.27
Comparison between the anisotropic waveguide theory and experimental data.

where m is the mode number. Combining Equations 7.5, 7.9, and the Bragg laser equation (Equation 7.3), the connection between the LC tilt angle θ and the lasing wavelength can be built. When the external electric field is applied, the LCs are aligned and θ is changed. Therefore, the lasing wavelength is turned. Figure 7.27 shows the comparison between the anisotropic waveguide calculations and experimental data [210]. The good agreement between the theoretical simulations and experimental data indicates the validity of the anisotropic waveguide theory. The period HPDLC grating structure, optical anisotropy of LC, and practical light propagation path in the HPDLC layer are all considered in this model. The emitted lasing wavelengths can be deduced from the dielectric anisotropy of LC, TM light wave propagation, and the abovementioned laser theory. Such anisotropic numerical analysis would be very useful when designing or optimizing tunable HPDLC DFB lasers.

7.5 Conclusions

In conclusion, the materials, morphologies, and applications of organic lasers of HPDLCs are introduced and discussed in this chapter. We catalog HPDLCs to three types according to their morphologies: (1) LC droplet-like morphology, (2) polymer scaffolding morphology, and (3) sliced polymer morphology. As LCs can be modulated by external fields, the alignments or configurations of LCs in the grating are important and elaborated for each type of HPDLC. HPDLCs can be used for various optical-electrical and opto-optical devices. The merits of simple configuration, commercially available materials, easy fabrication, and potentials to be integrated in all optical networks of HPDLCs will provide impetus to more research and practical applications.

References

1. P. G. de Gennes and J. Prost, *The Physics of Liquid Crystals* (2nd Edition), Oxford University Press, New York, 1993.
2. M. Kleman and O. D. Lavrentovich, *Soft Matter Physics: An Introduction*, Springer Verlag, New York, 2003.
3. F. W. Billmeyer, *Textbook of Polymer Science* (3rd Edition), Wiley, New York, 1984.
4. L. Vicari, *Optical Applications of Liquid Crystals*, IOP Publishing, Bristol and Philadelphia, 2003.
5. I. C. Khoo and S. T. Wu, *Optics and Nonlinear Optics of Liquid Crystals*, World Scientific, Singapore, 1993.
6. P. Yeh and C. Gu, *Optics of Liquid Crystal Displays*, Wiley, Hoboken, NJ, 2009.
7. I. Dierking, *Textures of Liquid Crystals*, Wiley-VCH, Weinheim, 2003.
8. S. Singh and D. A. Dunmur, *Liquid Crystals Fundamentals*, World Scientific, Singapore, 2002.
9. D. Demus, J. W. Goodby, G. W. Grey, H.-W. Spiess, and V. Vill, *Handbook of Liquid Crystals*, Wiley-VCH, Weinheim, 1998.
10. H. Yokoyama, Interfaces and Thin Films, In: *Handbook of Liquid Crystal Research*, P. J. Collings and J. S. Patel (Eds.), Oxford University Press, New York, 1997.
11. J. H. Kim, M. Yoneya, and H. Yokoyama, Tristable nematic liquid-crystal device using micropatterned surface alignment, *Nature* (2002), 420: 159–162.
12. J. S. Patel and H. Yokoyama, Continuous anchoring transition in liquid crystals, *Nature* (1993), 362: 525–527.
13. B. L. Feringa, *Molecular Switches*, Wiley-VCH, Weinheim, 2001.
14. Y. Zhao and T. Ikeda, *Smart Light-Responsive Materials: Azobenzene-Containing Polymers and Liquid Crystals*, Wiley-VCH, Boca Raton, FL, 2009.
15. J. Ma, X. Ye, and B. Jin, Structure and application of polarizer film for thin-film-transistor liquid crystal displays, *Displays* (2011), 32: 49–57.
16. J. Ma and L. Xuan, Invited review: Towards nanoscale molecular switch-based liquid crystal displays, *Displays* (2013), 34: 293–300.
17. J. Ma, Y. Li, T. White, A. Urbas, and Q. Li, Light-driven nanoscale chiral molecular switch: Full range color phototuning, *Chem. Commun.* (2010), 46: 3463–3465.
18. T. Kosa, L. Sukhomlinova, L. Su, B. Taheri, T. J. White, and T. J. Bunning, Light-induced liquid crystallinity, *Nature* (2012), 485: 347–349.
19. S.-T. Wu and D.-K. Yang, *Reflective Liquid Crystal Displays*, Wiley, New York, 2001.
20. D.-K. Yang and S.-T. Wu, *Fundamentals of Liquid Crystal Devices*, Wiley, New York, 2006.
21. P. S. Drzaic, *Liquid Crystal Dispersions*, World Scientific, Singapore, 1995.
22. I. C. Khoo, *Liquid Crystals* (2nd Ed.), Wiley, Hoboken, NJ, 2007.
23. J. W. Doane, N. A. Vaz, B. G. Wu, and S. Zumer, Field controlled light scattering from nematic microdroplets, *Appl. Phys. Lett.* (1986), 48: 269–271.
24. P. S. Drzaic, Polymer dispersed nematic liquid crystal for large area displays and light valves, *J. Appl. Phys.* (1986), 60: 2142–2148.
25. D.-K. Yang, L.-C. Chien, and J. W. Doane, Cholesteric liquid crystal/polymer dispersion for haze free light shutters, *Appl. Phys. Lett.* (1992), 60: 3102–3104.
26. D.-K. Yang, X.-Y. Huang, and Y.-M. Zhu, Bistable cholesteric reflective displays: materials and drive schemes, *Annu. Rev. Mater. Sci.* (1997), 27: 117–146.

27. H. Ren and S. T. Wu, Reflective reversed-mode polymer stabilized cholesteric texture light switches, *J. Appl. Phys.* (2002), 92: 797–800.

28. J. Ma, L. Shi, and D.-K. Yang, Bistable polymer stabilized cholesteric texture light shutter, *Appl. Phys. Express* (2010), 3: 021702.

29. R. L. Sutherland, L. V. Natarajan, and V. P. Tondiglia, Bragg gratings in an acrylate polymer consisting of periodic polymer-dispersed liquid-crystal planes, *Chem. Mater.* (1993), 5: 1533–1538.

30. R. L. Sutherland, V. P. Tondiglia, L. V. Natarajan, T. J. Bunning, and W. W. Adams, Electrically switchable volume gratings in polymer-dispersed liquid crystals, *Appl. Phys. Lett.* (1994), 64: 1074–1076.

31. T. J. Bunning, L. V. Natarajan, V. P. Tondiglia, and R. L. Sutherland, Holographic polymer-dispersed liquid crystals (H-PDLCs), *Annu. Rev. Mater. Sci.* (2000), 30: 83–115.

32. S. Y. Chou, P. R. Krauss, and P. J. Renstrom, Imprint lithography with 25-nanometer resolution, *Science* (1996), 272: 85–87.

33. M. Campbell, D. N. Sharp, M. T. Harrison, R. G. Denning, and A. J. Turberfield, Fabrication of photonic crystals for the visible spectrum by holographic lithography, *Nature* (2000), 404: 53–56.

34. C. Vieu, F. Carcenac, A. Pepin, Y. Chen, M. Mejias, A. Lebib, L. Manin-Ferlazzo, L. Couraud, and H. Launois, Electron beam lithography: Resolution limits and applications, *Appl. Surf. Sci.* (2000), 164: 111–117.

35. S. A. Jenekhe and X. L. Chen, Self-assembly of ordered microporous materials from rod-coil block copolymers, *Science* (1999), 283: 372–375.

36. Y. J. Liu and X. W. Sun, Holographic polymer-dispersed liquid crystals: Materials, formation, and applications, *Adv. OptoElectron.* (2008), 2008: 684349.

37. G. P. Crawford and S. Zumer, *Liquid Crystals in Complex Geometries: Formed by Polymer and Porous Networks*, CRC Press, London, 1996.

38. J. L. West, J. R. Kelly, K. Jewell, and Y. Ji, Effect of polymer matrix glass transition temperature on polymer dispersed liquid crystal electro-optics, *Appl. Phys. Lett.* (1992), 60: 3238–3240.

39. T. White, L. Natarajan, T. Bunning, and C. Guymon, Contribution of monomer functionality and additives to polymerization kinetics and liquid crystal phase separation in acrylate-based polymer-dispersed liquid crystals (PDLCs), *Liq. Cryst.* (2007), 34: 1377–1385.

40. J. Castellano, *Liquid Gold: The Story of Liquid Crystal Displays and the Creation of an Industry*, World Scientific, Singapore, 2005.

41. R. Caputo, A. V. Sukhov, N. V. Tabiryan, and C. Umeton, Efficiency dynamics of diffraction gratings recorded in liquid crystalline composite materials by a UV interference pattern, *Chem. Phys.* (1999), 245: 463–471.

42. R. Caputo, A. V. Sukhov, N. V. Tabirian, C. Umeton, and R. F. Ushakov, Mass transfer processes induced by inhomogeneous photo-polymerisation in a multicomponent medium, *Chem. Phys.* (2001), 271: 323–335.

43. A. Veltri, R. Caputo, C. Umeton, and A. V. Sukhov, Model for the photoinduced formation of diffraction gratings in liquid-crystalline composite materials, *Appl. Phys. Lett.* (2004), 84: 3492–3494.

44. S.-D. Wu and E. N. Glytsis, Holographic grating formation in photopolymers: analysis and experimental results based on a nonlocal diffusion model and rigorous coupled-wave analysis, *J. Opt. Soc. Am. B: Opt. Phys.* (2003), 20: 1177–1188.

45. M. R. Gleeson and J. T. Sheridan, A review of the modelling of free-radical photopolymerization in the formation of holographic gratings, *J. Opt. A: Pure Appl. Opt.* (2009), 11: 024008.

46. P. Hariharan, *Basics of Holography*, Cambridge University Press, New York, 2002.

47. M. L. Schilling, V. L. Colvin, L. Dhar, A. L. Harris, F. C. Schilling, H. E. Katz, T. Wysocki, A. Hale, L. L. Blyler, and C. Boyd, Acrylate oligomer-based photopolymers for optical storage applications, *Chem. Mater.* (1999), 11: 247–254.

48. K. Tanaka, K. Kato, and M. Date, Fabrication of holographic polymer dispersed liquid crystal (HPDLC) with high reflection efficiency, *Jpn. J. Appl. Phys.* (1999), 38: L277–L278.

49. A. Ogiwara, Effects of anisotropic diffractions on holographic polymer-dispersed liquid-crystal gratings, *Appl. Opt.* (2011), 50: 594–603.

50. M. A. Ellabban, I. Drevensek-Olenik, and R. A. Rupp, Huge retardation of grating formation in holographic polymer-dispersed liquid crystals, *Appl. Phys. B: Lasers Opt.* (2008), 91: 11–15.

51. Y. H. Cho and Y. Kawakami, A novel process for holographic polymer-dispersed liquid crystal system via simultaneous photo-polymerization and siloxane network formation, *Silicon Chem.* (2007), 3: 219–227.

52. T. J. White, W. B. Liechty, L. V. Natarajan, V. P. Tondiglia, T. J. Bunning, and C. A. Guymon, The influence of N-vinyl-2-pyrrolidinone in polymerization of holographic polymer dispersed liquid crystals (HPDLCs), *Polymer* (2006), 47: 2289–2298.

53. S. P. Gorkhali, J. Qi, and G. P. Crawford, Switchable quasi-crystal structures with five-, seven-, and ninefold symmetries, *J. Opt. Soc. Am. B: Opt. Phys.* (2006), 23: 149–158.

54. Y. H. Cho and Y. Kawakami, High performance holographic polymer dispersed liquid crystal systems using multi-functional acrylates and siloxane-containing epoxides as matrix components, *Appl. Phys. A: Mater. Sci. Process.* (2006), 83: 365–375.

55. G. Zharkova, I. Samsonova, S. Streltsov, V. Khachaturyan, A. Petrov, and N. Rudina, Electro-optical characterization of switchable Bragg gratings based on nematic liquid crystal-photopolymer composites with spatially ordered structure, *Microelectron. Eng.* (2005), 81: 281–287.

56. M. A. Ellabban, M. Fally, H. Ursic, and I. Drevensek-Olenik, Holographic scattering in photopolymer-dispersed liquid crystals, *Appl. Phys. Lett.* (2005), 87: 151101.

57. J. Qi, M. E. Sousa, A. K. Fontecchio, and G. P. Crawford, Temporally multiplexed holographic polymer-dispersed liquid crystals, *Appl. Phys. Lett.* (2003), 82: 1652–1654.

58. R. T. Pogue, L. V. Natarajan, V. P. Tondiglia, S. A. Siwecki, R. L. Sutherland, and T. J. Bunning, Controlling nano-scale morphology in switchable PDLC gratings, *Proc. SPIE* (1998) 0277-0786X.

59. Z. Zheng, J. Ma, W. Li, J. Song, Y. Liu, and L. Xuan, Improvements in morphological and electro-optical properties of polymer-dispersed liquid crystal grating using a highly fluorine-substituted acrylate monomer, *Liq. Cryst.* (2008), 35: 885–893.

60. Z. Zheng, J. Song, Y. Liu, F. Guo, J. Ma, and L. Xuan, Single-step exposure for two-dimensional electrically tuneable diffraction grating based on polymer dispersed liquid crystal, *Liq. Cryst.* (2008), 35: 489–499.

61. Z. Zheng, J. Ma, Y. Liu, and L. Xuan, Molecular dynamics of the interfacial properties of partially fluorinated polymer dispersed liquid crystal gratings, *J. Phys. D: Appl. Phys.* (2008), 41: 235302.

62. F. Vita, D. E. Lucchetta, R. Castagna, L. Criante, and F. Simoni, Large-area photonic structures in freestanding films, *Appl. Phys. Lett.* (2007), 91: 103114.

63. N. H. Nataj, E. Mohajerani, H. Jashnsaz, and A. Jannesari, Holographic polymer dispersed liquid crystal enhanced by introducing urethane trimethacrylate, *Appl. Opt.* (2012), 51: 697–703.

64. N. H. Nataj, A. Jannesari, E. Mohajerani, F. Najafi, and H. Jashnsaz, Photopolymerization behavior and phase separation effects in novel polymer dispersed liquid crystal mixture based on urethane trimethacrylate monomer, *J. Appl. Polym. Sci.* (2012), 126: 1676–1686.

65. H. Jashnsaz, E. Mohajerani, H. Nemati, S. H. Razavi, and I. A. Alidokht, Electrically switchable holographic liquid crystal/polymer Fresnel lens using a Michelson interferometer, *Appl. Opt.* (2011), 50: 2701–2707.

66. Y. Liu, B. Zhang, Y. Jia, and K. Xu, Improvement of the diffraction properties in holographic polymer dispersed liquid crystal Bragg gratings, *Opt. Commun.* (2003), 218: 27–32.

67. Y. J. Liu, X. W. Sun, H. T. Dai, J. H. Liu, and K. S. Xu, Effect of surfactant on the electro-optical properties of holographic polymer dispersed liquid crystal Bragg gratings, *Opt. Mater.* (2005), 27: 1451–1455.

68. Y. J. Liu and X. W. Sun, Electrically tunable three-dimensional holographic photonic crystal made of polymer-dispersed liquid crystals using a single prism, *Jpn. J. Appl. Phys.* (2007), 46: 6634–6638.

69. Y. J. Liu, Y. B. Zheng, J. Shi, H. Huang, T. R. Walker, and T. J. Huang, Optically switchable gratings based on azo-dye-doped, polymer-dispersed liquid crystals, *Opt. Lett.* (2009), 34: 2351–2353.

70. K. Kato, T. Hisaki, and M. Date, Alignment-controlled holographic polymer dispersed liquid crystal for reflective display devices, *Jpn. J. Appl. Phys.* (1999), 38: 805.

71. A. Redler and H. S. Kitzerow, Three-dimensional structure in holographic polymer-dispersed liquid crystals, *Polym. Adv. Technol.* (2013), 24: 7–9.

72. Y.-C. Su, C.-C. Chu, W.-T. Chang, and V. K. S. Hsiao, Characterization of optically switchable holographic polymer-dispersed liquid crystal transmission gratings, *Opt. Mater.* (2011), 34: 251–255.

73. M. W. Jang and B. K. Kim, Low driving voltage holographic polymer dispersed liquid crystals with chemically incorporated graphene oxide, *J. Mater. Chem.* (2011), 21: 19226–19232.

74. S. Gao, C. Zhang, Y. Liu, H. Su, L. Wei, T. Huang, N. Dellas, S. Shang, S. E. Mohney, and J. Wang, Lasing from colloidal InP/ZnS quantum dots, *Opt. Express* (2011), 19: 5528–5535.

75. A. Y.-G. Fuh, M. S. Li, and S. T. Wu, Transverse wave propagation in photonic crystal based on holographic polymer-dispersed liquid crystal, *Opt. Express* (2011), 19: 13428–13435.

76. A. Ogiwara and M. Watanabe, Optical reconfiguration by anisotropic diffraction in holographic polymer-dispersed liquid crystal memory, *Appl. Opt.* (2012), 51: 5168–5177.

77. A. Ogiwara, M. Watanabe, and R. Moriwaki, Formation of temperature dependable holographic memory using holographic polymer-dispersed liquid crystal, *Opt. Lett.* (2013), 38: 1158–1160.

78. J. Klepp, C. H. Pruner, Y. Tomita, C. Plonka-Spehr, P. Geltenbort, S. Ivanov, G. Manzin, K. H. Andersen, J. Kohlbrecher, and M. A. Ellabban, Diffraction of slow neutrons by holographic SiO_2 nanoparticle-polymer composite gratings, *Phys. Rev. A* (2011), 84: 013621.

79. J. D. Busbee, L. V. Natarajan, V. P. Tongdilia, T. J. Bunning, R. A. Vaia, and P. V. Braun, SiO_2 nanoparticle sequestration via reactive functionalization in holographic polymer-dispersed liquid crystals, *Adv. Mater.* (2009), 21: 3659–3662.

80. K. Pavani, I. Naydenova, J. Raghavendra, S. Martin, and V. Toal, Electro-optical switching of the holographic polymer-dispersed liquid crystal diffraction gratings, *J. Opt. A: Pure Appl. Opt.* (2009), 11: 024023.

81. M. Fally, M. Bichler, M. A. Ellabban, I. D. Olenik, C. Pruner, H. Eckerlebe, and K. Pranzas, Diffraction gratings for neutrons from polymers and holographic polymer-dispersed liquid crystals, *J. Opt. A: Pure Appl. Opt.* (2009), 11: 024019.

82. S. C. Sharma, A review of the electro-optical properties and their modification by radiation in polymer-dispersed liquid crystals and thin films containing CdSe/ZnS quantum dots, *Mater. Sci. Eng. B* (2010), 168: 5–15.

83. H. Zhang, H. Xianyu, J. Liang, Y. Bétrémieux, G. P. Crawford, J. Noto, and R. Kerr, Switchable circular-to-point converter based on holographic polymer-dispersed liquid-crystal technology, *Appl. Opt.* (2007), 46: 161–166.

84. E. Kim, J. Woo, and B. Kim, Diffraction grating in a holographic polymer dispersed liquid crystal based on polyurethane acrylate, *Liq. Cryst.* (2007), 34: 79–85.

85. R. L. Sutherland, Polarization and switching properties of holographic polymer-dispersed liquid-crystal gratings. I. Theoretical model, *J. Opt. Soc. Am. B: Opt. Phys.* (2002), 19: 2995–3003.

86. R. L. Sutherland, L. V. Natarajan, V. P. Tondiglia, S. Chandra, C. K. Shepherd, D. M. Brandelik, S. A. Siwecki, and T. J. Bunning, Polarization and switching properties of holographic polymer-dispersed liquid-crystal gratings. II. Experimental investigations, *J. Opt. Soc. Am. B: Opt. Phys.* (2002), 19: 3004–3012.

87. K. K. Vardanyan, J. Qi, J. N. Eakin, M. D. Sarkar, and G. P. Crawford, Polymer scaffolding model for holographic polymer-dispersed liquid crystals, *Appl. Phys. Lett.* (2002), 81: 4736–4738.

88. M. D. Sarkar, J. Qi, and G. P. Crawford, Influence of partial matrix fluorination on morphology and performance of HPDLC transmission gratings, *Polymer* (2002), 43: 7335–7344.

89. M. D. Sarkar, N. L. Gill, J. B. Whitehead, and G. P. Crawford, Effect of monomer functionality on the morphology and performance of the holographic transmission gratings recorded on polymer dispersed liquid crystals, *Macromolecules* (2003), 36: 630–638.

90. B. K. Kim, M. W. Jang, H. C. Park, H. M. Jeong, and E. Y. Kim, Effect of graphene doping of holographic polymer-dispersed liquid crystals, *J. Polym. Sci. A: Polym. Chem.* (2012), 50: 1418–1423.

91. Y. Liu, H. Dai, E. Leong, J. Teng, and X. Sun, Electrically switchable two-dimensional photonic crystals made of polymer-dispersed liquid crystals based on the Talbot self-imaging effect, *Appl. Phys. B: Lasers Opt.* (2011), 104: 659–663.

92. I. Drevensek-Olenik, I. Urbancic, M. Copic, M. E. Sousa, and G. P. Crawford, In-plane switching of holographic polymer-dispersed liquid crystal transmission gratings, *Mol. Cryst. Liq. Cryst.* (2008), 495: 177–185.

93. S. P. Gorkhali, J. Qi, and G. P. Crawford, Electrically switchable mesoscale Penrose quasicrystal structure, *Appl. Phys. Lett.* (2005), 86: 01110.

94. J. Qi, M. D. Sarkar, G. T. Warren, and G. P. Crawford, In situ shrinkage measurement of holographic polymer dispersed liquid crystals, *J. Appl. Phys.* (2002), 91: 4795–4800.

95. C. C. Bowley, G. P. Crawford, and H. Yuan, Reflection from dual-domains in a holographically-formed polymer-dispersed liquid crystal material, *Appl. Phys. Lett.* (1999), 74: 3096–3098.

96. M. S. Park, Y. H. Cho, B. K. Kim, and J. S. Jang, Fabrication of reflective holographic gratings with polyurethane acrylate (PUA), *Curr. Appl. Phys.* (2002), 2: 249–252.

97. M. S. Park, B. K. Kim, and J. C. Kim, Reflective mode of HPDLC with various structures of polyurethane acrylates, *Polymer* (2003), 44: 1595–1602.

98. M. S. Park and B. K. Kim, Transmission holographic gratings produced using networked polyurethane acrylates with various functionalities, *Nanotechnology* (2006), 17: 2012–2017.

99. Z. Zheng, L. Yao, R. Zhang, Z. Zou, Y. Liu, Y. Liu, and L. Xuan, Thermo-stability of acrylate based holographic polymer dispersed liquid crystal gratings, *J. Phys. D: Appl. Phys.* (2009), 42: 115504.

100. W. Huang, S. Deng, W. Li, Z. Peng, Y. Liu, L. Hu, and L. Xuan, A polarization-independent and low scattering transmission grating for a distributed feedback cavity based on holographic polymer dispersed liquid crystal, *J. Opt.* (2011), 13: 085501.

101. L. V. Natarajan, C. K. Shepherd, D. M. Brandelik, R. L. Sutherland, S. Chandra, V. P. Tondiglia, D. Timlin, and T. J. Bunning, Switchable holographic polymer-dispersed liquid crystal reflection gratings based on thiol-ene photopolymerization, *Chem. Mater.* (2003), 15: 2477–2484.

102. R. Caputo, L. De Sio, A. Veltri, C. Umeton, and A. V. Sukhov, Development of a new kind of switchable holographic grating made of liquid-crystal films separated by slices of polymeric material, *Opt. Lett.* (2004), 29: 1261–1263.

103. I. Drevensek-Olenik, M. Jazbinsek, M. E. Sousa, A. K. Fontecchio, G. P. Crawford, and M. Eopie, Structural transitions in holographic polymer-dispersed liquid crystals, *Phys. Rev. E* (2004), 69: 051703.

104. L. V. Natarajan, D. P. Brown, J. M. Wofford, V. P. Tondiglia, R. L. Sutherland, P. Lloyd, R. Jakubiak, R. Vaia, and T. J. Bunning, Visible light initiated thiol-ene based reflection H-PDLCs, *Proc. SPIE* (2005), 59360F.

105. J. M. Wofford, L. V. Natarajan, V. P. Tondiglia, R. L. Sutherland, P. F. Lloyd, S. A. Siwecki, and T. J. Bunning, Holographic polymer dispersed liquid crystal (HPDLC) transmission gratings formed by visible light initiated thiol-ene photopolymerization, *Proc. SPIE* (2006), 63320Q.

106. M. J. Birnkrant, Combining holographic patterning and block copolymer self-assembly to fabricate hierarchical volume gratings, Thesis of Drexel University, Philadelphia, PA, 2009.

107. R. A. Ramsey and S. C. Sharma, Switchable holographic gratings formed in polymer-dispersed liquid-crystal cells by use of a He-Ne laser, *Opt. Lett.* (2005), 30: 592–594.
108. R. A. Ramsey, S. C. Sharma, and K. Vaghela, Holographically formed Bragg reflection gratings recorded in polymer-dispersed liquid crystal cells using a He-Ne laser, *Appl. Phys. Lett.* (2006), 88: 051121.
109. P. Pilot, Y. B. Boiko, and T. V. Galstian, Near-IR (800 to 855 nm) sensitive holographic photopolymer-dispersed liquid crystal materials, *Proc. SPIE* (1999), 143–150.
110. P. Nagtegaele and T. V. Galstian, Holographic characterization of near infrared photopolymerizable materials, *Synth. Met.* (2002), 127: 85–87.
111. M. Jazbinsek, I. D. Olenik, M. Zgonik, A. K. Fontecchio, and G. P. Crawford, Characterization of holographic polymer dispersed liquid crystal transmission gratings, *J. Appl. Phys.* (2001), 90: 3831–3837.
112. J. J. Butler, M. S. Malcuit, and M. A. Rodriguez, Diffractive properties of highly birefringent volume gratings: investigation, *J. Opt. Soc. Am. B: Opt. Phys.* (2002), 19: 183–189.
113. M. Yemtsova, A. Kirilyuk, A. F. van Etteger, T. Rasing, Determination of liquid crystal orientation in holographic polymer dispersed liquid crystals by linear and nonlinear optics, *J. Appl. Phys.* (2008), 104: 073115.
114. R. Jakubiak, D. P. Brown, L. V. Natarajan, V. Tondiglia, P. Lloyd, R. L. Sutherland, T. J. Bunning, and R. A. Vaia, Influence of morphology on the lasing behavior of pyrromethene 597 in a holographic polymer dispersed liquid crystal reflection grating, *Proc. SPIE* (2006), 63220A.
115. R. Jakubiak, T. J. Bunning, R. A. Vaia, L. V. Natarajan, and V. P. Tondiglia, Electrically switchable, one-dimensional polymeric resonators from holographic photopolymerization: a new approach for active photonic bandgap materials, *Adv. Mater.* (2003), 15: 241–244.
116. L. V. Natarajan, D. P. Brown, J. M. Wofford, V. P. Tondiglia, R. L. Sutherland, P. F. Loyd, and T. J. Bunning, Holographic polymer dispersed liquid crystal reflection gratings formed by visible light initiated thiol-ene polymerization, *Polymer* (2006), 47: 4411–4420.
117. H. Kogelnik, Coupled wave theory for thick hologram gratings, *Bell Syst. Tech. J.* (1969), 69: 2909–2946.
118. R. L. Sutherland, V. P. Tondiglia, L. V. Natarajan, and T. J. Bunning, Phenomenological model of anisotropic volume hologram formation in liquid-crystal-photopolymer mixtures, *J. Appl. Phys.* (2004), 96: 951–965.
119. J. Qi, L. Li, M. D. Sarkar, and G. P. Crawford, Nonlocal photopolymerization effect in the formation of reflective holographic polymer-dispersed liquid crystals, *J. Appl. Phys.* (2004), 96: 2443–2450.
120. R. L. Sutherland, V. P. Tondiglia, L. V. Natarajan, and T. J. Bunning, Evolution of anisotropic reflection gratings formed in holographic polymer-dispersed liquid crystals, *Appl. Phys. Lett.* (2001), 79: 1420–1422.
121. R. L. Sutherland, V. P. Tondiglia, L. V. Natarajan, P. F. Lloyd, and T. J. Bunning, Coherent diffraction and random scattering in thiol-ene-based holographic polymer-dispersed liquid crystal reflection gratings, *J. Appl. Phys.* (2006), 99: 123104.
122. V. P. Tondiglia, R. L. Sutherland, L. V. Natarajan, P. F. Lloyd, and T. J. Bunning, Droplet deformation and alignment for high efficiency polarization-dependent

holographic polymer-dispersed liquid-crystal reflection gratings, *Opt. Lett.* (2008), 33: 1890–1892.

123. R. L. Sutherland, V. P. Tondiglia, L. V. Natarajan, P. F. Lloyd, and T. J. Bunning, Enhancing the electro-optical properties of liquid crystal nanodroplets for switchable Bragg gratings, *Proc. SPIE* (2008): 705003.

124. F. Vita, D. E. Lucchetta, R. Castagna, O. Francescangeli, L. Criante, and F. Simoni, Detailed investigation of high-resolution reflection gratings through angular-selectivity measurements, *J. Opt. Soc. Am. B: Opt. Phys.* (2007), 24: 471–476.

125. D. Lucchetta, O. Francescangeli, L. Criante, F. Simoni, L. Pierantoni, T. Rozzi, M. Scoponi, and S. Rossetti, Optical and mechanical shrinkage effects in dye-doped photonic bandgap structures based on organic materials, *Phys. Rev. E* (2006), 73: 011708.

126. J. Qi, Holographic polymer-dispersed liquid crystals: Physics and applications, Thesis of Brown University, Providence, RI, 2003.

127. H. Pu, D. Yin, B. Gao, H. Gao, H. Dai, and J. Liu, Dynamic characterizations of high diffraction efficiency in volume Bragg grating formed by holographic photopolymerization, *J. Appl. Phys.* (2009), 106: 083111.

128. W. Huang, Y. Liu, Z. Diao, C. Yang, L. Yao, J. Ma, and L. Xuan, Theory and characteristics of holographic polymer dispersed liquid crystal transmission grating with scaffolding morphology, *Appl. Opt.* (2012), 51: 4013–4020.

129. L. D. Sio, N. Tabiryan, R. Caputo, A. Veltri, and C. Umeton, POLICRYPS structures as switchable optical phase modulators, *Opt. Express* (2008), 16: 7619–7624.

130. G. Montemezzani and M. Zgonik, Light diffraction at mixed phase and absorption gratings in anisotropic media for arbitrary geometries, *Phys. Rev. E* (1997) 55, 1035–1047.

131. L. D. Sio, A. Veltria, R. Caputo, A. D. Luca, G. Strangi, R. Bartolino, and C. P. Umeton, POLICRYPS composite structures: realization, characterization and exploitation for electro-optical and all-optical applications, *Liq. Cryst. Rev.* (2013), 1: 2–19.

132. R. Caputo, A. D. Luca, L. D. Sio, L. Pezzi, G. Strangi, C. Umeton, A. Veltri et al., POLICRYPS: A liquid crystal composed nano/microstructure with a wide range of optical and electro-optical applications, *J. Opt.* (2009), 11: 024017.

133. R. Caputo, A. Veltri, C. Umeton, and A. V. Sukhov, Kogelnik-like model for the diffraction efficiency of POLICRYPS gratings, *J. Opt. Soc. Am. B: Opt. Phys.* (2005), 22: 735–742.

134. R. Caputo, L. D. Sio, A. Veltri, C. Umeton, and A. V. Sukhov, Realization of POLICRYPS gratings: Optical and electro-optical properties, *Mol. Cryst. Liq. Cryst.* (2005), 441: 111–129.

135. R. Caputo, I. Trebisacce, L. D. Sio, and C. P. Umeton, Phase modulator behavior of a wedge-shaped POLICRYPS diffraction grating, *Mol. Cryst. Liq. Cryst.* (2011), 549: 29–36.

136. G. Strangi, V. Barna, R. Caputo, A. De Luca, C. Versace, N. Scaramuzza, C. Umeton, R. Bartolino, and G. N. Price, Color-tunable organic microcavity laser array using distributed feedback, *Phys. Rev. Lett.* (2005), 94: 063903.

137. R. Caputo, A. Veltri, C. P. Umeton, and A. V. Sukhov, Characterization of the diffraction efficiency of new holographic gratings with a nematic film–polymer-slice sequence structure, *J. Opt. Soc. Am. B: Opt. Phys.* (2004), 21: 1939–1947.

138. I. Drevensek-Olenik, M. Fally, and M. A. Ellabban, Temperature dependence of optical anisotropy of holographic polymer-dispersed liquid crystal transmission gratings, *Phys. Rev. E* (2006), 74: 021707.

139. A. Y.-G. Fuh, M.-S. Tsai, T.-C. Liu, and L.-C. Chien, Temperature dependence of the dynamical behavior of holographic gratings formed in polymer dispersed liquid crystals films, *Jpn. J. Appl. Phys.* (1997), 36: 6839–6846.

140. G. P. Crawford, Electrically switchable Bragg gratings, *Opt. Photonics News* (2003), 14: 54–49.

141. L. Domash, Y.-M. Chen, C. Gozewski, P. Haugsjaa, and M. Oren, Electronically switchable Bragg gratings for large scale NXN fiber optic crossconnects, *Proc. SPIE* (1997): 214–228.

142. R. L. Sutherland, L. V. Natarajan, V. P. Tondiglia, S. A. Siwecki, S. Chandra, and T. J. Bunning, Switchable holograms for displays and telecommunications, *Proc. SPIE* (2001): 1–10.

143. Y. J. Liu, X. W. Sun, J. H. Liu, H. T. Dai, and K. S. Xu, A polarization insensitive 2x2 optical switch fabricated by liquid crystal-polymer composite, *Appl. Phys. Lett.* (2005), 86: 041115.

144. J. Zheng, G. Sun, Y. Jiang, T. Wang, A. Huang, Y. Zhang, P. Tang, S. Zhuang, Y. Liu and S. Yin, H-PDLC based waveform controllable optical choppers for FDMF microscopy, *Opt. Express* (2011), 19: 2216–2224.

145. L. D. Sio, A. Tedesco, N. Tabirian, and C. Umeton, Optically controlled holographic beam splitter, *Appl. Phys. Lett.* (2010), 97: 183507.

146. G. P. Crawford, T. G. Fiske, and L. D. Silverstein, Reflective color displays based on PSCT and H-PDLC technologies, *SID'96.* (1996), 99–102.

147. G. P. Crawford, T. G. Fiske, and L. D. Silverstein, Reflective color LCDs based on H-PDLC and PSCT technologies, *J. Soc. Inf. Display* (1997), 5: 45–48.

148. H. Yuan, J. Colegrove, G. Hu, T. Fiske, A. Lewis, J. Gunther, L. Silverstein, C. Bowley, G. Grawford, L. Chien, and J. Kelly, HPDLC color reflective displays, *Proc. SPIE* (1999): 196–206.

149. L. Domash, G. Grawford, A. Ashmead, R. Smith, M. Popovich, and J. Storey, Holographic PDLC for photonic applications, *Proc. SPIE* (2000): 46–58.

150. M. J. Escuti, P. Kossyrev, G. P. Crawford, T. Fiske, J. Colegrove, and L. D. Silverstein, Expanded viewing-angle reflection from diffuse holographic-polymer dispersed liquid crystal films, *Appl. Phys. Lett.* (2000), 77: 4262–4264.

151. J. Qi and G. P. Crawford, Holographically formed polymer dispersed liquid crystal displays, *Displays* (2004), 25: 177–186.

152. M. S. Park, E. H. Kim, and B. K. Kim, Applications of holographic PDLC for full color display, *J. Polym. Eng.* (2008), 28: 169–178.

153. D. R. Cairns, C. C. Bowley, S. Danworaphong, A. K. Fontecchio, G. P. Crawford, L. Li, and S. M. Faris, Optical strain characteristics holographically formed polymer-dispersed liquid crystal films, *Appl. Phys. Lett.* (2000), 77: 2677–2679.

154. M. L. Ermold, K. Rai, and A. K. Fontecchio, Hydrostatic pressure response of polymer-dispersed liquid crystal gratings, *J. Appl. Phys.* (2005), 97: 104905.

155. K. Rai and A. K. Fontecchio, Optimization of pressure response in HPDLC gratings based on polymer composition, *Mol. Cryst. Liq. Cryst.* (2006), 450: 183–190.

156. V. K. S. Hsiao, W. D. Kirkey, F. Chen, A. N. Cartwright, P. N. Prasad, and T. J. Bunning, Organic solvent vapor detection using holographic photopolymer reflection gratings, *Adv. Mater.* (2005) 17: 2211–2214.

157. A. E. Siegman, *Lasers*, University Science, Mill Valley, CA, 1986.

158. G. S. He, T.-C. Lin, V. K. S. Hsiao, A. N. Cartwright, P. N. Prasad, L. V. Natarajan, V. P. Tondiglia, R. Jakubiak, R. A. Vaia, and T. J. Bunning, Tunable two-photon pumped lasing using a holographic polymer-dispersed liquid-crystal grating as a distributed feedback element, *Appl. Phys. Lett.* (2003), 83: 2733–2735.

159. D. E. Lucchetta, L. Criante, O. Francescangeli, and F. Simoni, Wavelength flipping in laser emission driven by a switchable holographic grating, *Appl. Phys. Lett.* (2004), 84: 837–839.

160. H. Kogelnik and C. V. Shank, Stimulated emission in a periodic structure, *Appl. Phys. Lett.* (1971), 18: 152–154.

161. R. Jakubiak, L. V. Natarajan, V. Tondiglia, G. S. He, P. N. Prasad, T. J. Bunning, and R. A. Vaia, Electrically switchable lasing from pyrromethene 597 embedded holographic-polymer dispersed liquid crystals, *Appl. Phys. Lett.* (2004), 85: 6095–6097.

162. D. E. Lucchetta, L. Criante, O. Francescangeli, and F. Simoni, Light amplification by dye-doped holographic polymer dispersed liquid crystals, *Appl. Phys. Lett.* (2004), 84: 4893–4895.

163. L. Criante, D. E. Lucchetta, F. Vita, R. Castagna, and F. Simoni, Distributed feedback all-organic microlaser based on holographic polymer dispersed liquid crystals, *Appl. Phys. Lett.* (2009), 94: 111114.

164. V. K. S. Hisao, C. Lu, G. S. He, M. Pan, A. N. Cartwright, P. N. Prasad, R. Jakubiak, R. A. Vaia, and T. J. Bunning, High contrast switching of distributed-feedback lasing in dye-doped H-PDLC transmission grating structures, *Opt. Express* (2005), 13: 3787–3794.

165. W. Cao, A. Munoz, P. Palffy-Muhoray, and B. Taheri, Lasing in a three-dimensional photonic crystal of the liquid crystal blue phase II, *Nat. Mater.* (2002), 1: 111.

166. W. Cao, P. Palffy-Muhoray, B. Taheri, A. Marino, and G. Abbate, Lasing thresholds of cholesteric liquid crystals lasers, *Mol. Cryst. Liq. Cryst.* (2005), 429: 101–110.

167. M. F. Moreira, I. C. S. Carvalho, W. Cao, C. Bailey, B. Taheri, and P. Palffy-Muhoray, Cholesteric liquid-crystal laser as an optic fiber-based temperature sensor, *Appl. Phys. Lett.* (2004), 85: 2691–2693.

168. Y. J. Liu, X. W. Sun, P. Shum, H. P. Li, J. Mi, and W. Ji, Low-threshold and narrow-linewidth lasing from dye-doped holographic polymer-dispersed liquid crystal transmission gratings, *Appl. Phys. Lett.* (2006), 88: 061107.

169. S. Riechel, U. Lemmer, J. Feldmann, T. Benstem, W. Kowalsky, U. Scherf, A. Gombert, and V. Wittwer, Laser modes in organic solid-state distributed feedback lasers, *Appl. Phys. B: Lasers Opt.* (2000), 71: 897–900.

170. C. Ye, L. Shi, J. Wang, and D. Lo, Simultaneous generation of multiple pairs of transverse electric and transverse magnetic output modes from titania zirconia organically modified silicate distributed feedback waveguide lasers, *Appl. Phys. Lett.* (2003), 83: 4101–4103.

171. J. Wang, F. Chen, R. Li, H. Dong, J. Fan, L. Zhang, L. Shi, and K. Y. Wong, Optically pumped distributed feedback thin film waveguide lasers with multiwavelength and polarized emissions, *Appl. Phys. B: Lasers Opt.* (2012), 107: 163–169.

172. R. Jakubiak, V. P. Tondiglia, L. V. Natarajan, R. L. Sutherland, P. Lloyd, T. J. Bunning, and R. A. Vaia, Dynamic lasing from all-organic two-dimensional photonic crystals, *Adv. Mater.* (2005), 17: 2807–2811.

173. R. Jakubiak, V. P. Tondiglia, L. V. Natarajan, R. L. Sutherland, P. Lloyd, and T. J. Bunning, Stimulated emission from pyrromethene 597 in holographic polymer dispersed liquid crystal structures, *Proc. SPIE* (2005): 202–207.

174. R. Jakubiak, V. P. Tondiglia, L. V. Natarajan, P. F. Lloyd, R. L. Sutherland, R. A. Vaia, and T. J. Bunning, Lasing of pyrromethene 597 in 2D holographic polymer dispersed liquid crystals: influence of columnar conformation, *Proc. SPIE* (2009): 72320K.

175. D. Luo, X. W. Sun, H. T. Dai, Y. J. Liu, H. Z. Yang, and W. Ji, Two-directional lasing form a dye-doped two-dimensional hexagonal photonic crystal made of holographic polymer-dispersed liquid crystals, *Appl. Phys. Lett.* (2009), 95: 151115.

176. D. Luo, X. W. Sun, H. T. Dai, H. V. Demir, H. Z. Yang, and W. Ji, Temperature effect on the lasing from a dye-doped two-dimensional hexagonal photonic crystal made of holographic polymer-dispersed liquid crystals, *J. Appl. Phys.* (2010), 108: 013106–013103.

177. H. Coles and S. Morris, Liquid-crystal lasers, *Nat. Photonics* (2010), 4: 676–685.

178. K. Sakoda, Enhanced light amplification due to group-velocity anomaly peculiar to two- and three-dimensional photonic crystals, *Opt. Express* (1999), 4: 167–176.

179. N. Susa, Threshold gain and gain-enhancement due to distributed-feedback in two-dimensional photonic-crystal lasers, *J. Appl. Phys.* (2001), 89: 815–823.

180. G. A. Turnbull, P. Andrew, W. L. Barnes, and I. D. W. Samuel, Operating characteristics of a semiconducting polymer laser pumped by a microchip laser, *Appl. Phys. Lett.* 82 (2003) 313–315.

181. V. P. Tondiglia, L. V. Natarajan, R. L. Sutherland, D. Tomlin, and T. J. Bunning, Holographic formation of electro-optical polymer-liquid crystal photonic crystals, *Adv. Mater.* (2002), 14: 187–191.

182. M. J. Escuti, J. Qi, and G. P. Crawford, Tunable face centered-cubic photonic crystal formed in holographic polymer dispersed liquid crystals, *Opt. Lett.* (2003), 28 522–524.

183. M. J. Escuti, J. Qi, and G. P. Crawford, Two-dimensional tunable photonic crystal formed in a liquid-crystal/polymer composite: threshold behavior and morphology, *Appl. Phys. Lett.* (2003), 83: 1331–1333.

184. M. J. Escuti and G. P. Crawford, Mesoscale three dimensional lattices formed in polymer dispersed liquid crystals: a diamond-like face centered cubic, *Mol. Cryst. Liq. Cryst.* (2004), 421: 23–26.

185. M. Devetak, J. Milavee, R. A. Rupp, B. Yao, and I. Drevebsek-Olenik, Two-dimensional photonic lattices in polymer-dispersed liquid crystal composites, *J. Opt. A: Pure Appl. Opt.* (2009), 11: 024020.

186. W. Huang, Z. Diao, Y. Liu, Z. Peng, C. Yang, J. Ma, and L. Xuan, Distributed feedback polymer laser with an external feedback structure fabricated by holographic polymerization technique, *Org. Electron.* (2012), 13: 2307–2311.

187. W. Huang, Y. Liu, L. Hu, Q. Mu, Z. Peng, C. Yang, and L. Xuan, Second-order distributed feedback polymer laser based on holographic polymer dispersed liquid crystal grating, *Org. Electron.* (2013), 14: 2299–2305.

188. Z. Diao, S. Deng, W. Huang, L. Xuan, L. Hu, Y. Liu, and J. Ma, Organic dual-wavelength distributed feedback laser empowered by dye-doped holography, *J. Mater. Chem.* (2012), 22: 23331–23334.

189. Z. Diao, L. Xuan, L. Liu, M. Xia, L. Hu, Y. Liu, and J. Ma, A dual-wavelength surface-emitting distributed feedback laser from a holographic grating with

an organic semiconducting gain and a doped dye, *J. Mater. Chem. C.* (2014), 2: 6177–6182.

190. F. J. Duarte, *Tunable Laser Applications* (2nd Ed.), CRC Press, New York, 2009.

191. D. Bruneau, H. Cazeneuve, C. Loth, and J. Pelon, Double-pulse dual-wavelength alexandrite laser for atmospheric water vapor measurement, *Appl. Opt.* (1991), 30: 3930–3937.

192. X. He, X. Fang, C. Liao, D. Wang, and J. Sun, A tunable and switchable single-longitudinal-mode dual-wavelength fiber laser with a simple linear cavity, *Opt. Express* (2009), 17: 21773–21781.

193. D. Dragoman and M. Dragoman, *Advanced Optoelectronic Devices*, Springer, New York, 1998.

194. J. Clark and G. Lanzani, Organic photonics for communications, *Nat. Photonics* (2010), 4: 438–446.

195. R. Wu, Y. Li, J. Wu, J. Ma, and Q. Dai, A study of lasing wavelength by DOS in the temperature-tunable cholesteric liquid crystal lasers, *Opt. Commun.* (2013), 300: 1–4.

196. Y. J. Liu, X. W. Sun, H. I. Elim, and W. Ji, Gain narrowing and random lasing from dye-doped polymer-dispersed liquid crystals with nanoscale liquid crystal droplets, *Appl. Phys. Lett.* (2006), 89: 011111.

197. Y. J. Liu, X. W. Sun, H. I. Elim, and W. Ji, Effect of liquid crystal concentration on the lasing properties of dye-doped holographic polymer-dispersed liquid crystal transmission gratings, *Appl. Phys. Lett.* (2007), 90: 011109.

198. R. Ozaki, T. Matsui, M. Ozaki, and K. Yoshino, Electrically color-tunable defect mode lasing in one-dimensional photonic-band-gap system containing liquid crystal, *Appl. Phys. Lett.* (2003), 82: 3593–3595.

199. M. Ozaki, M. Kasano, T. Kitasho, D. Ganzke, W. Haase, and K. Yoshino, Electro-tunable liquid-crystal laser, *Adv. Mater.* (2003), 15: 974–977.

200. R. Ozaki, Y. Matsuhisa, M. Ozaki, and K. Yoshino, Electrically tunable lasing based on defect mode in one-dimensional photonic crystal with conducting polymer and liquid crystal defect layer, *Appl. Phys. Lett.* (2004), 84: 1844–1846.

201. R. Ozaki, T. Shinpo, K. Yoshino, M. Ozaki, and H. Moritake, Tunable liquid crystal laser using distributed feedback cavity fabricated by nanoimprint lithography, *Appl. Phys. Express* (2008), 1: 012003.

202. V. I. Kopp, B. Fan, H. K. M. Vithana, and A. Z. Genack, Low-threshold lasing at the edge of a photonic stop band in cholesteric liquid crystals, *Opt. Lett.* (1998), 23: 1707–1709.

203. S. Furumi, S. Yokoyama, A. Otomo, and S. Mashiko, Electrical control of the structure and lasing in chiral photonic band-gap liquid crystals, *Appl. Phys. Lett.* (2003), 82: 16–18.

204. B. Maune, M. Loncar, J. Witzens, M. Hochberg, T. Baehr-Jones, D. Psaltis, A. Scherer, and Y. M. Qiu, Liquid-crystal electric tuning of a photonic crystal laser, *Appl. Phys. Lett.* (2004), 85: 360–362.

205. H. P. Yu, B. Y. Tang, J. H. Li, and L. Li, Electrically tunable lasers made from electro-optically active photonics band gap materials, *Opt. Express* (2005), 13: 7243–7249.

206. A. D. Ford, S. M. Morris, and H. J. Coles, Photonics and lasing in liquid crystals, *Mater. Today* (2006), 9: 36–42.

207. L. Chen, F. Gao, Y. Bu, F. Jia, C. Liu, and Z. Cai, Tunable distributed feedback lasing from leaky waveguides based on gel-glass dispersed liquid crystal thin films, *Mater. Lett.* (2011), 65: 3476–3478.

208. S. Klinkhammer, N. Heussner, K. Huska, T. Bocksrocker, F. G. Ringer, C. Vannahme, T. Mappes, and U. Lemmer, Voltage-controlled tuning of an organic semiconductor distributed feedback laser using liquid crystals, *Appl. Phys. Lett.* (2011), 99: 023307.

209. W. Huang, Z. Diao, L. Yao, Z. Cao, Y. Liu, J. Ma, and L. Xuan, Electrically tunable distributed feedback laser emission from scaffolding morphologic holographic polymer dispersed liquid crystal grating, *Appl. Phys. Express* (2013), 6: 022702.

210. Z. Diao, W. Huang, Z. Peng, Q. Mu, Y. Liu, J. Ma, and L. Xuan, Anisotropic waveguide theory for electrically tunable distributed feedback laser from dye-doped holographic polymer dispersed liquid crystal, *Liq. Cryst.* (2013), 41: 239–246.

211. G. Abbate, L. DeStefano, and E. Santamato, Transverse-magnetic nonlinear modes in a nematic liquid-crystal slab waveguide, *J. Opt. Soc. Am. B: Opt. Phys.* (1996), 13: 1536–1541.

212. S. Yamamoto, Y. Koyamada, and T. Makimoto, Normal-mode analysis of anisotropic and gyrotropic thin-film waveguides for integrated optics, *J. Appl. Phys.* (1972), 43: 5090–5097.

213. K. Okamoto, *Fundamentals of Optical Waveguides* (2nd Ed.), Academic Press, Burlington, MA, 2010.

8

Organic and Organic–Inorganic Hybrid Electrochromic Materials and Devices

Prakash R. Somani

CONTENTS

8.1 Introduction to Chromism and Chromic Materials

The word *chromism* means *color change*. In the modern sense, it represents *change in absorption, reflection, or refraction* of the material/coatings. When the color change is in the visible region (400–800 nm) of the spectrum, it is directly observable by our naked eye. However, this need not always be the case. Today, photodetectors have replaced the job of the eye, and the color change can be in ultraviolet, visible, or infrared regions. Materials that show such color change are known as *chromic materials*. Color change in these chromic materials is due to external stimuli such as change in temperature, illumination by light, pressure, and so on.

Color change can be reversible or irreversible. In other words, certain materials display new color/new optical properties as long as external stimuli are present and regain their original color/optical properties once the external stimuli are removed. In such a case, the material is said to possess *reversible chromism*. The most common example is that of photochromic sunglasses, which change color as long as they are held under sunlight and regain their original color in shadow. On the other hand, certain materials display new color/optical properties even when the external stimuli are removed and do not regain their old color/optical properties. Such materials are said to possess *irreversible chromism*. There might be different reasons for such irreversible color change, including but not limited to conversion of the material/molecule from one form to another (e.g., irreversible *cis-* to *trans-* transformation), degradation (partial or complete) of the material itself, and so on. In general, both reversible and irreversible chromic materials are useful; more useful are materials that display *reversible chromism*. This is due to the simple fact that the same material/device can be used multiple times. Chromism can be classified into different types depending on the stimuli/cause. Table 8.1 displays important types of chromism and its causes/stimuli.

TABLE 8.1

Important Types of Chromism and Their Causes/Stimuli

No.	Chromism	Cause
1.	Electrochromism	Application of a voltage
2.	Thermochromism	Change in temperature
3.	Photochromism	Illumination by light
4.	Piezochromism	Application of pressure/change in pressure applied
5.	Vapochromism	Due to vapor/gas contact and/or its concentration
6.	Humidochromism	Change in humidity
7.	Ionochromism	Due to ions
8.	Halochromism	Due to change in pH
9.	Tribochromism	Due to mechanical friction

Chromic materials and devices are finding increasing applications in diverse areas. Photochromic sunglasses have been in use for a long time. Certain dyes and pigments that show solvatochromism and thermochromism are in use as indicators and in textiles, respectively. Recently, electrochromic (EC) smart windows (a combination of an EC window and solar cell) and electrochromic displays (ECDs) have been commercialized. Such applications are important from the viewpoint of energy saving in buildings and reducing pollution and should be developed and commercialized further. In other words, the technology of chromic materials and devices is an *eco-friendly* and *green technology*. The market for EC materials and devices is predicted to be about a few billion dollars, which is significant.

Chromism (change in color) is also present in nature. Certain plants and animals do change their color with respect to the surrounding environment and/or for their protection.

8.2 Electrochromism and Electrochromic Materials

Theoretical studies have predicted that the absorption and emission spectra of certain dyes may be shifted by several nanometers on application of a strong electric field. This effect is called *electrochromism*, in analogy to *thermochromism* and *photochromism*, which represents change of color due to heat (thermal energy) and light, respectively. Currently, this definition does not represent electrochromism very well.

An EC material changes color in a persistent but reversible manner by an electrochemical reaction called electrochromism. Electrochromism is a reversible and visible (i.e., differentiable) change in transmittance and/or reflectance that is associated with an electrochemically induced (or driven) oxidation–reduction reaction. It results from the generation of different visible-region electronic absorption bands on switching between redox states. The color change is commonly between a transparent (bleached) and a colored state, or between two colored states. If more than two redox states are electrochemically available, the EC material may exhibit different colors and may be called *polyelectrochromic* or can be said to possess *multicolor electrochromism.* This optical change is effected by a small electric current at low dc potential of few volts.

Several materials show EC color change. For example, transition metal oxides, Prussian blue (PB), viologens, phthalocyanines (Pcs), conducting polymers (CPs), and so on.

8.3 Origin of Electrochromic Color Change

An EC material changes color in a persistent but reversible manner by an electrochemical reaction, which usually involves the insertion or removal of ions in/from an EC material (i.e., oxidation or reduction reaction). This causes changes in the optical bandgap (or energy gap between highest occupied molecular orbital [HOMO] and lowest unoccupied molecular orbital [LUMO] levels) and hence the color of the material. In certain materials, the electrochemical reaction only involves interaction of ions with the material to change the oxidation/reduction state and hence the optical properties and color.

An EC device is essentially a rechargeable battery in which the EC electrode is separated by a suitable solid or liquid electrolyte from a charge balancing counter electrode, and the color change occurs by charging or discharging the electrochemical cell with an applied potential of a few volts. After the resulting pulse of the current has decayed and the color change has been effected, the new redox state persists, with little or no input of power, due to the so-called *memory effect*. Figure 8.1 illustrates a schematic of an EC device. The EC electrodes can work in either reflective or transmissive mode. Such electrodes are made using a transparent conducting electrode coated with an EC material. The counter electrode can be of any material that provides a reversible electrochemical reaction in devices. In variable light transmissive EC devices, the counter electrode also has to be transparent. This is due to the fact that light should be transmitted through the device. The counter electrode can be colorless in both of its redox forms or EC in a complementary mode to the primary EC electrode/material, because the entire system is in the optical path. Hence, transparent electrolytes are required in transmissive devices.

Transparent conducting electrodes made from indium-tin-oxide (ITO)-coated glass/plastic substrates, aluminum-doped ZnO, fluorine-doped tin oxide electrodes (FTO), semitransparent metal films, transparent CP films, and so on are generally used. For applications that are designed to operate in a reflective mode (e.g., displays), the counter electrode can be of any material with a suitable reversible redox reaction.

FIGURE 8.1
Schematic diagram of an electrochromic device.

8.4 Important Characterization Techniques

EC phenomena involve color change and electrochemistry. Ultraviolet-visible optical absorption spectroscopy or diffuse reflectance spectroscopy is generally employed to characterize the changes in color/optical properties of the material. EC materials are generally first studied at a single working electrode, under potentiostatic or galvanostatic control, using three-electrode circuitry. Electrochemical techniques such as cyclic voltammetry, coulometry, and chronoamperometry, all with, as appropriate, *in situ* spectroscopic measurements, are employed for characterization. X-ray photoelectron spectroscopy (XPS) is commonly used to determine the oxidation/reduction state(s) of the materials. Other physiochemical characterization techniques such as X-ray diffraction, scanning electron microscopy, transmission electron microscopy, Fourier transform infrared spectroscopy, energy dispersive X-ray analysis, atomic force microscopy, and so on are used to get detailed information about the various physicochemical properties of the material(s) used.

EC materials are characterized either *in situ* (when the electrochemical reaction is occurring/color change is happening) or *ex situ* (after the electrochemical reaction has occurred). *In situ* characterization gives information about real-time measurements, whereas *ex situ* measurements give information about the material(s) after the color change/electrochemical reaction has occurred.

The purpose of this chapter is to introduce EC phenomena and materials used, with a special emphasis on organic and organic–inorganic hybrid materials, new concepts and applications, current status, and future prospects for the technology.

8.5 Inorganic Electrochromic Materials

8.5.1 Transition Metal Oxides

Oxides of many transition metals (in the film form), for example, iridium [1,2], rhodium [3], ruthenium [4], tungsten [5], manganese [6], cobalt [6], and so on have been shown to possess EC properties. This class of EC material has been classified under inorganic EC materials. Transition metal oxide (TMO) films have been deposited by several techniques, such as vacuum evaporation [7], sputtering [7], spray deposition [8], electrodeposition [9], electrochemical oxidation of tungsten metal [10], chemical vapor deposition [10], sol–gel method [10], and so on. TMO films can be electrochemically switched to a nonstoichiometric redox state, which has an intense EC absorption band due to optical intervalence charge transfer (CT) [10–13]. A typical and widely studied example is the tungsten trioxide (WO_3) system; its electrochromism was first reported in 1969 [10–13]. Tungsten oxide has a nearly cubic structure, which may be simply described as an *empty-perovskite* type formed by WO_6 octahedra that share corners. The empty space inside the cubes is considerable and this provides the availability of a large number of interstitial sites where the guest ions can be inserted.

WO_3, with all tungsten sites as oxidation state W^{VI}, is a transparent thin film. On electrochemical reduction, W^V sites are generated to give the EC (blue coloration to the film) effect. Although there is still controversy about the detailed coloration mechanism, it is generally accepted that the injection and extraction of electrons and metal cations (Li^+, H^+, etc.) play an important role. WO_3 is a cathodically ion insertion material. The blue coloration in the thin film of WO_3 can be erased by electrochemical oxidation. In the case of Li^+ cations, the electrochemical reaction can be written as Equation 8.1 and the generalized equation can be written as Equation 8.2:

$$WO_3 + x(Li^+ + e^-) \rightarrow Li_x W_{(1-x)}{}^{VI} W_x^V O_3$$
$$(\text{transparent}) \qquad\qquad (\text{blue})$$

(8.1)

$$WO_3 + xM^+ + xe^- \rightarrow M_x WO_3$$
$$(\text{transparent}) \qquad\qquad (\text{blue})$$

(8.2)

The fractional number of sites that are filled in the WO_3 lattice is indicated by the subscript x in the general formula $M_x WO_3$. At low x, the films have an intense blue color caused by photo-effected intervalence charge transfer between adjacent W^V and W^{VI} sites. At higher x, insertion irreversibly forms a metallic bronze, which is red or golden in color. The process is promoted by cathodic polarization, which induces ion insertion and electron injection. The inserted ions expand the lattice of the guest oxide while the compensating electrons modify its electronic structure and in turn its optical properties. It can be stated simply that the injected electrons are trapped by a W^{6+} forming a W^{5+}, while M^+ remains ionized in the interstitial sites of the WO_3 lattice. This gives rise to the formation of tungsten bronze having electrical and optical properties different from those of the pristine oxide. In fact, the pristine state WO_3 is pale yellow and a poor electrical conductor, while in the intercalated $M_x WO_3$ state, it becomes highly conducting and blue in color with absorption spectra around 0.5–0.6 μm.

The aforementioned W^{6+}/W^{5+} intervalance transition model implies a certain delocalization of electrons, which is consistent with the enhancement of the conductivity that accompanies the insertion processes. However, although this is the most accepted theory, other models, including a nonlocalized electron model, are proposed to explain the EC mechanism of tungsten oxide, in particular, and of inorganic ion insertion compounds in general.

The characteristics of the EC process can be conventionally examined by directly comparing the optical and electrochemical response of the WO_3 electrode. Figure 8.2 shows the cyclic voltammetry and the optical transmittance of a WO_3 electrode measured in a cell having a Li^+ ion conducting electrolyte and a Li metal counter electrode. The cathodic scan promotes the development of the process: current flows in the WO_3 electrode and the blue color L_xWO_3 bronze develops. Accordingly, the optical transmittance decays to reach a minimum at the voltammetric peak. Reversing the scan results in the removal of the Li^+ ions with the restoration of the pale-yellow-colored WO_3 pristine oxide and the transmittance increases to reach nearly 100%. Although the process is readily reversible and can be repeated a large number of times, a careful observation of the voltammograms reveals some slight differences between the initial cycle and the subsequent cycling. Figure 8.3 shows that the development of the EC effect requires a sort of preliminary activation process. Such a process, which is commonly encountered in insertion compounds, may be described as an initial and permanent uptake of M^+ ions, which somewhat opens the route for the following fast and reversible insertion–withdrawal reactions. The extent of the process may vary from material to material and for some insertion ECMs it appears crucial in establishing prolonged cyclability [14]. A very good, informative review has been written by Granqvist [15].

The EC properties of WO_3 films have also been studied by combining it with other EC materials such as Nb_2O_5, CPs, and so on, and the ECDs were made and studied [16–21]. WO_3 research has always been driven by the many possible commercial applications and prototype alphanumeric displays, and EC mirrors were also reported. The so-called amorphous or

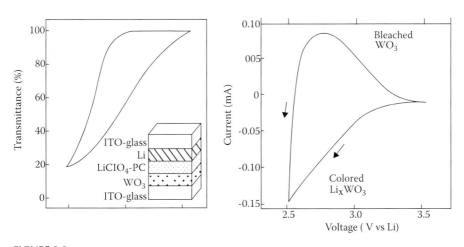

FIGURE 8.2
Cyclic voltammetry and optical transmittance of a WO_3 electrode in a $LiClO_4$-PC solution. Light source: He–Ne laser (6328 Å); scan rate: 20 mV/s; sample thickness: 600 Å. (From S. Passerini and B. Scrosati, *Solid State Ion.*, 55, 56, 1992, 520.)

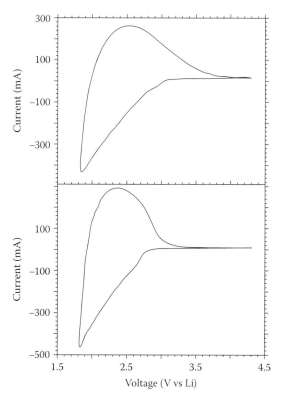

FIGURE 8.3
Initial (upper curve) and stabilized (lower curve) voltammetric response of a WO_3 electrode in a $LiClO_4$-PC solution. Scan rate: 20 mV/s; sample thickness: 2500 Å. (From S. Passerini and B. Scrosati, *Solid State Ion.*, 55, 56, 1992, 520.)

poorly crystallized tungsten oxide thin films have developed as the most important EC material. EC material, such as the viologens, is of great interest because of its reversible clear to deep blue coloration in transmission and high coloration efficiency (CE). An important way to assess the many known insertion/extraction films is to compare spectral coloration efficiencies, CE (λ), for the visible region:

$$CE\ (\lambda) = \Delta\ OD\ (\lambda)/q \qquad (8.3)$$

where $\Delta\ OD\ (\lambda)$ represents the change in single-pass, transmitted optical density at the wavelength of interest λ, because of transfer of charge q, as $cm^2 \cdot C^{-1}$. The CE, also called electrochromic efficiency (EE), depends strongly on the sample preparation. The EE may vary from 40 to 100 $cm^2 \cdot C^{-1}$ and accordingly the response time (i.e., the coloration–decoloration switching time) may vary from milliseconds up to few seconds. The cycle life (i.e., the number of coloration–decoloration cycles) depends on the electrolyte used and can reach up to 10^6 cycles. Adherence to Lambert's law must either be assumed or tested to avoid pitfalls with thin films; cathodic coloration is well described by CE (λ) at moderate values of q for many inorganic amorphous films [22]. The CE is determined spectroelectrochemically, using a cell that employs a cathode–anode pair in a liquid electrolyte. The cell is operated such that only the electrode of interest is in the light path of the spectrophotometer.

At present, the major aim is the development of Smart Windows for control of thermal conditions within a building, thereby reducing winter heating and summer cooling requirements. Glass manufacturers have recognized this opportunity for a long time. Pilkington Technology Center has produced a prototype EC window (dimensions: 0.7 m × 1 m), which when colored is capable of reducing light transmittance by a factor of four [12]. Also, it is worthwhile to mention here that numerous patents were granted to American Cyanamid as a result of its display-oriented work. Such EC systems are of a simple two-electrode sandwich device construction as described in Section 8.3. In the development of such variable transmission windows, glass companies favor the sputtering technique because it is already in place for the production of a range of coatings for architectural glazing [23].

Many other TMOs in the thin film form are EC, for example, the oxides of molybdenum and vanadium given in Table 8.2. All the previous materials of TMOs given in Table 8.2 are cathodically ion insertion materials, that is, a more intensely absorbing redox state is produced on reduction. In contrast, group VIII metal oxides become colored on electrochemical oxidation (anodic ion insertion materials). The typical examples are given in Table 8.2, Part B [24–32].

8.5.2 Prussian Blue

PB, iron(III) hexacyanoferrate(II), is the prototype of a number of polynuclear transition metal hexacyanometallates, which form an important class

TABLE 8.2

Electrochromic Transition Metal Oxides

S. No.	Cathodically Ion Insertion Materials	Anodically Ion Insertion Materials	Color	
			Oxidized State	Reduced State
Part A				
1.	MoO_3		Transparent	Blue
2.	V_2O_5		Yellow	Blue-black
3.	Nb_2O_5		Yellow	Blue
4.	WO_3		Transparent	Blue
Part B				
1.		$Ir(OH)_3$	Blue-black	Transparent
2.		$Ni(OH)_2$	Brown-bronze	Transparent or pale green

of insoluble mixed valance compounds. They have the general formula M'_k $[M''(CN)_6]_l$ (l, k integral), where M' and M'' are transition metals with different formal oxidation numbers. In these compounds, a given element exists in two oxidation states with the possibility of charge/electron transfer between them. PB is ferric ferrocyanide. A high spin ferric ion is coupled to a low spin ferrous ion by a cyanide bridge. The coupling is Fe(III)–N≡C≡Fe(II), that is, the first ferric ion couples to the nitrogen and the ferrous ion to the carbon. Robin [33] has shown that the electrons have a 99% probability of being on the low spin ion coordinated to the carbon.

Although PB has been an important inorganic pigment and has been manufactured on a large scale for use in paints, lacquers, printing inks, and other color uses, not much was known about the electrochemistry of PB. This might be due to the lack of knowledge for the preparation of the thin films of PB. Neff et al. [34,35] have shown a method for preparation of thin films of PB on platinum (Pt) and gold electrodes and demonstrated the redox behavior of PB. The first report [34] concerning the electrochemistry and electrochromism of PB promoted numerous investigations into the properties of PB thin films. We have used PB as a sensitizer to improve the EC response of CPs (particularly polypyrrole [PPy] and polyaniline [PANI]) and for photosensitivity [36,37]. Also, we have constructed ECDs (laboratory scale) and studied their EC responses in liquid electrolytes, which are explained in Section 8.6.4.2.

As is well known, there are two proposed formulae of PB, *water-insoluble PB* $[Fe^{III}_4\{Fe^{II}(CN)_6\}_3]$ and *water-soluble PB* $[KFe^{III}Fe^{II}(CN)_6]$ [38–40]. The electrochemical reduction and oxidation of PB can lead to *Prussian white* (PW) (Everitt's salt) and *Prussian green* (PG) (Berlin green), respectively. PB thin films are generally formed by electrochemical reduction of solutions containing iron (III) and hexacyanoferrate (III) ions [41–44]. Reduction of the brown-yellow soluble complex Prussian brown [PX, iron (III) hexacyanoferrate (III), present in equilibrium with the iron (III) and hexacyanoferrate (III) ions] is the principal electron-transfer process in PB electrodeposition, as shown in the equation below:

$$[Fe^{III}Fe^{III}(CN)_6] + e^- \rightarrow [Fe^{III}Fe^{II}(CN)_6]^-$$

$$\text{(PX)} \qquad\qquad\qquad \text{(PB)}$$

$$(8.4)$$

Charge compensating cations (initially Fe^{3+}, then K^+ on potential cycling in K^+-containing supporting electrolyte) are present in the PB film for electroneutrality [45]. Partial electrochemical oxidation of PB in pure supporting electrolyte yields PG, a type historically known as Berlin green (BG), as can be presented by

$$3[Fe^{III}Fe^{II}(CN)_6]^- \rightarrow [Fe^{III}_3\{Fe^{III}(CN)_6\}_2\{Fe^{II}(CN)_6\}]^- + 2e^-$$

$$\text{(PB)} \qquad\qquad\qquad \text{(PG)}$$

$$(8.5)$$

Although in bulk form PG is believed to have a fixed composition with anion composition, for thin films there is a continuous composition range between PB and PX, which becomes golden yellow in the fully oxidized form [45]. The latter may be obtained by electrochemical oxidation of a particularly pure form of PB [41,45]:

$$[Fe^{III}Fe^{II}(CN)_6]^- \rightarrow [Fe^{III}Fe^{III}(CN)_6] + e^- \qquad (8.6)$$
$$\text{(PX)} \qquad\qquad\qquad \text{(PB)}$$

Reduction of PB yields PW, also known as Everitt's salt, which appears transparent as a thin film:

$$[Fe^{III}Fe^{II}(CN)_6]^- + e^- \rightarrow [Fe^{II}Fe^{II}(CN)_6]^{2-} \qquad (8.7)$$
$$\text{(PW)} \qquad\qquad\qquad \text{(PB)}$$

For all of the aforementioned EC redox reactions, Equations 8.4–8.7, there is a concomitant ion ingress/egress in the films for electroneutrality. The PB film deposited electrochemically on an ITO electrode shows a color change from blue to transparent when reduced by applying a potential of 0.5 V versus saturated calomel electrode (SCE) in an aqueous (0.01 M) KCl electrolyte for 60 seconds. Our XPS results shown in Figure 8.4 confirm Equation 8.7.

Early PB–ECD employed PB as a sole EC material, for example a seven segment display using PB-modified SnO_2 working and counter electrode at 1 mm separation and an ITO/PB–Nafion/ITO solid state device [46,47]. For the solid state system, device fabrication involved chemical rather than electrochemical formation of PB on immersion of a membrane of the solid polymer electrolyte (SPE) Nafion in aqueous solutions of $FeCl_3$, then $K_3Fe(CN)_6$. The resulting PB-containing Nafion composite film was sandwiched between two ITO electrodes. The construction and optical behavior of an ECD utilizing a single film of PB, without addition of conventional electrolyte, have also been described [48]. In such designs, a film of PB is sandwiched between two optically transparent electrodes (Figure 8.5a). On application of an appropriate potential across the film, oxidation occurs near the positive electrode and the reduction occurs near the negative electrode to yield PX and PW, respectively (Figure 8.5b). The conversion of the outer portions of the film results in a net bleaching of the device. The functioning of the device relies on the fact that PB can be bleached both anodically (to the yellow state) and cathodically (to the transparent state) and that it is a mixed conductor through which the potassium ions can move to provide the charge compensation required for the EC redox reactions.

Since PB and WO_3 are anodically and cathodically coloring EC materials, respectively, they are used together in a single device so that their EC

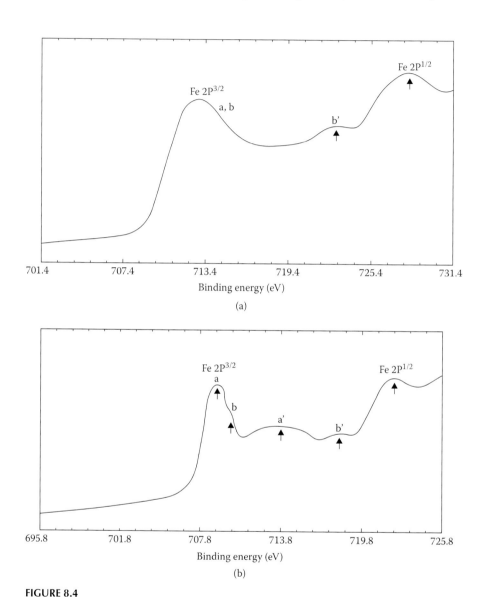

FIGURE 8.4
(a) X-ray photoelectron spectroscopy spectrum of the electrochemically deposited PB film, and
(b) the reduced film obtained by applying a reducing potential of −0.5 V versus SCE for 60
seconds in 0.01 M aqueous KCl electrolyte. (From P. Somani and S. Radhakrishnan, *Chem. Phys.
Lett.*, 292, 1998, 218.)

reactions are complementary [49]. Our results on PPy/PB and PANI/PB composite films show that the use of PB not only enhances the EC response of
the CPs but also extends the EC response to a wider region of the visible
spectrum, thus acting as a sensitizer.

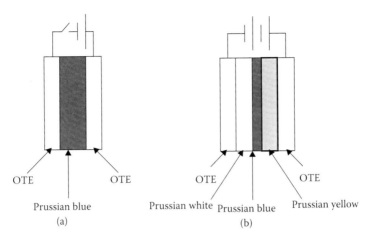

FIGURE 8.5
Electrochromic display using PB film only (a) before and (b) after application of the voltage.
(From M. K. Carpenter and R. S. Conell, *J. Electrochem. Soc.*, 137, 1990, 2464.)

8.6 Organic Electrochromic Materials

8.6.1 Phthalocyanines

Since the first report of the multicolor electrochromism of thin films of lutetium bis(phthalocyanine) ([Lu(Pc)$_2$]) in 1970, numerous metal Pcs have been investigated for their EC properties. Such compounds have a metal ion either at the center of a single Pc ring or between two rings in a sandwich-type compound. Solid state electrochromism has been observed in Pc compounds of more than 30 metals, predominantly among the sandwich-type diphthalocyanine (bisphthalocyanine) of the lanthanide and actinide rare earths and the related Group III elements lanthanide, yttrium, and scandium. It also occurs in diphthalocyanines of the Group IV elements tin, zirconium, and hafnium. Single-ring EC Pcs include the metal-free and magnesium compounds and those of transition elements, including divalent iron, cobalt, nickel, copper, zinc, and molybdenum [50–56].

Because lutetium diphthalocyanine [Lu(Pc)$_2$] provides a full range of colors from orange or reddish orange to violet and the films are easily prepared by vacuum sublimation, this compound has been most extensively studied among the lanthanide complexes. [Lu(Pc)$_2$] is generally formed as a vivid green film on vacuum sublimation. Moskalev and Kirin [50] were the first to report the [Lu(Pc)$_2$] EC spectra, using an aqueous 0.1 M KCl electrolyte. Typical curves for the cycled green, light blue, and orange forms in 0.1 M KCl are shown in Figure 8.6. Other spectra representing further reduction of the film to dark purple-blue and violet are given in Nicholson and Weismuller [57]. [Lu(Pc)$_2$] films undergo a series of ring-based redox processes. On oxidation,

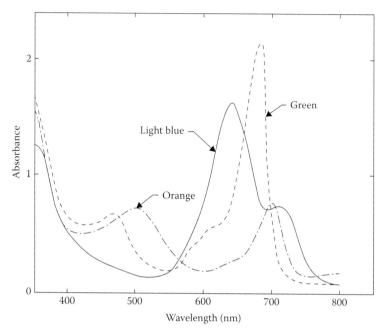

FIGURE 8.6
Absorption spectra of a lutetium diphthalocyanine film in aqueous 1 M KCl. (From P. N. Moskalev and I. S. Kirin, *Opt. Spektrosk.*, 29, 1970, 414.).

films can be switched first to a yellow-tan form and then to a red form. On reduction, the green state can be switched first to a blue redox form and then to a violet blue form. Although [Lu(Pc)$_2$] films can exhibit five colors, only the blue-green transition is utilized in most prototype ECDs. Mechanical problems such as film fracture and/or loss of adhesion to the electrode substrate arise from anion ingress/egress during color switching. Despite such difficulties, [Lu(Pc)$_2$]-based ECDs with good reversibility, fast response time, and little degradation over >5 × 10^6 cycles have been described.

[Lu(Pc)$_2$] films are usually prepared by vacuum sublimation, and the practical problems encountered with extending this technology to the manufacture of practical devices have been recognized. These include the slow rate of deposition, the partial decomposition of [Lu(Pc)$_2$] under sublimation conditions, and the presence of electrochemically inaccessible sites in the resulting films. Electro-polymerization of [Lu(T4APc)$_2$] (T4APc = 4,4′,4″,4‴-tetraaminophthalocyanine) has been reported as a potentially viable route to the fabrication of practical ECDs. The oxidative film-forming mechanism is assumed to be analogous to that described for PANI, although the poly[Lu(T4APc)$_2$] films are believed to be oligomeric in nature rather than truly polymeric.

Although loss of electroactivity in such films was found on sweeping to positive potentials, the electrochemistry in Dimethyl sulfoxide at negative potentials is well behaved, with the observation of two broad quasi-reversible one-electron redox couples. Spectroelectrochemical measurements revealed switching times of less than two seconds for the observed green–gray–blue color transitions in this region. The electro-polymerization method is also applicable to single-ring transition metal Pc complexes, and the preparation of films with reasonable optical densities required minutes compared to the several hours potential cycling that is needed for [Lu(T4APc)$_2$]. The redox reactions and the color changes of two of the metal complexes studied are as follows:

$$Poly[Co^{II}T4APc] + n\,e^- \rightarrow Poly[Co^{I}T4APc]^-$$
$$\text{(blue-green)} \qquad\qquad \text{(yellow-brown)} \tag{8.8}$$

$$Poly[Co^{I}T4APc]^- + ne^- \rightarrow Poly[Co^{I}T4APc]^{2-}$$
$$\text{(yellow-brown)} \qquad \begin{pmatrix} \text{red-brown as thick film and} \\ \text{deep pink as thin film} \end{pmatrix} \tag{8.9}$$

$$Poly[Ni^{II}T4APc] + ne^- \rightarrow Poly[Ni^{II}T4APc]^-$$
$$\text{(green)} \qquad\qquad \text{(blue)} \tag{8.10}$$

$$Poly[Ni^{II}T4APc]^- + n\,e^- \rightarrow Poly[Ni^{II}T4APc]^{2-}$$
$$\text{(blue)} \qquad\qquad \text{(purple)} \tag{8.11}$$

The first reduction in the cobalt polymer is assigned to a metal-centered redox process, resulting in the appearance of a new metal to ligant charge transfer transition, with the second reduction being ligand-centered. In the case of nickel polymer, both redox processes are ligand-based.

A further alternative to vacuum deposition method is the Langmuir–Blodgett (LB) technique, which is suitable for both monolayer and multilayer film formation. The electrochemical properties of a variety of substituted and unsubstituted Pc metal complexes as multilayer LB films have been studied. The first paper on this subject reported the electrochemical study of alkoxy-substituted [Lu(Pc)$_2$]-LB films exhibited a one-electron reversible oxidation corresponding to a transition from green to orange and blue forms, respectively, with the electron transport through the multilayer being, at least in parts, diffusion controlled. An explanation of the relatively facile redox reaction in such multilayers is that the Pc ring is large compared with the alkyl tail projected area, with enough space and channels present in the LB films to allow ion transport. High-quality LB films of M^{II} tetrakis [(3,3′-dimethyl-1-butoxy) carbonyl] phthalcyanine (M=Cu, Ni) have been reported. Ellipsometric and polarized optical absorption measurement suggest that the Pc molecules are oriented with their large faces perpendicular to the dipping direction and to the substrate plane.

That the LB technique is amenable to the fabrication of ECDs is supported by a recent report of a new thin film display based on LB films of the prascodymium bisphthalocyanine complex. The EC electrode in the display was

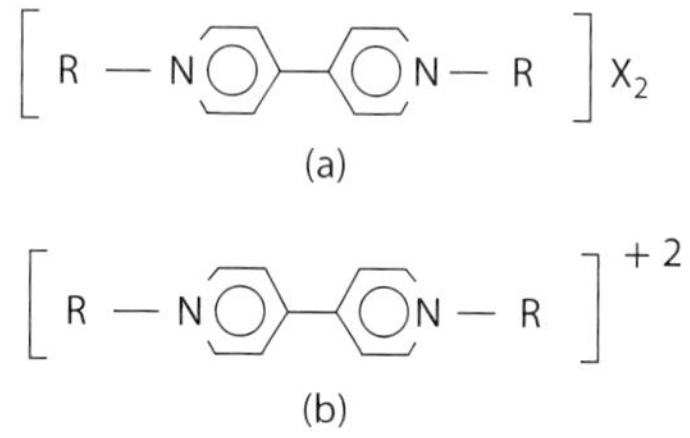

$$\left[R - N\bigcirc\!\!-\!\!\bigcirc N - R \right] X_2$$

(a)

$$\left[R - N\bigcirc\!\!-\!\!\bigcirc N - R \right]^{+2}$$

(b)

FIGURE 8.7
(a) General chemical formulae of viologen and (b) viologen ion.

fabricated by the deposition of multilayers (10–20 layers, approximately 100–200 Å) of praseodymium bisphthalocyanine onto ITO-coated glass (7 cm × 4 cm) slides. The display exhibited blue-green-yellow-red polyelectrochromicity when a potential ranging from –2 to +2 V was applied. After, 10^5 cycles, no significant changes were observed in the spectra of this color state. The high stability of the device was ascribed to the preparation of well-ordered monolayers (by LB technique), which seems to allow better diffusion of the counterions into the films and improve the reversibility and stability of the system [58].

8.6.2 Viologens

The viologens family (4,4′-dipyridinium compounds) has the general chemical formula shown in Figure 8.7a, where R may be an alkyl, cyclo-alkyl, or other substitutes and X is a halogen. The name *viologen* was coined by Michaelis and Hills for 4,4′-dipyridinium compounds because they become deeply blue-purple on reduction. The viologen ion (as shown in Figure 8.7b) can have a two-step reduction, that is, a one-electron or two-electron reduction. The most extensively studied material from the viologen family is the methyl viologen (MV). Elofson and Edsberg reported the first electron reduction of MV at –0.6 V versus SCE. It is reversible (the coplanarity of the two heterocyclic nuclei facilitates the reversible reduction) and independent of pH. The second electron reduction is not electrochemically reversible, but MV can be reoxidised by air. The second reduction was reported at –1.038 V versus SCE in the range of pH = 5–13 and slightly pH dependent below pH = 5. A polarograph of MV dichloride in an aqueous solution with pH = 5.3 clearly exhibits a two-step reduction with two half-wave potentials located at –0.68 and –1.07 V versus SCE, respectively, confirming the results of Elofson and Edsberg and the cyclic voltammetry results.

The scope for viologen systems is indicated by the following observations [59–61a]:

- The color of the precipitate is a function of the nitrogen substituent. For example, if the substituent is *N*-heptyl, then a purple deposit is obtained; if it is *p*-Cyano phenyl, then a green deposit results.

- The electrical properties of the deposit can be tailored by selection of the anion. With *n*-heptyl viologen dication and bromide anions, the deposit is conducting; substitution of the bromide by dihydrogen phosphate anions results in an insulating deposit.
- Electrodes can be functionalized with electrochemically active polymeric materials derived from, for example, the hydrolytically unstable viologen *N*,*N'*-bis[-3-(trimethoxysilyl)propyl]-4,4' bipyridinium dibromide, potentially realizing solid state EC devices.
- Conventionally, ECD achieves contrast change by modulating light using the optical absorption of the electrodeposited film.

8.6.3 Carbon Materials

Fullerenes are finding an increasing number of applications, such as in organic solar cells, dye-sensitized solar cells (DSSCs), light-emitting diodes, and so on. Many of these devices are based on electrochemistry and are optoelectronic devices. It is thus interesting and important to know if fullerenes show EC properties. Important members of the fullerene family are C_{60} and C_{70} (shown in Figure 8.8) due to their relative abundance in soot during production. A change in the color of C_{60} thin films from yellow-brown (undoped state) to silver-black when doped with alkali metals has been observed. Cordoba de Torresi et al. [61b,61c] report the reversible color change of C_{60} thin films produced by electrochemical Li^+ insertion. Undoped film is light brown; doped is dark brown. It will be interesting to know if the thin films of higher fullerenes such as C_{70} show electrochromism or not. Such studies are not yet reported (to the best of my knowledge).

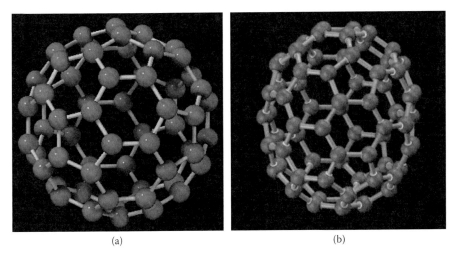

(a)   (b)

FIGURE 8.8
(a) Fullerene (C_{60}) and (b) Fullerene (C_{70}).

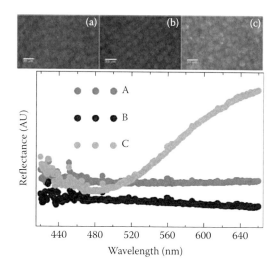

FIGURE 8.9

(See color insert.) Optical micrographs of the same graphite particles at different stages of lithium content: (a) no lithium, (b) $LiCl_2$, and (c) LiC_6. The scale bars are 20 µm in all pictures. Plot of the corresponding reflectance versus wavelength spectra. (From A. Timmons et al., Quantification of color changes in graphite upon the electrochemical intercalation of Li, *212th ECS Meeting*, Abstract 755.)

Graphite electrodes are commonly used in lithium-ion batteries. Charging and discharging of the battery involves insertion and removal of lithium ions into/from graphite electrodes. This electrochemical process is well studied, and also involves changes in lattice (lattice expansion due to insertion of Li ions and contraction due to removal of Li ions). Graphite film also exhibits electrochromism (color change) due to insertion of lithium ions. Figure 8.9 displays the EC color change in graphite thin film and the corresponding reflectance spectra for various Li ion insertion stages [61d]. However, optical EC contrast in most of the carbon materials is poor and hence such pure carbon materials are of little use for EC device applications.

8.6.4 Conducting Polymers

Chemical and electrochemical polymerization of various organic aromatic molecules such as pyrrole, aniline, thiophenes, furan, carbazole, and so on produces novel electronically CPs. In the oxidized forms, such CPs are *doped* with counter anions (p-doping) and possess a delocalized π-electron band structure. The energy gap (E_g) between the highest occupied π-electron band (valence band) and the lowest unoccupied band (conduction band) determines the intrinsic optical (and of course electrical) properties of these materials.

It is well known that we can tune the optical properties of these materials by controlled doping (and/or dedoping). The doping process (oxidation) introduces polarons, which are the major charge carriers. Other charge carriers such as solitons, bipolarons, and so on are also very important in such

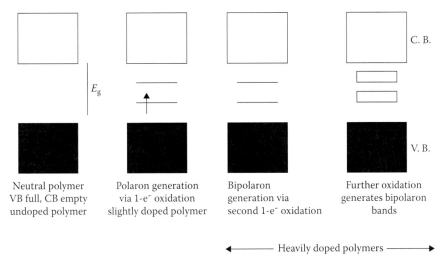

FIGURE 8.10
Schematic of the energy level diagrams of polarons and bipolarons in conducting polymers.

systems (as shown in Figure 8.10). The introduction of new states in the band-gap (or forbidden gap) with the doping process is the main reason for the changes in the optical properties of CPs and is the main cause for the electro-chromism of such materials. Reduction of CPs with concurrent counter anion exit removes the electronic conjugation, to give the *undoped* (neutral) electrically insulating form. CPs can also undergo cathode doping with cation insertion (n-doping). However, n-doped forms are less stable than p-doped forms. There are few reports on the electrochromism of n-doped CPs.

In recent years, CPs have gained a lot of attention for ECDs. This is due to the fact that all electroactive polymers and CPs are potentially EC materials, and are more processable than inorganic EC materials and offer the advantage of a high degree of color tailorability. This tailorability has been achieved through the modification of various polymer systems via monomer functionalization and copolymerization as well as with the use of blends, laminates, and composites. Complex colors are achieved by mixing two existing colors in a dual polymer device.

In CPs, EC changes are induced by redox processes, which are accompanied by ion insertion/expulsion and result in modification of the polymer's electronic properties, giving rise to changes in color of the materials. The color exhibited by the polymer is closely related to the bandgap and the dopant ions. A major focus in the study of EC polymeric materials has been controlling their colors by main chain and pendant group structural modification. Polyheterocycles have proven to be of special interest for this due to their environmental stability under ambient and in-use conditions.

Three major strategies for color control are used with EC properties. The polymer's bandgap (defined as the onset of the π–π* transition) is directly related to the relative energies of the HOMO and the LUMO. By judiciously substituting the polymer's repeat unit, the electronic properties can be controlled by the induced steric and electronic effects. These substituents determine the effective conjugation length and the electron density of the polymer backbone. Copolymers offer a second means of controlling the EC properties of the CPs. Copolymerization of distinct monomers or homopolymerization of hybrid monomers containing several distinct units can lead to an interesting combination of the properties observed in the corresponding homopolymers. Indeed, it has been observed that the color of copolymers based on carbazole, thiophenes, and pyrrole derivatives can be controlled by altering the ratio of the respective monomers. Blends, laminates, and composites offer a third method, similar to copolymers, for combining the EC properties of several systems. The use of two polymers covering different color regions is the simplest way to achieve multicolor electrochromism [62–65].

8.6.4.1 Polypyrrole

The EC properties of PPy were investigated in thin film forms. PPy thin films are usually prepared by electrochemical polymerization of pyrrole (either in aqueous or organic solvent solution such as acetonitrile together with a suitable electrolyte). The mechanism for electrochemical polymerization of pyrrole is shown in Scheme 8.1. It involves oxidation and dimerization of pyrrole followed by aromatization and oxidation of the dimer, the aromatic dimer, and the higher molecular weight oligomers, which are known to oxidize more easily than the monomer. By extension of these reactions, the steps for the polymerization process can be described by reactions 4 and 5 in Scheme 8.1. The polymerization reaction has been shown to involve two electrons/molecules of pyrrole and the resulting polymer is produced in the oxidized state with 0.25–0.33 cation centers per pyrrole unit depending on the electrolyte anion used. PPy thickness is controlled through the charge passed.

Doped (oxidized) PPy film is blue-violet (λ_{max} = 670 nm). Electrochemical reduction yields the yellow-green (λ_{max} = 420 nm) undoped form. The schematic of the doping/dedoping process can be given as:

$$PPy_{ox} + n\, e^- \Leftrightarrow PPy_{red} \tag{8.12}$$

Removal of all dopant anions from PPy yields a pale-yellow film. However, complete dedoping is only achieved if the PPy films are extremely thin. This means that PPy of thickness commensurate with device construction (>1 μm) has a low contrast ratio. The electrochromism of PPy is unlikely to be exploited, mainly due to the degradation of the film on repetitive color switching. CPs with improved EC properties are, however, formed on electrochemical polymerization of 3,4-disubstituted pyrrole.

We have combined PB films with PPy films and constructed ECDs. Our results show that the use of PB not only enhances the EC response (contrast) but also extends the response to a wider region of the visible spectrum, thus

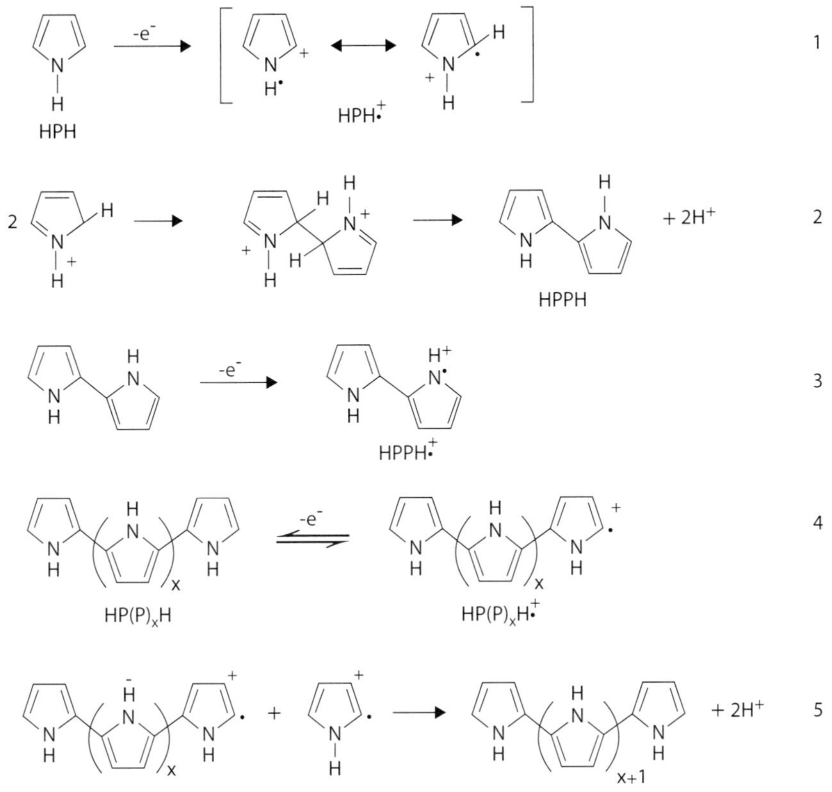

SCHEME 8.1
Mechanism for the electrochemical polymerization of the polypyrrole.

acting as a sensitizer for improving the EC response of PPy (Figure 8.11). We have also found that the EC contrast of such composite films is strongly dependent on the size of the counterions, electronegativity, and thickness of the films.

8.6.4.2 Polyaniline

The electrical and EC properties of PANI depend not only on its oxidation state but also on its protonation state, and hence the pH value of the electrolyte used. The electrochemistry of PANI thin films has been extensively investigated in aqueous acid solutions and in organic media, and several redox mechanisms involving protonation–deprotonation and/or anion ingress/egress have been proposed. Figure 8.12 shows various PANI forms and their interconversions. PANI films are polyelectrochromic (transparent yellow to green, dark blue, and black), the yellow–green transition being durable to repetitive color switching.

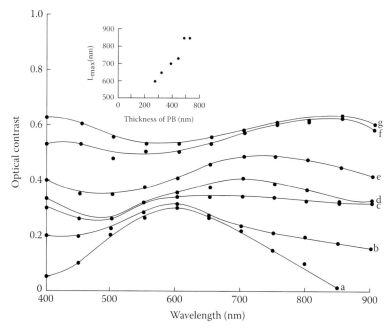

FIGURE 8.11

Effect of PB films on the electrochromic contrast of polypyrrole films in 0.01 M aqueous LiClO$_4$ electrolyte. Curves (a)–(g) correspond to the PB film thickness of 0, 280, 315, 380, 435, 475, and 520 nm, respectively. The inset shows the shift in wavelength of maximum contrast L_{max} with the PB film layer thickness. (From Somani et al., *Acta. Mater.*, 11, 48, 2000, 2859.)

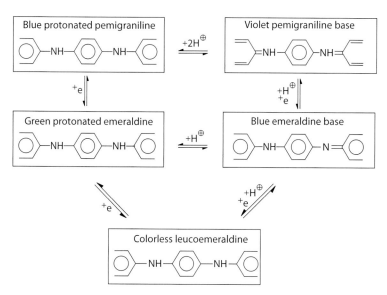

FIGURE 8.12

Polyaniline forms and their interconversions.

Kaneko et al. have prepared the PANI films by electro-polymerization of aniline on a Pt or Nesa glass at 1 V versus Ag–AgCl using 1 mol/dm³ aniline dissolved in 2 M aqueous HCl solution for 15 seconds. The results show that under neutral to basic conditions, the PANI film prepared was black and did not show any electrochromism studied by visible spectroscopy. The PANI film prepared under acidic conditions showed reversible multicolor electrochromism.

The cyclic voltammogram and the color changes under continuous scanning of the applied potential from −0.7 V to 0.6 V versus Ag–AgCl are shown in Figure 8.13a. The figure shows reversible color changes from yellow, yellowish green, green, dark green, bluish green, blue, and violet to brown, depending on the applied potential. Figure 8.13b shows the visible absorption spectrum measured after applying a potential to a PANI film coated on a Nesa glass. The spectra show two major absorption regions, around 300–500 nm and above 500 nm. The absorption at the wavelength above 500 nm shifts to a shorter wavelength when the applied potential is changed from −0.7 to 0.7 V versus Ag–AgCl electrode. This blue shift of the absorption corresponds to color changes of yellow (−0.7 V), green (0 V), blue (0.2–0.3 V), violet (0.5 V), and brown (0.7 V). If the absorption maximum shifts to around 500 nm, as is expected when the applied potential is above 0.7 V versus Ag–AgCl, the PANI film must exhibit a red color. However, application of a potential above this value indicates an irreversible color change of the PANI film.

The EC colors shown by the PANI films are very close to the three primary colors (i.e., green, blue, and red) except for red. If the absorption maximum could be shifted to a shorter wavelength up to 500 nm, it would show all three primary colors. Introduction of a substituent or another monomeric unit might lead to an EC device that can show all colors.

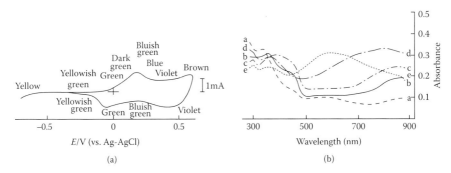

FIGURE 8.13
(a) Cyclic voltammogram and the color changes in the polyaniline (PANI) film coated on Pt and dipped in a 10–2 M aqueous HCl solution containing 0.2 mol/dm³ LiCl, scan rate: 10 mV/s. (b) Absorption spectra of PANI film coated on a Nesa glass dipped in a 10–2 M aqueous HCl solution containing 0.2 mol/dm³ LiCl measured after applying potentials of (a) −0.7 V, (b) 0 V, (c) 0.2 V, (d) 0.5 V, and (e) 0.7 V versus Ag–AgCl electrode. (From Kaneko et al., *Makromol. Rapid. Commun.*, 8, 1987, 179.)

PANI has been combined with PB in complementary ECDs that exhibit deep blue ↔ green electrochromism. The EC compatibility is obtained by combining the colored oxidized state of the polymer with the blue PB and the bleached reduced state of the polymer with PG; both liquid electrolyte and solid state configurations have been described.

$$\text{Oxidized PANI} + \text{PB} \rightarrow \text{Emeraldine PANI} + \text{PG}$$

(Deep blue)   (Green) $\qquad$ (8.13)

While electrochemical polymerization is a suitable method for the preparation of relatively low surface area EC CP films, it may not be suitable for fabricating large area ECDs. Efforts have been made to synthesize soluble CPs such as poly(o-methoxyaniline), which can then be deposited as thin films by casting from solution, by relatively easier techniques such as dip coating, spin coating, and so on. In a novel approach, large area EC coatings have been prepared by incorporating PANI into polyacrylate–silica hybrid sol–gel networks using suspended particles or solutions and then spray or brush coating onto ITO surfaces [66–72].

8.6.4.3 Polythiophene

Like PPy, polythiophene thin films can be prepared by electrochemical polymerization of thiophene. EC properties of polythiophene and of the polymers of several substituted thiophenes are reported in Table 8.3. Tuning of color states is possible by suitable choice of thiophene monomer. For example, the EC properties of polymer films prepared from 3-methylthiophene-based oligomers are strongly dependent on the relative positions of the methyl groups on the polymer backbone.

When a multicolor ECD is constructed by combination of different types of EC CPs, an EC CP is often combined with a second EC polymer showing a different specific color at a quite different operation potential. Poly(aniline)–poly(sodium acrylate)–poly(thiophene) and poly(aniline)–poly(sodium acrylate)–poly(3-methylthiophene) combined films are constructed by combining poly(aniline)–poly(sodium acrylate), which shows its specific color in the oxidized state, with poly(thiophene) and poly(3-methyl thiophene), which show

TABLE 8.3

Polythiophenes

		Polymer λ_{max} (nm) and Color	
S. No.	**Monomer**	**Oxidized State**	**Reduced State**
1.	Thiophene	730 (blue)	470 (red)
2.	3-Methylthiophene	750 (deep blue)	480 (red)
3.	3,4-Dimethylthiophene	750 (dark blue)	620 (pale brown)
4.	2,2'-Bithiophene	680 (blue-gray)	460 (red-orange)

TABLE 8.4

Electrochromism Observed in Conducting Polymers

			Color of the Films		
S. No.	Polymer	Anion	Oxidized State (Doped)	Reduced State (Undoped)	Response Time
1.	Polypyrrole	ClO_4^-	Brown	Yellow	20
2.	Polythiophene	ClO_4^-	Brown	Green	45
3.	Polymethylthiophene	BF_4^-	Blue	Red	12
4.	Poly-3,4 dimethylthiophene	ClO_4^-	Dark blue	Blue	60
5.	Poly-2,2'–bithiophene	CF_3^-, SO_3^-	Blue/gray	Red	40
6.	Octacyanophthalocyanine	H^+	Green	Blue	1
7.	Polyaniline (PANI)	Cl^-	Green/blue	Yellow	100

their specific colors in the reduced state. An electroactive polymer laminate for use in an ECD comprises a conductive substrate, a first layer of an electroactive polymer, and a second layer of an electroactive polymer prepared from a second monomer having an oxidation potential higher than that of the first monomer and adhering to the first layer [73–75]. Table 8.4 gives the comparison of electrochromism observed in most studied CPs.

8.6.4.4 Other Conducting Polymers and Gels

Polyisothianaphthene (PITN) is the first example of the transparent CP. It exhibits reversible p-type electrochemical doping with an associated high contrast color change with an excellent stability. Doped thin films of PITN have very low optical density. Reversible electrochromism with a characteristic switching time of a few hundred milliseconds has been found. The experimental results demonstrate that the rate-limiting step in the EC switching process involves diffusion of dopant counterions into the polymer structure. Also, it has been shown that the polymer morphology plays an important role in determining the EC switching. By controlling the morphology, so as to minimize the diffusion distance in CPs, it should be possible to achieve sufficient speed to be of interest for a variety of technological applications [76].

Poly(o-aminophenol) (PAP)—a ladder polymer with phenoxazine rings as electroactive sites—can be prepared by electrooxidative polymerization of o-aminophenol in an acidic solution. The PAP film is strongly adhesive to the electrode surfaces, with very uniform, semiconducting properties, and displays EC properties. The oxidative form is bronze-brown and the reduced form is pale green (almost colorless). The maximum absorption peak of the oxidized form was observed at 440 nm and that of the reduced form at approximately 600 nm. A reversible color change of the film between brown and pale green (almost colorless) was observed when the electrode potential was cycled between −0.7 V and +0.7 V versus SCE. The brown color of the oxidized form remains substantially unchanged even after standing it in air at open circuit for a long period (about six months); the film possessed the memory effect. On the other hand, the reduced form was gradually oxidized by oxygen in air and consequently becomes colored. The colorlessness of the reduced form can be held under an atmosphere of nitrogen gas. The response time is somewhat high, that is, several seconds were required to complete the coloring of the film and then to bleach it completely [77].

It has been observed that, under certain experimental conditions, the electrochemical reduction of *o*-chloranil (*o*-CA) yields electrode adherent blue films, which can be bleached by oxidation. The writing/erasing lifetime of the *o*-CA/*o*-CA⁻ system on a Pt electrode, greater than 10^5 cycles, deteriorates if the reduction potential allows the formation of *o*-CA^{2-} dianions. The electrochemical reduction of *p*-chloranil under similar experimental conditions produces a yellow color at the electrode, which diffuses into the solution. Hence, ortho-quinones seem better suited for EC applications than *para*-compounds [78].

Polymer gel films [poly(1-vinyl-2-pyrrolidin-one-co-*N*, *N*′ methylenebisacrylamide] that contain simple organic ECs [*p*-diacetylbenzene, dimethyl- or diethyl terephthalate] have been prepared and their EC response is being investigated. A single film ECD has been prepared and studied. Its simple construction is suitable for large area displays [79].

8.7 Comparison between Inorganic and Polymeric EC Materials

Table 8.5 gives a comparison between inorganic and polymeric EC materials.

8.8 Nanomaterials in Electrochromic Devices and Applications

Nanomaterials offer several interesting properties and advantages for EC applications because the properties of the materials are different from their bulk counterparts when the dimensions are in the nanometer region. The ratio of atoms or molecules on the surface and in the bulk (i.e., surface-to-volume ratio) is very high and the properties of the materials are mainly governed by the surface atoms or molecules. Nanotechnology has changed many of our notions. The earlier notion that the bandgap of a material has a fixed value is no longer valid. It is now a common knowledge that the bandgap of the material is particle-size dependent and increases as the size decreases (particularly in the nanometer region). It is now possible to tune

TABLE 8.5

Comparison between Inorganic and Polymeric Electrochromic Materials

S. No.	Property	Inorganic Materials	Polymers
1.	Method of preparation	Needs sophisticated techniques such as vacuum evaporation, spray pyrolysis, and sputtering	The material can be easily prepared by simple chemical, electrochemical polymerization and the films can be obtained by simple techniques such as dip coating and spin coating
2.	Processibility of the materials	The materials are poor in processibility	The material can be processed very easily
3.	Cost for making the final product (device)	High as compared to the polymer-based devices	Low cost as compared to the inorganic materials
4.	Colors obtainable	Limited numbers of colors are available from a given material	Colors depend on the doping percentage, choice of the monomer, operating potential, etc.; hence, a large number of colors are available with polymeric materials
5.	Contrast	Contrast is moderate	Very high contrast can be obtained
6.	Switching time (milliseconds)	10–750	10–120
7.	Lifetime	10^3–10^5 cycles	10^4–10^6 cycles

the optical properties (such as bandgap, optical absorption, color, and luminescence) of materials by controlling their size (bandgap engineering). Nano forms of almost all known materials have been synthesized and studied. Today, there is hardly any material of which a nano form is not made and studied. Nanomaterials have also enabled the synthesis of highly porous structures and films used in applications like DSSC, supercapacitor electrodes, battery electrodes, fuel cell electrodes, sensors, gas adsorption and storage, and so on. The application of nanomaterials in EC devices is mainly due to the following:

- Ability to tune the optical properties of the materials and hence the color (i.e., bandgap engineering)
- High surface area of nanomaterials
- Possibility of high-porosity films and structures in which counterions can move easily

Several studies have been published on use of nanomaterials and thin films for EC applications. A simple Google search can easily provide more than a few hundred research papers and patents in the area of EC materials and devices using nanomaterials. Studies are mostly related to synthesis and the study of nanostructured EC films by different methods (ink-jet printing, vacuum evaporation, sol–gel thin films, etc.), different nano forms (nanoparticles, nanowires, nanotubes, etc.) and EC properties, and different device structures. Although the basic science of electrochromism also remains the same in nanomaterials/nanostructures, their use brings some added advantages, such as fast switching, better contrast, better penetration of counterions, and color tunability in different parts of the spectrum. A detailed account of the application of nanomaterials in EC devices and applications is outside the scope of this chapter [80–116].

8.9 Current Status and Use of Electrochromic Devices

8.9.1 Moving Electrochromic Pixels

A microfabricated, movable EC pixel based on conducting PPy has been prepared and demonstrated by Smela. It uses the fact that the conjugated polymers undergo a volume change of several percent and a color change when their oxidation state is changed electrochemically [117]. This demonstration shows that, in principle, many different kinds of electronic, optical, and micromechanical devices could be put on movable platforms. One can envision moving light-emitting diodes or semiconductor lasers, for example, or a comb-drive microgrippers being moved into position to grasp a small object [118].

8.9.2 All-Plastic and Flexible Electrochromic Displays and Devices

Segmented alphanumeric ECDs are already demonstrated and commercialized by several companies. Figure 8.14 shows a photograph of the segmented alphanumeric ECDs by NTera. Such ECDs have many advantages over other nonemissive displays such as liquid crystal displays (LCDs), which are listed in Table 8.6.

Efforts for production of all-plastic and flexible EC devices are in progress, which facilitates new technological applications. Two factors prompted the

FIGURE 8.14
Segmented alphanumeric display prototypes prepared by NTera.

TABLE 8.6

Comparison of Electrochromic Displays with Other Nonemissive Displays

S. No.	Property	Electrochromic Display	Electrophoretic Display	Field Effect Liquid Crystal Display	Dynamic Scattering Liquid Crystal Display	Dipole Suspension Display
1.	Viewing angle	Wide	Wide	Narrow	Narrow	Narrow
2.	Optical mode	T^a, R^b, P^c	R	T, R, P	T, R, P	T, R
3.	Color	Two or more	Two	B/W or dye	B/W	B/W
4.	Resolution	Electrode limited	Electrode limited	Electrode limited	Electrode limited	Electrode limited
5.	Operating mode	dc pulse	dc pulse	ac	ac	ac
6.	Voltage (V)	0.25–20	30–80	2–10	10–30	2–30
7.	Power (μW/cm^2)	1–10	15	<0.1	1–10	1–10
8.	Energy (mJ/cm^2)	10–100	6×10^{-4}			
9.	Memory	Yes	Yes	No/Yes	No/Yes	No
10.	Threshold	Poor	No	Poor	Poor	Poor
11.	Contrast					
12.	Write time (ms)	100–1000	60	20	20	20
13.	Erase time (ms)	100–500	30	1–500	100	30
14.	Operating life	10^5–10^6 cycles		>2 × 10^4 h	>1 × 10^4 h	

ᵃ Transmission.
ᵇ Reflection.
ᶜ Projection.

assembly of all-plastic EC devices: (1) commercial production of transparent electrodes with thermoplastics, for example, ITO on PET, where PET is poly(ethylene terephthalate); and (2) synthesis of new CPs with low energy bandgap. Most of the so-called all-plastic EC devices described in the literature comprise not only plastic components, but also inorganic compounds as one of the EC materials. Moreover, most of these materials, including those polymers with low energy bandgap, do not show the high flexibility and elasticity inherent to polymeric materials. One exception is PANI doped with dodecylbenzene sulfonic acid or phosphoric acid diesters, which shows plastic characteristics. Hence, to assemble an all-plastic and flexible EC device, it is necessary to improve the mechanical properties of the CPs used as EC materials. This can be done by mixing the CPs with thermoplastics or elastomers to produce polymeric blends, which show the EC properties of the CPs associated with the mechanical properties of common polymers.

Several studies on plastic and flexible EC devices have been reported. All-plastic and flexible EC devices have been fabricated and described by De Paoli et al. using two optically complementary conductive polymer blends deposited on ITO–PET and a polymeric electrolyte. Two polymer blends were prepared by mixing the CPs poly(ET2) and poly(4,4′-dipentoxy-2,2′-bithiophene) (poly(NNDMBP)) with the elastomer poly(epichlorohydrin-co-ethylene oxide) (Hydrin-C). The SPE was a mixture of Hydrin-C and $LiClO_4$. The results show that blending with Hydrin-C improves not only the mechanical properties of the CPs, but also their EC properties. Such devices show good optical characteristics, such as stability to successive EC cycling, short optical response time, and a good optical contrast. Scale-up of such devices is in progress [117].

8.9.3 ECDs for Time–Temperature Integration

An attempt has been made to use an EC device for time–temperature integration. The demonstrated device could be used for smart labeling of frozen goods. A use by date is valid only if the product has been stored and transported under the correct temperature conditions. Demand for sensors that could be incorporated into packaging and provide a visual indication if an item has experienced conditions that cause deterioration is very high.

The rate of deterioration is expressed by an Arrhenius equation and the cumulative effect over time is given by an expression of the form:

$$\text{Degree of deterioration} = \int A\exp(-E_a / RT)dt \qquad (8.14)$$

When an EC device is operating under a constant potential across the (polymer) electrolyte, the current depends on the temperature according to the Arrhenius expression (at least over a small temperature range). The total charge passed in a given time is given by an expression similar to the integral of Equation 8.14. It is therefore feasible to tailor the device configuration

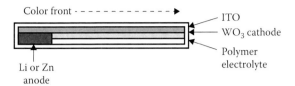

FIGURE 8.15
Schematic diagram of the electrochromic display used for time–temperature monitoring (From Colley et al., *Polym. Int.*, 49, 2000, 371.)

to match the time–temperature profile for decay of a particular foodstuff. Budd et al. have demonstrated an EC device like that in Figure 8.15 for time–temperature monitoring with an active EC layer of WO_3, which becomes blue on intercalation with lithium or hydrogen:

$$yLi^+ + ye^- + WO_3 \rightarrow Li_yWO_3 \tag{8.15}$$

Metal foil is used as an anode and the polymer electrolyte incorporates an appropriate salt. ITO-coated glass is used as a top transparent conducting electrode. Coloration occurs initially in the region closest to the anode. As the region near to the anode becomes saturated, the color front progresses along the device length at a rate dependent on the conductivity of the electrolyte, and consequently on the temperature. Poly[oxymethylene-oligo(oxyethylene)] was used in this prototype device [119,120].

8.9.4 Electrochromic (Smart) Windows

An electrochromic window (EW) is an ECD that allows electrochemically driven tuning of light transmission and reflections. Such EC optical switching devices are usually called *smart windows*. They can be used for a variety of applications where the optical modulation effect can be used significantly, for example, regulation of incident solar energy and glare for the improvement of energy efficiency of buildings, vehicles, aircraft, spacecraft, and ships. The potential market for smart windows is very large and includes (in the automobile sector) rearview mirrors (which are already in the marketplace), sunroofs, visors, and side and smart windows. Other uses are large area information displays to be used in airports and railway stations as well as EC eyeglasses and sunglasses.

Unlike ECD, in EW, the entire system is in the optical path, which puts some restrictions on the electrolyte to be used. Either it should be transparent or EC in a complementary mode with respect to the primary EC electrode. These requirements are not easy to fulfill and hence most of the efforts have been devoted to the characterization of materials capable of assuring the desired complementary switchable optical function. Such EWs can be combined with solar cells so that the power required to bring the EC color change can be obtained from the solar cells.

(a)   (b)

FIGURE 8.16
(See color insert.) Electrochromic materials in architectural pilot projects: automatically dimming smart windows from a Berkeley lab (a) 10:30 a.m. and (b) 10:50 a.m. (Adapted with permission from Windows and Daylighting Group, Lawrence Berkeley National Laboratory, Berkeley, California.)

Figure 8.16 shows a photograph of the automatically dimming smart windows developed at a Berkeley laboratory. The effect of light modulation by the windows is easily visible. Pilot scale production and commercialization of such smart windows have been undertaken by few companies. Gentex Corporation (http://www.gentex.com) has developed and demonstrated a dimmable aircraft window system, which can be used for glare reduction or can darken the window completely so as to block the incoming light fully.

8.9.5 Cathode Ray Tubes with Variable Transmittance

An alternative to the common brilliance adjustments of TV tubes, when room illumination alters, is an electrochromically darkening cathode ray tube (CRT) screen employing oxides, which is being studied at Philips [121]. EC darkening is preferable to direct electrical control as color values are thereby better preserved. Other applications of EC materials such as for entry-ticket security [122,123] and so on, are quite interesting.

8.9.6 Electrochromic Mirror for Automobiles for Glare Reduction

One of the successful products developed and under commercialization utilizing EC materials is electrochromic rearview mirrors for automobiles. Light reflected from the rearview mirrors of automobiles during the nighttime produces glare. Content/information in the mirror almost becomes invisible due to such glare, which needs to be reduced so that other content becomes visible. This is smartly done with the help of EC materials. The EC coatings coupled with the mirror darken and reduce the amount of light reflected, and hence reduce the glare. Gentex has started commercial production of such EC mirrors. EC mirrors demonstrated by Gentex and their effects during day- and nighttime are shown in Figures 8.17 and 8.18.

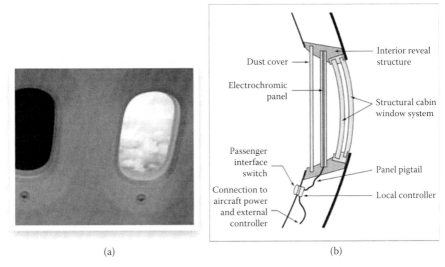

(a)   (b)

FIGURE 8.17
(See color insert.) (a) Electrochromic dimmable aircraft window developed by Gentex Corporation; (b) schematic of the electrochromic window fitting inside the aircraft. (Adapted from Gentex Corporation, http://www.gentex.com.)

FIGURE 8.18
(See color insert.) Electrochromic mirror for automobiles for glare reduction: automatically dimming rearview mirror from Gentex (http://www.gentex.com/automotive_products2.html).

8.10 Comparison of ECDs with Other Nonemissive Displays

A comparison of ECDs with other nonemissive displays is given in Table 8.6. It can be seen that ECDs are comparable in properties with those of LCDs with the added advantages that they can be made into different colors without the addition of external dye, and that the window does not depend on the viewing angle and can be easily prepared in the form of large-area windows.

8.11 Industries Working on Electrochromic Materials and Products

Some of the important industries working in the area of EC materials, devices, and products are the following:

SAGE Glass (http://www.sageglass.com)

Anoop Agrawal, Electrochromix, Inc. (Colorado)

Chromogenics (http://www.chromogenics.com)

Gentex (http://www.gentex.com)

Saint Gobain (http://www.saint-gobain-recherche.fr), PPG, Ashai, Pilkington (all glass manufacturers)

e-Chromic (San Jose startup) (http://www.echromic.com)

Varitint (startup by Jaideep Raje and Gayatri Swaminathan)

NTera (ECDs)

8.12 Drawback of Electrochromic Technology

One of the major drawbacks (as per the author's opinion) of EC technology is that it consumes electrical power. Electric power—by and large—is mainly produced from coal, natural gas, hydrothermal, and nuclear power. Many of the conventional energy sources are depleting, and nuclear energy has safety and environmental issues. Almost all the methods for electricity production are at the cost of the environment. Consumption of electric power by EC materials and devices could be a serious setback for certain applications. For example, instead of supplying electric power to an EC window, thereby controlling the amount of sunlight coming into an apartment, I would prefer to simply use curtains for the same purpose. Even though such an EC window is driven by solar cell output, it is just fine. For large area applications, power requirements are going to matter in the future. Other applications, like antiglare rearview mirrors for automobiles, ECDs, and so on, are expected to take off.

8.13 Conclusions and Future Prospects

In the last two decades, the preliminary focus was related to understanding the physical and chemical properties of various EC materials. Of particular interest were organic, polymeric, and inorganic materials. Efforts were also made to demonstrate prototype devices and their commercialization. Various products, such as segmented alphanumeric ECDs, rearview mirrors for automobiles, sunroofs and visors, and smart windows, are under prototype production. Widespread application of ECDs, particularly for architectural applications, will depend on reducing costs, increasing device lifetime, and overcoming device stability issues. Further, existing EC smart windows require an external power source for their operation. Photoelectrochromic systems that change color electrochemically but only on being illuminated may be more appropriate candidates for smart windows. The study and development of photoelectrochromic materials and devices for large-area window applications are in progress and need to be focused further. Commercial production of all-plastic EC devices, smart windows, and moving EC images (such as moving pixels and alphanumeric displays) has already been started.

This chapter did not discuss nonvisible EC response, which is important in both solar screening and optical fiber systems. Recent interest in EC devices for multispectral energy modulation by reflectance and absorbance has extended the working definition. EC devices are now being studied for modulation of radiation in the near-infrared, thermal infrared, and microwave regions, and *color* can mean response of detectors for these wavelengths, not just the human eye.

References

1. D. N. Buckley and L. D. Burke, *J. Chem Soc. Faraday* 72, 1 (1975) 1447.
2. D. N. Buckley, L. D. Burke, and J. K. Mukahy, *J. Chem. Soc. Faraday* 72, 1 (1976) 1896.
3. L. D. Burke and E. J. M. O'Sakan, *J. Electroanal. Chem.* 93 (1978) 11.
4. L. D. Burke and D. P. Whelan, *J. Electroanal. Chem.* 103 (1979) 179.

5. L. D. Burke, T. A. M. Thomey, and D. P. Whelan, *J. Electroanal. Chem.* 107 (1980) 201.

6. L. D. Burke and O. J. Murphy, *J. Electroanal. Chem.* 109 (1980) 199–212.

7. B. W. Faughnan, R. S. Crandall and P. H. Heyman, *RCA Rev.* 36 (1975) 177.

8. P. S. Patil, L. D. Kadam, and C. D. Lokhande, *Solar Ener. Mater. Solar Cells* 53 (1998) 229–234; D. Craigen, A. Mackintosh, J. Hickman, and K. Calbow, *J. Electrochem. Soc.* 133(7), (1986) 1529.

9. L. D. Burke and O. J. Murphy, *J. Electroanal. Chem.* 112 (1980) 379.

10. P. M. S. Monk, R. J. Mortimer, and D. R. Rosseinsky, *Electrochromism: Fundamentals and Applications*, VCH, Weinheim, Germany, 1995.

11. H. Byker, in: K. C. Ho, D. A. MacArthur (Eds.), *Electrochromic Materials II*, PV 94–2, Electrochemical Society Proceeding Series, Pennington, NJ, 1994, pp. 3–13.

12. M. Green, *Chem. Ind.* 17 (1996) 641.

13. C. G. Granqvist, *Handbook of Inorganic Electrochromic Materials*, Elsevier, Amsterdam, the Netherlands, 1995.

14. M. S. Whittingham, *Prog. Solid State Chem.* 12 (1970) 41 (a); S. Passerini and B. Scrosati, *Solid State Ion.* 55, 56 (1992) 520 (b).

15. C. G. Granqvist, *Solar Ener. Mater. Solar Cells* 60 (2000) 201.

16. N. Machida, M. Tatsumisago, and T. Minami, *J. Electrochem. Soc.* 9, 133 (1986) 1963.

17. M. T. Nguyen and L. H. Dao, *J. Electrochem. Soc.* 7, 136 (1989) 2131.

18. K. Kuwabara, S. Ichikawa, and K. Sugiyama, *J. Electrochem. Soc.* 10, 135 (1988) 2432.

19. D. Craigen, A. Mackintash, J. Hickman, and K. Calbow, *J. Electrochem. Soc.* (July 1986) 1529.

20. K. Hirochi, M. Kitabatake, and O. Yamazaki, *J. Electrochem. Soc.* (September 1986) 1973.

21. M. Akhtar, R. M. Paiste, and H. A. Weakliem, *J. Electrochem. Soc.* (June 1988) 1597.

22. S. A. Roberts, D. R. Bloomquist, R. D. Willett, and H. W. Dodgen, *J. Am. Chem. Soc.* 103 (1981) 2603.

23. R. A. Batchelor, M. S. Burdis, and J. R. Siddle, *J. Electrochem. Soc.* 143 (1996) 1050.

24. G. Gottfield, J. D. E. McIntyre, G. Ben, and J. L. Shay, *Appl. Phys. Lett.* 33 (1978) 208.

25. L. M. Schiavone, S. Dautremont, G. Beni, and J. L. Shay, *J. Electrochem. Soc.* 128 (1981) 1339.

26. G. Beni, in: *Proceedings of the 159th Meeting of Electrochemical Society*, Minneapolis, MN, May 1981, Abstr. No. 154.

27. M. Fantini and A. Gorenstein, *Solar Ener. Mater.* 16 (1987) 487.

28. J. S. E. M. Stevensson and C. G. Granqvist, *Appl. Phys. Lett.* 23, 49 (1986) 1566.

29. A. Gorenstein, F. Decker, W. Estrada et al., *J. Electroanal. Chem.* 277 (1990) 277.

30. S. Passerini, B. Scrosati, and A. Gorenstein, *J. Electrochem. Soc.* 137 (1990) 3297.

31. R. Pileggi, B. Scrosati, and S. Passerini, in: D. F. Shriver, R. A. Huggins, and M. Balkanski (Eds.), *Solid State Ionics, Part II*, Vol. 210, Warrendale, PA, Materials Research Society, 1991, p. 249.

32. F. Decker, S. Passerini, R. Pilleggi, and B. Scrosati, *Electrochem. Acta* 37 (1992) 1033.

33. M. B. Robin, *Inorg. Chem.* 1 (1962) 337.

34. V. D. Neff, *J. Electrochem. Soc.* 125 (1978) 886.

35. D. Ellis, M. Eckhoff, and V. D. Neff, *J. Phys. Chem.* 85 (1981) 1225.
36. P. Somani, D. P. Amalnerkar, and S. Radhakrishnan, *Synth. Met.* 110 (2000) 181.
37. P. Somani and S. Radhakrishnan, *Mater. Chem. Phys.* 2, 70 (2001) 150.
38. A. Ludi and H. U. Gudel, *Struct. Bond.* (Berlin) 14 (1973) 1.
39. G. C. Allen and N. S. Hush, *Prog. Inorg. Chem.* 8 (1967) 357.
40. M. B. Robin and P. Day, *Adv. Inorg. Chem. Radiochem.* 10 (1967) 247.
41. R. J. Mortimer and D. R. Rosseinsky, *J. Electroanal. Chem.* 151 (1983) 133.
42. P. Somani, and S. Radhakrishnan, *Chem. Phys. Lett.* 292 (1998) 218.
43. P. Somani and S. Radhakrishnan, Charge transport processes in conducting polypyrrole sensitized with Prussian blue, *Mater. Chem. Phys.*, 76 (2002) 15–19.
44. P. Somani, A. B. Mandale, and S. Radhakrishnan, *Acta Mater.* 11, 48 (2000) 2859.
45. R. J. Mortimer and D. R. Rosseinsky, *J. Chem. Soc., Dalton Trans.* (1984) 2059.
46. K. Itaya, K. Shibayama, H. Akahoshi, and S. Toshima, *J. Appl. Phys.* 53 (1982) 804.
47. K. Honda, J. Ochiai, and H. Hayashi, *J. Chem. Soc. Chem. Commun.* 1 (1986) 168.
48. M. K. Carpenter and R. S. Conell, *J. Electrochem. Soc.* 137 (1990) 2464.
49. K. C. Ho, T. G. Rukavina, and C. B. Greenberg, in: K. C. Ho and D. A. MacArthur (Eds.), *Electrochromic Materials II*, PV 94–2, Electrochemical Society Proceeding Series, Pennington, NJ, 1994, p. 252.
50. P. N. Moskalev and I. S. Kirin, *Opt. Spektrosk.* 29 (1970) 414.
51. P. N. Moskalev and I. S. Kirin, *Russ. J. Phys. Chem.* 46 (1972) 1019.
52. G. C. S. Collins and D. J. Schiffrin, *J. Electroanal. Chem.* 139 (1982) 335.
53. L. G. Tomilova, N. A. Ovchinnikova, E. A. Lukyanets, *Zh. Obsheh. Khim.* 57 (1987) 2100.
54. T. Sugiyama, T. Okamoto, and H. Yamamoto, *Jpn. Appl. Phys. Symp.* (1984) 14 p-D-15.
55. J. L. Kahl, L. R. Faulkner, K. Dwarakanath, and H. Tachikawa, *J. Am. Chem. Soc.* 108 (1986) 5434.
56. J. M. Green and L. R. Faulkner, *J. Am. Chem. Soc.* 105 (1983) 2950.
57. M. M. Nicholson and T. P. Weismuller, *A Study of Colors in Lutetium Diphthalocyanine Electrochromic Displays*, Final Report, Contract N00014-81-C-0264, C82-268/201, AD-A12083, October 1982, Rockwell International, Anaheim, CA.
58. M. M. Nicholson, Electrochromism and display devices, in: C. C. Leznoff and A. B. P. Lever (Eds.), *Phthalocyanines: Properties and Applications*, Vol. 3, VCH, Weinheim, Germany.
59. D. J. Barclay and D. H. Martin, Electrochromic displays, Chapter 15 in: E. R. Howells (Ed.), *Technology of Chemicals and Materials for Electronics*, Society of Chemical Industry, London, United Kingdom, p. 266.
60. J. Bruinink, Electrochromic display devices, in: A. R. Kmetz and F. K. VonWillisen (Eds.), *Non-emissive Electro-optic Displays*, Plenum Press, New York, 1976, p. 201.
61. H. R. Zeller, Principles of electrochromism as related to display applications, in: A. R. Kmetz and F. K. Von Willisen (Eds.), *Non-emissive Electro-optic Display Devices*, Plenum Press, New York, 1976, p. 149 (a); R. C. Haddon, *Acc. Chem. Res.* 25 (1992) 127 (b); S. I. Cordoba de Torresi, R. M. Torresi, G. Ciampi, and C. A. Luengo, *J. Electroanal. Chem.* 377 (1994) 283 (c); A. Timmons, I. G. Hill, and J. R. Dahn, Quantification of color changes in graphite upon the electrochemical intercalation of Li, *212th ECS Meeting*, Abstract 755 (d).
62. B. Scrosati, *Applications of Electroactive Polymers*, Chapman & Hall, London, United Kingdom, 1998.

63. A. H. Fawcett (Ed.), *High Value Polymers*, The Royal Society of Chemistry, Cambridge, United Kingdom, 1996.
64. H. G. Kiss (Ed.), *Conjugated Conducting Polymers*, Springer Series in Solid State Physics, Springer, Berlin, Germany, 1992.
65. D. B. Cotts and Z. Reyes (Eds.), *Electrically Conductive Organic Polymers for Advanced Applications,* Noyes Data Corporation, USA, 1986.
66. M. T. Nguyen and L. H. Dao, *J. Electrochem. Soc.* 136 (1989) 2131.
67. T. Kobayashi, H. Yoneyama, and H. Tamura, *J. Electroanal. Chem.* 161 (1984) 419.
68. M. Morita, *J. Polym. Sci. B* 32 (1994) 231.
69. A. Watanabe, K. Mori, Y. Iwasaki, Y. Nakamura, and S. Niizuma, *Macromolecules* 20 (1987) 1793.
70. S. Gottesfeld, A. Redondo, and S. W. Feldberg, *J. Electrochem. Soc.* 1 (1987) 271.
71. M. C. Bernard, A. H. L. Goff, and W. Zeng, *Electrochem. Acta* 44 (1998) 781.
72. M. Kaneko, H. Nakarmura, and T. Shimomura, *Macromol. Rapid. Commun.* 8 (1987) 179.
73. G. Schopf and G. Kobmehl, *Polythiophenes: Electrically Conductive Polymers Advances in Polymer Science*, Springer, Berlin, Germany, 1998.
74. A. Corradini, A. M. Marinangeli, and M. Mastragostino, *Solid State Ion.* 28–30 (1988) 1738.
75. D. M. Welsh, A. Kumar, E. W. Meijer, and J. R. Reynolds, *Adv. Mater.* 11 (1999) 1379.
76. D. S. K. Mudigonda, J. L. Boehme, I. D. Brotherston, D. L. Meeker, and J. P. Ferraris, *Chem. Mater* 12 (2000) 1508–1509.
77. H. Yashima, M. Kobayashi, K. B. Lee, D. Chung, A. J. Heeger, and F. Wudi, *J. Electrochem. Soc.* 134 (1987) 46.
78. T. Ohsaka, S. Kunimura, and N. Oyama, *Electrochimica Acta* 33 (1988) 639.
79. A. D. Monvernay, A. Cherigui, P. C. Lacaze, and J. E. Dubois, *J. Electroanal. Chem.* 169 (1984) 157.
80. M. Layani, P. Darmawan, W. L. Foo, L. Liu, A. Kamyshny, D. Mandler, S. Magdassi, and P. S. Lee, *Nanoscale* 6 (2014) 4572–4576.
81. J. W. Liu, J. Zheng, J. L. Wang, J. Xu, H. H. Li, and S. H. Yu, *Nano Lett.* 13, 8 (2013) 3589–3593.
82. K. W. Park and M. F. Toney, *Electrochem. Comm.* 7 (2005) 151–155.
83. V. Jain, M. Khiterer, R. Montazami, H. M. Yochum, K. J. Shea, and J. R. Heflin, *Applied Materials and Interfaces* 1,1 (2009) 83–89.
84. V. Jain, H. M. Yochum, R. Montazami, and J. R. Heflin, *Appl. Phys. Lett.* 92 (2008) 033304.
85. V. Jain, R. Sahoo, J. R. Jinschek, R. Montazami, H. M. Yocjum, F. L. Beyer, A. Kumar, and J. R. Heflin, *Chem. Comm.* (2008) 3663–3665.
86. J. Wang, X. W. Sun, and Z. Jiao, *Materials* 3 (2010) 5029–5053.
87. J. W. Liu, J. Zheng, J. L. Wang, J. Xu, H. H. Li, and S. H. Yu, Ultrathin $W_{18}O_{49}$ nanowire assemblies for electrochromic devices, *Nano Letters* 13/8 (2013) 3589–3593.
88. S. F. Hong and L. C. Chen, Nano-Prussian Blue analogue/PEDOT:PSS composites for electrochromic windows, *Solar Energy Material Solar Cells* 104 (2012) 64–74.
89. J. Wang, X. W. Sun, and Z. Jiao, Application of nanostructures in electrochromic materials and devices: Recent progress, *Materials* 3 (2010) 5029–5053.
90. Y. Mizukoshi and K. Yamada, Electrochromic characteristic of NiO/Au composites nano-rod array membrane, *Bull. Soc. Photogr. Imag. Japan* 24/1 (2014) 12–17.

91. M. R. J. Scherer and U. Steiner, Efficient electrochromic devices made from 3D nanotubular gyroid networks, *Nano Letters* 13 (2013) 3005–3010.
92. A. C. Dillon, A. H. Mahan, R. Deshpande, P. A. Parilla, K. M. Jones, and S. H. Lee, Metal oxide nanoparticles for improved electrochromic and lithium ion battery technologies, *Thin Solid Films* 516 (2008) 794–797.
93. C. G. Kuo and Y. C. Tung, The electrochromic properties of nanostructured bilayer WO3 thin films prepared by DC sputter and electrochemical method, *Journal of Science and Innovation* 4/1 (2014) 21–26.
94. X. Yang, G. Zhu, S. Wang, R. Zhang, L. Lin, W. Wu, and Z. L. Wang, A self powered electrochromic device driven by a nanogenerator, *Energy and Environmental Science* (2012), DOI: 10.1039/c2ee23194h.
95. Z. Jiao, J. Wang, L. Ke, X. Liu, H. V. Demir, M. F. Yang, and X. W. Sun, Electrochromic properties of nanostructured tungsten trioxide (hydrate) films and their applications in a complementary electrochromic device, *Electrochimica Acta* 63 (2012) 153–160.
96. V. Jain, R. Sahoo, J. R. Jinschek, R. Montazami, H. M. Yochum, F. L. Beyer, A. Kumar, and J. R. Heflin, High contrast solid state electrochromic devices based on ruthenium purple nanocomposites fabricated by layer-by-layer assembly, *Chem. Comm.* (2008) 3663–3665.
97. H. M. A. Soliman, A. B. Kashyout, M. S. E. Nouby, and A. M. Abosehly, Effect of hydrogen peroxide and oxalic acid on electrochromic nanostructured tungsten oxide thin films, *Int. J. Electrochem. Sci.* 7 (2012) 258–271.
98. M. Vidotti and S. I. C. de Torresi, Nanochromics: Old materials, new structures and architectures for high performance devices, *J. Braz. Chem. Soc.* 19/7 (2008) 1248–1257.
99. J. P. Coleman, J. J. Freeman, P. Madhukar, and J. H. Wagenknecht, *Displays* 20 (1999) 145–154.
100. C. W. Kung, T. C. Wang, J. E. Mondloch, D. F. Jimenez, D. M. Gardner, W. Bury et al., *Chem. Mater.* 25 (2013) 5012–5017.
101. N. N. Dinh, D. H. Ninh, T. T. Thao, and T. V. Van, *J. Nanomat.* 2012 (2012), doi:10.1155/2012/781236.
102. M. Vidotti and S. I. C. de Torresi, *J. Braz. Chem. Soc.* 19, 7 (2008) 1248–1257.
103. A. Verma and P. K. Singh, *Indian J. Chem.* 52A (2013) 593–598.
104. S. H. Lee, R. Deshpande, P. A. Parilla, K. M. Jones, B. To, A. H. Mahan, and A. C. Dillon, *Adv. Mater.* 18 (2006) 763–766.
105. Z. Jiao, J. Wang, L. Ke, X. Liu, H. V. Demir, M. F. Yang, and X. W. Sun, *Electrochemica Acta* 63 (2012) 153–160.
106. C. M. White, D. T. Gillaspie, E. Whitney, S. H. Lee, and A. C. Dillon, *Thin Solid Films* 517 (2009) 3596–3599.
107. R. A. Patil, R. S. Devan, J. H. Lin, Y. R. Ma, P. S. Patil, and Y. Liou, *Sol. Energy Mater. Sol. Cells* 112 (2013) 91–96.
108. C. G. Kuo, C. Y. Chou, Y. C. Tung, J. H. Chen, *J. Marine Sci. Technol.* 20/4 (2012) 365–368.
109. O. Lavi, G. L. Frey, A. Siegmann, and Y. E. Eli, *Electrochem. Comm.* 10 (2008) 1210–1213.
110. I. Shiyanovskaya, M. Hepel, and E. Tewksburry, *J. New Mater. Electrochem. Systems* 3 (2000) 241–247.
111. M. Kateb, V. Ahmadi, and M. Mohseni, Proceedings of 4th International Conference on Nanostructures (ICNS4), March 12–14, 2012, Kish Island, I. R. Iran, pp. 876–878.

112. K. J. Patel, M. S. Desai, C. J. Panchal, H. N. Deota, and U. B. Trivedi, *J. Nano Electro. Phys.* 5, 2 (2013) 02023.

113. S. Y. Li, G. A. Niklasson, and C. G. Granqvist, *Appl. Phys. Lett.* 101 (2012) 071903.

114. H. Wei, J. Zhu, S. Wu, S. Wei, and Z. Guo, *Polymer* 54 (2013) 1820–1831.

115. A. Llordes, G. Garcia, J. Gazquez, and D. J. Milliron, *Nature* 500 (2013) 323–32. doi:10.1038/nature12398.

116. M. C. Rao, *J. Non-oxide Glasses* 5, 1 (2013) 1–8.

117. E. Smela, *Adv. Mater.* 11 (1999) 1343.

118. H. Tsutsumi, Y. Nakagawa, K. Miyazaki, M. Morita, and Y. Matsuda, *J. Polym. Sci. A* 30 (1992) 1725.

119. W. A. Gazotti, G. Casalbore-Miceli, and A. Geri, *Adv. Mater.* 10 (1998) 1522.

120. R. A. Colley, P. M. Budd, J. R. Owen, and S. Balderson, *Polym. Int.* 49 (2000) 371.

121. G. C. de Vries, *Electrochim. Acta* 44 (1999) 3185.

122. P. M. S. Monk, C. Turner, and S. P. Akhtar, *Electrochim. Acta* 44 (1999) 4817.

123. D. R. Rosseinsky and R. J. Mortimer, *Adv. Mater.* 13 (2001) 783.

9

Polymer Nanostructures through Packing of Spheres

Lawrence A. Renna, Timothy S. Gehan, and D. Venkataraman

CONTENTS

9.1 Introduction

Nanoscale structures and morphologies impart desired functions to many polymer-based materials (Aida et al. 2012; Bates et al. 2012; Hammond 2004, 2012; Hawker and Russell 2005; Ikkala and ten Brinke 2004; Palmer and Stupp 2008; Ramanathan and Darling 2011). Thus far, amazing arrays of structures, morphologies, and materials have been created using either lithography or molecular self-assembly (Aida et al. 2012; Bates et al. 2012; Hawker and Russell 2005; Palmer and Stupp 2008; Park et al. 2003; Xia and Whitesides 1998). Yet, there are structures or morphologies that are difficult to create using known strategies or with specific materials of choice. As an example, consider the bicontinuous structure (Figure 9.1a), which has been targeted for many applications, such as porous materials (Velev and Lenhoff 2000) or materials that require charge transfer and charge transport like organic photovoltaic devices (OPVs) (Hentze and Antonietti 2001; Jones and Lodge 2012; Simon et al. 2001; Szymanski et al. 2005; Wiesenauer and Gin 2012). Although the bicontinuous structure is well known in coil–coil diblock copolymer thin films and in amphiphilic molecules (Matsen and Bates 1996), this structure is rare in rod–rod and rod–coil block copolymers (Cheng et al. 2002; Kimber et al. 2010; Lee et al. 2000; Lin et al. 2012; Liu and Fredrickson 1996; Ryu and Lee 2008). Moreover, the bicontinuous structure will not form if the constituent blocks of the diblock copolymer are miscible.

Although the phase-segregated morphologies have been well established for coil–coil copolymers (Fredrickson and Bates 1996; Matsen and Bates 1996; Zhao and Brittain 2000), such morphologies for rod–coil or rod–rod copolymers are not well established (Chen et al. 1996; de Boer et al. 2001; Gutacker et al. 2010; Halperin 1990; Klok and Lecommandoux 2001; Lecommandoux et al. 2003; Lee et al. 2001; Liu et al. 2011; Olsen et al. 2007, 2008; Olsen and Segalman 2005, 2008; Scherf et al. 2009, 2008; Segalman et al. 2009; Shen and Eisenberg 1999; Tschierske 1998). Therefore, the block copolymer strategy is limited to coil–coil polymers. Bicontinuous structures with domain sizes <50 nm cannot be easily created by lithography because the size features are too small (Pease and Chou 2008; Xia et al. 2011; Yang et al. 2009). This example illustrates the need for new strategies to create nanoscale morphologies through self-assembly, which provides access to structures that are not obtainable using current methods.

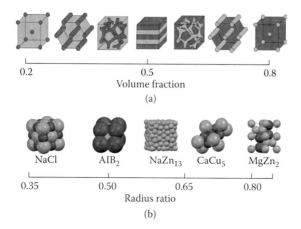

FIGURE 9.1
(a) Morphologies (cubic, cylindrical, gyroid, and lamellar) obtained in diblock polymers as a function of the volume fraction, and (b) morphologies that can be obtained by the assembly of nanoparticles with different radius ratios.

9.2 Concept of Sphere Packing

The radius ratio (R_{cation}/R_{anion}) rules enunciated by Pauling in 1929 can be used to predict the structure of ionic crystals and the coordination number of the ions from the sizes of the constituent anion and cation (Pauling 1929). Akin to the formation of binary ionic crystals, two types of nanoparticles can self-assemble into ordered *superlattices* or *supraparticle assemblies* at a much larger scale (Chen and O'Brien 2008; Li et al. 2011; Sharma et al. 2011; Shevchenko et al. 2006a; Talapin 2008; Yang et al. 2008; Zhang and Wang 2009; Zhang et al. 2006). Similar to ionic crystals, the packing of the nanoparticles depends on the ratio of the radii of the particles ($\gamma = R_{small}/R_{large}$). Unlike in ionic crystals, the self-assembly of nanoparticle mixtures occurs even if the nanoparticles do not have electrostatic attraction (Eldridge et al. 1993; Shevchenko et al. 2006a).

The primary mechanism that drives assembly of spherical particles into crystalline arrays (or superlattices) is entropy maximization by adopting the most efficient packing (Barry and Dogic 2010; Eldridge et al. 1993; Frenkel 1993; Grzelczak et al. 2010). Recent computational studies indicate a rich structural diversity in the assembly of spherical and nonspherical nanoparticles (Damasceno et al. 2012; Phillips and Glotzer 2012). Until a few years ago, the focus of this area was the underlying physics of assemblies of hard spheres. The past five years has witnessed a surging interest in the area of sphere assemblies within the chemistry community because of the pioneering works of Murray, Talapin, and coworkers on the assembly of inorganic nanoparticles (Dong et al. 2010, 2011; Shevchenko et al. 2006a, 2006b, 2007; Ye et al. 2011). The growing attention and importance of supraparticle assemblies is aptly captured by the following quote in a recent issue of *Nano Letters*: "The simplicity of supraparticle assemblies, their multifunctionality, structural versatility, and similarity with viral particles represent interesting directions for future research in nanostructures" (Pelaz et al. 2012).

Extending the concept of nanoparticle packing to create polymer-based nanostructures will allow us to (a) obtain stable nanoscale structures in a single step, through self-assembly from *any* two or more spherical moieties; (b) independently preassemble the polymer domains with the required molecular assembly *within* the domain size dictated by the size of the particle; (c) systematically alter nanoscale morphology through changes in the size and shape of the nanoparticles or interparticle interactions or both; and (d) rationally design nanostructured materials using multiple polymer components in the assembly. The concept of spherical packing is a powerful bottom-up approach that provides hierarchal control and tunability in each step of the self-assembly process to reliably obtain stable nanoscale morphologies.

What are the roadblocks in obtaining functional polymer-based nanostructures through the self-assembly of spheres? Unlike the synthesis of inorganic nanoparticles, the synthesis of organic nanoparticles is not as well established. In the field of inorganic nanoparticles there are extensive protocols to obtain nanoparticles with narrow size dispersity, an essential feature needed to obtain ordered structures. Most inorganic nanoparticles are dispersed in organic solvents; however, organic nanoparticles are typically dispersed in water. Therefore, protocols for the formation of ordered assemblies from aqueous dispersions need to be developed. In this chapter, we review the current protocols for the synthesis of organic nanoparticles and methods of self-assembly of nanoparticles, which may serve as a starting point for researchers interested in this field. In terms of synthesis, we focus on conjugated polymer nanoparticles (CPNs); however, their synthetic methods are quite versatile and can be readily applied to other organic materials. For assemblies, we focus on the various methods to assemble inorganic and organic nanoparticles.

9.3 Synthesis of Conjugated Polymer Nanoparticles

CPNs have been shown to be very useful materials for a wide variety of applications, including optoelectronic devices (Kietzke et al. 2003; Labastide et al. 2011; Millstone et al. 2010; Park et al. 2011), biological sensing/imaging (Chong et al. 2012; Fernando et al. 2010; Wu et al. 2008), and gene/drug delivery (Chong et al. 2012; Silva et al. 2010). There are two common methods for synthesizing CPNs: the reprecipitation method and the miniemulsion method (Feng et al. 2013; Pecher and Mecking 2010; Tuncel and Demir 2010).

In the reprecipitation method, the desired polymer is dissolved in a *good* solvent (e.g., tetrahydrofuran [THF]), and then added into a *bad* solvent (e.g., water) for the polymer, while vigorously stirring or sonicating. In the miniemulsion method, a polymer or a monomer is dissolved in the organic (oil) phase, which is then added to the water phase containing a surfactant. The mixture is sonicated and then the solvent is either evaporated or the monomer is polymerized to form the nanoparticle dispersion. The miniemulsion method allows polymer nanoparticles to be prepared from either monomers or presynthesized polymers. However, the reprecipitation method is limited to forming CPNs from only presynthesized polymers.

9.3.1 Reprecipitation Method

The reprecipitation method for forming polymer nanoparticles starts with dissolving the desired polymer in a good solvent (e.g., THF) that is miscible with water. The polymer solution is then quickly added to water with vigorous agitation. The good solvent is removed from the solution either by heating or under reduced pressure, to result in polymer nanoparticles dispersed in water. Surprisingly, these nanoparticle dispersions are reported to be stable for weeks to months without the use of a stabilizing agent. In this method, the size of the nanoparticle is controlled by the polymer concentration and the initial solvent choice (Potai and Traiphol 2013).

One of the first papers reporting the synthesis of CPNs using the reprecipitation method was by Masuhara and coworkers (Kurokawa et al. 2004). They reported the synthesis of nanoparticles of the functionalized polythiophene (P3DDUT) seen in Figure 9.2. A 0.25 mL solution of P3DDUT in THF was injected into 10 mL of water and vigorously stirred, forming a P3DDUT nanoparticle dispersion. The nanoparticle diameter was tuned from 40 to 400 nm by changing the weight percent of P3DDUT in THF. The authors observed that the absorption, emission, and thermochromic properties of the nanoparticles changed with size. As the nanoparticle size decreased, the ultraviolet-visible (UV-Vis) absorption maxima blue shifted with reduced vibronic structure indicating changes to polymer morphology. The emission spectra followed a similar trend where the emission maxima blue also shifted with less fine structure as the particle size decreased. The authors proposed that this difference in spectroscopic properties was due to lattice softening from changes in the molecular conformation of the polymer. For nanoparticles smaller than 140 nm, the authors reported that the nanoparticle synthesis protocol was inconsistent and had large size dispersity.

McNeill's group (Szymanski et al. 2005) reported the preparation of poly[2-methoxy-5-(2-ethylhexyloxy)-1,4-phenylenevinylene] (MEH-PPV) nanoparticles containing single-polymer chains. The nanoparticles showed a size range of mostly between 5 and 10 nm, and the dispersion was reported to be stable for weeks. A transmission electron microscope (TEM) image of these nanoparticles is shown in Figure 9.3. As the size of the particles was comparable to the radius of gyration of a single polymer chain, these particles are sometimes referred as polymer nanodots. Preparing such small nanoparticles requires a dilute initial solution of polymer (0.005 wt% polymer in THF) and a high THF to water volume ratio. From near-field scanning optical microscopy (NSOM), UV-Vis spectroscopy, and fluorescence spectroscopy, the authors concluded that the nanoparticles contained individual polymer chains, and had a smaller mean conjugation length that was attributed to bending or kinking of the polymer within the nanoparticle.

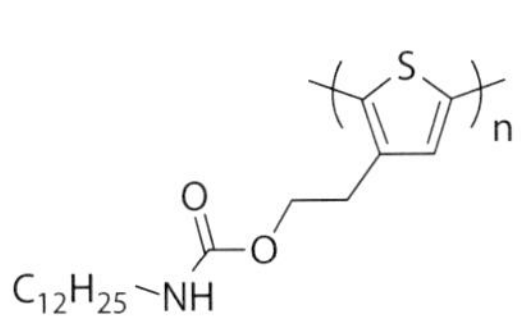

FIGURE 9.2
Chemical structure of P3DDUT.

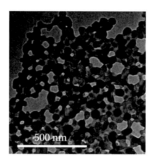

FIGURE 9.3
TEM image of MEH-PPV nanoparticles prepared with the reprecipitation method. (Reprinted with permission from Szymanski et al. 2005, 8543–8546. Copyright 2005 *American Chemical Society.*)

The spectroscopic signatures of CPNs affect their use in optoelectronic (Kietzke et al. 2003; Labastide et al. 2011; Millstone et al. 2010; Park et al. 2011) and biosensing applications (Chong et al. 2012; Fernando et al. 2010; Wu et al. 2008). Therefore, it is important to understand the impact of size and internal morphology on the photophysics of CPNs.

Traiphol and coworkers (Potai and Traiphol 2013) demonstrated that the choice of initial solvent the polymer is dissolved in affects the internal morphology of CPNs on reprecipitation. This was demonstrated with two conjugated polymers, MEH-PPV and poly(3-octylthiophene) (P3OT), using THF or dichloromethane (DCM) as initial solvents. The reprecipitation procedure used was very similar to those reported in Kurokawa et al. (2004) and Szymanski et al. (2005), and it should be noted that the volume ratio of organic solvent to water was below the water solubility limit for both THF and DCM. The authors demonstrated that when switching from THF to DCM, the MEH-PPV average nanoparticle diameter increases from 40 to 120 nm. This trend is consistent for P3OT nanoparticles as well. The UV-Vis absorption maxima of the MEH-PPV nanoparticle dispersion, compared to solution, red shifted when THF was used as the initial solvent and blue shifted when using DCM. The photoluminescence (Yodh et al. 2001) emission maxima of the nanoparticles using THF is red shifted by 50 nm in comparison to the CPNs prepared using DCM. The authors state:

> When the water solubility of initial solvent is relatively low, the collapse of conjugated chains occurs prior to the assembling process. The resultant CPNs contain large fraction of collapsed coils. Therefore, the photophysical properties of collapsed coils, which absorb and emit light at high-energy region, are dominant in this type of CPNs. When the initial solvent is miscible with water, the mixing process is relatively fast. The conjugated polymers are driven to assemble into CPNs with minimal change of individual chain conformation. This type of CPNs contains

large fraction of aggregates and exhibits smaller size. Their absorption
and PL spectra occur at low-energy region.

Figure 9.4 shows this difference when using each of the initial solvents.

Unlike the previous example where the internal structure of the CPNs
was tuned during the synthetic process, Joo and coworkers (Lee et al. 2011)
developed a method to tune the internal morphology of nanoparticles
after synthesizing them using the reprecipitation method. Joo and coworkers prepared poly(3-hexylthiophene) (P3HT) nanoparticle dispersions by
adding a 1 mg/mL solution of P3HT in THF dropwise to vigorously stirred
water. The P3HT nanoparticle dispersions were then hydrothermally
annealed in a Parr bomb for 10 hours at a temperature ranging from 60°C
to 150°C and allowed to cool slowly. There was a red shift of the peak max
in the UV-Vis absorption spectra with increasing annealing temperature.
This red shift in absorbance is attributed to a planarization of the thiophene rings, which is induced by the high pressure during the annealing process. The 0-0 and 0-1 peak ratio of the absorption spectra starts to
increase at 90°C and further increases with increasing temperature. The
emission spectra slightly red shifted with increasing annealing temperature and a slight increase in the emission intensity at 710 and 810 nm. The
authors conclude that hydrothermal annealing of the nanoparticles leads
to more interchain interactions and ring deformations within the polymer
chains because of the high pressure on the nanoparticles.

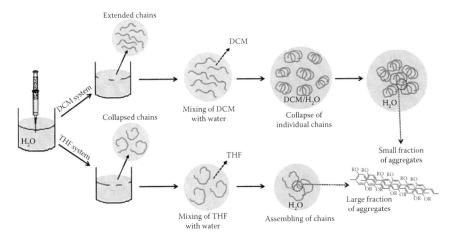

FIGURE 9.4
(See color insert.) Proposed schematic for nanoparticle formation when changing initial
solvents. (Reprinted from *Journal of Colloid and Interface Science*, 403, Potai and Traiphol,
Controlling chain organization and photophysical properties of conjugated polymer nanoparticles prepared by reprecipitation method: The effect of initial solvent, 58–66, Copyright 2013,
with permission from Elsevier.)

9.3.2 Miniemulsion Method for Synthesizing Conjugated Polymer Nanoparticles

9.3.2.1 Miniemulsion Method of Presynthesized Conjugated Polymer Nanoparticles

In the miniemulsion method, the desired polymer is dissolved in an organic solvent that is not miscible with water (e.g., chloroform, toluene, xylene, and chlorobenzene) and injected into an aqueous solution containing surfactants. The biphase mixture is stirred for about one hour to form a macroemulsion, where the organic solvent droplets are larger than 1 μm. This macroemulsion is sonicated using an ultrasonicator. The sonication breaks up the macroemulsion droplets, forming a miniemulsion of smaller droplets with sizes ranging from 30–500 nm (Landfester 2001). These droplets are claimed to be monodisperse and are stabilized by the surfactant in solution. The droplet size is dependent on multiple variables: surfactant concentration, organic solvent to water volume ratio, organic solvent chosen (Landfester et al. 2002), and sonication power (Keum et al. 2011). To our knowledge, the effect each of these variables has on CPNs has not been fully explored to date. Coalescence of droplets can occur due to Ostwald ripening, where the organic solvent diffuses from the droplet. Ostwald ripening can be prevented by using an ultrahydrophobe, typically a long linear hydrocarbon or even the polymer in solution. After forming the miniemulsion, the solution is stirred and heated to remove the organic solvent, forcing the polymers in the solution to aggregate, and leaving a surfactant-stabilized CPN dispersion (Figure 9.5).

Ionic small molecule surfactants are typically used to stabilize and achieve nanometer-sized droplets. Polymeric surfactants have also been used but are not as common (Wu et al. 2009). Sodium dodecyl sulfate (SDS) is the most common surfactant used for synthesizing CPNs. Antonietti and coworkers have shown that cationic surfactants such as cetyltrimethylammonium bromide (CTAB) can also be used as the surfactant (Landfester et al. 1999).

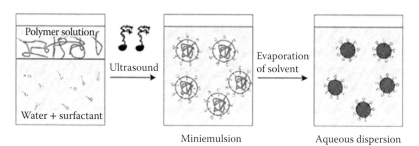

FIGURE 9.5
Graphical depiction of the miniemulsion method for preparing polymer nanoparticles from preformed polymers. (Reprinted by permission from Macmillan Publishers Ltd. *Nature Materials*, Kietzke et al. 2003, Copyright 2003.)

Kietzke and coworkers (Landfester et al. 2002) reported the synthesis of CPNs using poly(fluorene)- and poly(cyclopentadithiophene)-based polymers. The nanoparticles were synthesized through the miniemulsion process using chloroform as the organic solvent and SDS as the surfactant in water. The authors demonstrated that the size of the nanoparticles could be controlled by the SDS concentration and the polymer concentration, with sizes ranging from 75 to 250 nm. Interestingly, they reported small-sized nanoparticles even with a large weight percent of polymer in solution. The authors also demonstrated the formation of nanoparticle films on glass and showed that the films were fluorescent. Therefore, they suggested that it is possible to prepare multilayered films with CPNs for optoelectronic devices. Upon annealing the nanoparticle film above the polymer's glass transition temperature, the authors observed coalescence of the particles. This work represents a very important step to realizing the application of CPNs for optoelectronic devices.

Venkataraman and coworkers (Nagarjuna et al. 2012) prepared P3HT nanoparticles using the miniemulsion method to determine the effect of the initial solvent on the nanoparticles' morphology. Using SDS as the surfactant in water, P3HT nanoparticles were prepared from three solvents: chloroform (good solvent for P3HT), toluene (marginal solvent for P3HT), and a 1:4 volume mixture of toluene to chloroform. A TEM image of P3HT nanoparticles prepared using chloroform as the initial solvent can be seen in Figure 9.6. The authors used UV-Vis absorption data and single-particle photoluminescence decay data to probe the internal morphology. The authors concluded that when using chloroform as the solvent, the polymer is conformationally distorted and rapidly aggregates on removal of the solvent

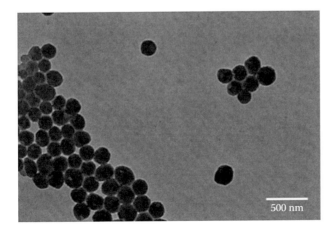

500 nm

FIGURE 9.6
TEM image of P3HT nanoparticles prepared using the miniemulsion method with chloroform as the initial solvent. (Reprinted with permission from Nagarjuna et al. 2012. Copyright 2012 American Chemical Society.)

at 80°C. Because the temperature is above the boiling point of chloroform, the polymer cannot be solvent-annealed and is consistent with the observation of higher dispersity in the aggregate structure within the nanoparticle. When toluene is used as the solvent for P3HT, the polymer aggregates before all of the toluene is evaporated, and as it is heated below toluene's boiling point the aggregates have time to solvent-anneal, which could cause tighter packing of the polymer chains. When using the mixed system, as it is mostly chloroform, it is assumed that the polymers have greater chain mobility and toluene allows the aggregates to slowly solvent-anneal, leading to a higher degree of structural order within the particle. This work is an in-depth example of how to tune the CPNs' morphology during synthesis using the miniemulsion method.

Scherf and coworkers (Kietzke et al. 2003) attempted to obtain nanoscale morphologies in conjugated polymer films using single-component CPNs or using two-component conjugated polymer blend nanoparticles. The authors prepared single-component CPNs of two fluorescent polymers, m-LPPP and PF11112 (Figure 9.7), and spin coated thin films following their previous procedure (Landfester et al. 2002). The authors spin coated a film from a mixture of these nanoparticles, as they have been shown to exhibit energy transfer between the polymers. Upon annealing the film of the mixture of nanoparticles and exciting it at 380 nm (absorption max of PF11112), the emission intensity of m-LPPP increases fourfold and the emission of PF11112 is quenched, indicating energy transfer between the polymers. When preparing a film of just m-LPPP nanoparticles and exciting it under the same conditions, the authors observed very weak emission, proving that emission from the mixed particle film was due to energy transfer between polymers. The authors then conclude "that the decrease in PF11112 emission intensity is simultaneously accompanied by an increase of the m-LPPP emission is further proof that annealing does not lead to any significant changes in the electronic properties of one or both components."

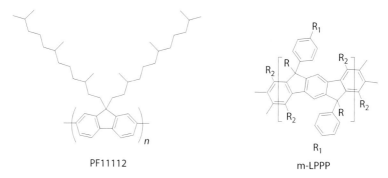

PF11112   m-LPPP

FIGURE 9.7
Chemical structures for PF11112 and m-LPPP; where R = methyl, R1 = dodecyl, and R2 = hexyl.

In an attempt to apply this method to OPVs, the authors prepared mixed nanoparticle blends of a hole-transporting polymer poly(9,9′-dioctylfluorene-co-bis (*N*,*N*′-(4,butylphenyl))bis(*N*,*N*′-phenyl-1,4-phenylene)diamine) (PFB) and an electron transporting polymer poly(9,9′-dioctylfluorene-cobenzothiadiazole) (F8BT). These two polymers were mixed in chloroform in a 1:1 weight ratio and nanoparticles were prepared using the miniemulsion method; they also tried using xylene instead of chloroform to see if solvent has an effect on nanoparticle morphology/size. PFB and F8BT have been shown to have an external quantum efficiency (EQE) between 0.25% and 4% depending on the choice of initial solvent (Arias et al. 2001). The nanoparticles' mean diameter was around 50 nm for both chloroform and xylenes; they were spin coated onto indium tin oxide and then Ca/Al was evaporated on top. Whether using xylene or chloroform, the authors report a peak EQE of about 1.7%, which is attributed to the same size of the nanoparticles. Therefore, the EQE in nanoparticle-based systems are mainly attributed to the size of each component's domain, assuming a bicontinuous network. Although they do not report a power conversion efficiency, this is a very important step in the development of CPN–based OPVs.

9.3.2.2 *Miniemulsion Polymerization of Conjugated Polymer Nanoparticles*

One main advantage that the miniemulsion method has over the reprecipitation method is that one can start either from presynthesized polymers or from monomers and then synthesize the polymer within the emulsion droplets. Originally, for CPNs, this was done by oxidative polymerization with a conjugated monomer: pyrrole (Jang et al. 2002), aniline (Vincent 1995), acetylene (Huber and Mecking 2006), and 3,4-ethylenedioxythiophene (EDOT) (Henderson et al. 2001). The initial reason to disperse these conducting polymers as nanoparticles using the miniemulsion method is that the parent polymers are typically not very soluble in any solvent.

Mecking and coworkers (Baier et al. 2009) demonstrated that the miniemulsion polymerization method can be extended using Glaser's catalytic step-growth polymerization to prepare CPNs from conjugated monomers. They report the synthesis of fluorescent nanoparticles of poly(phenylene diethynylene) and poly(fluorene diethynylene). To alter the emission spectra of the nanoparticle dispersions, either a functionalized fluorenone or a functionalized perylenediimide was added in small molar ratios to each polymer solution before preparing the CPNs. The nanoparticles had an emission maximum ranging from 496 to 691 nm. The nanoparticles were prepared with tetradecyltrimethylammonium bromide as the surfactant and were around 30 nm in diameter.

With the intention of preparing red- and near-infrared-emitting CPNs, Mecking and coworkers (Huber et al. 2012) prepared low-bandgap CPNs from fluorene and electron-withdrawing monomers using palladium-catalyzed Sonogashira cross-coupling within the miniemulsion. The nanoparticles were prepared using SDS as the surfactant, resulting in particle sizes ranging from 42 to 62 nm. The nanoparticles' emission was controlled by changing the monomers used in the polymerization and the molar ratios of monomers used, with emission up to the 1000-nm region. Low-bandgap polymers are typically prepared using palladium-catalyzed coupling and have been shown to be particularly effective in organic optoelectronic devices. Therefore, this work shows that even these reactions can be utilized within miniemulsions to prepare low-bandgap polymer nanoparticles.

Recently, Sprakel and coworkers (Kuehne et al. 2012) reported the synthesis of fluorene- and benzothiadiazole-based copolymer nanoparticles using miniemulsion Suzuki–Miyaura polymerization. The authors used a modified method from the typical miniemulsion polymerization, using both sonication and mechanical stirring to form the emulsion. Using this technique, they were able to form monodisperse CPNs. The authors demonstrate that the size and spectroscopic properties can be tuned by changing the monomers used in the synthesis. The sizes attained were all quite large, ranging from 167 to 1842 nm. The authors demonstrate that these nanoparticles do readily assemble to form photonic crystals. Preparing monodisperse CPNs has been a challenge; here Sprakel and coworkers provide a way using palladium cross-coupling miniemulsion polymerization. Some drawbacks of this method are that the nanoparticles are very large and need to be much smaller for applications in optoelectronic devices (<167 nm). Also the miniemulsion polymerization method can trap catalysts and other impurities within the CPNs.

9.4 Assembly of Nanoparticles

Inorganic nanoparticles/nanocrystals have been the focus of the assembly of nanoparticles into specific morphologies. The low size dispersity of inorganic nanoparticle samples is an important factor in obtaining ordered structures. A diverse number of techniques to obtain these structures have been established utilizing inorganic nanoparticles. Therefore, many of the techniques that will be discussed herein to self-assemble nanoparticles have only been applied using inorganic nanoparticles. It is the expectation that the same principles can be extended to organic nanoparticles.

There are numerous methods for the self-assembly of nanoparticles into colloidal crystals, glasses, and superlattices (Grzelczak et al. 2010; Murray et al. 2000; Shevchenko et al. 2006a; Xu et al. 2013; Zhang et al. 2010). The methods in this section have been broken down into the following categories: self-assembly by evaporation, self-assembly at the liquid–air interface, self-assembly at the liquid–liquid interface, and other types of self-assembly that do not necessarily fit into the aforementioned classifications.

9.4.1 Self-Assembly by Evaporation

In this method, the slow evaporation of the liquid phase of a colloidal dispersion on a substrate produces a self-assembled structure. Although the technique is relatively simple, there are myriad attractive, repulsive, and disruptive forces that complicate the self-assembly process. The solvent evaporation method has been used to obtain ordered arrays of spherical particles (Bartlett et al. 1992; Puntes et al. 2001; Shevchenko et al. 2006a; Sun et al. 2000; Zeng et al. 2002), nanorods (Zanella et al. 2011), and nanotriangles (Walker et al. 2010).

One of the earliest examples using solvent evaporation to obtain an ordered binary nanoparticle assembly was reported by Bartlett et al. (1992). They utilized the radius ratio of the particles and partial volume fraction to tune the morphology of the nanoparticle assembly. The authors demonstrated that poly(methyl methacrylate) (PMMA) nanospheres (186 and 321 nm) dispersed in decalin/carbon disulfide self-assembled into four different crystal structures: A, AB_2, AB_{13}, or B, at a radius ratio of 0.62 and varying particle volume fraction. The dispersant (also known as dispersing phase) was chosen to have the same refractive index as the nanoparticles, and therefore reduces the van der Waals interactions between nanoparticles. When such enthalpic forces are screened, as in this case, the system behaves similarly to the age-old physics problem of assembling hard spheres where the entropy gain in maximizing particle free volume is the dominant factor. SiO_2 nanoparticles (200–700 nm) dispersed in ethanol have also been shown to self-assemble on a vertically aligned substrate into close-packed structures. The authors demonstrated that the film's thickness could be controlled from a single monolayer of nanoparticles to hundreds of nanoparticle thick films by tuning the volume fraction and particle size (Jiang et al. 1999).

When a colloidal solution is allowed to dry unperturbed on a horizontal substrate, the *coffee ring effect* is often observed (Yunker et al. 2011). This effect is due to the nonequilibrium state of a drying droplet where different evaporation rates throughout the droplet and small temperature fluctuations result in capillary flows within the droplet. These forces move nanoparticles outward toward the droplet's contact line, causing a concentration gradient of nanoparticles in the film after drying is complete (Figure 9.8b and c). Recent work has shown that by simply changing the shape of the nanospheres to nanoellipsoids, the coffee ring effect can be suppressed (Figure 9.8a). This is due to the long-range interactions between particles that are generated by the anisotropic shape of the ellipsoids, resulting in non-close-packed structures.

Another interesting solution to avoid the coffee ring effect is to control the evaporation rate by placing the sample in an environment that contains some degree of saturation of the solvent. The saturation can be changed to tune the evaporation rate, providing an additional tool to self-assemble nanoparticles (Bian et al. 2011; Hanrath et al. 2009; Singh et al. 2011; Xie et al. 2013). By slowing the evaporation rate, the evaporation gradient in the droplet is

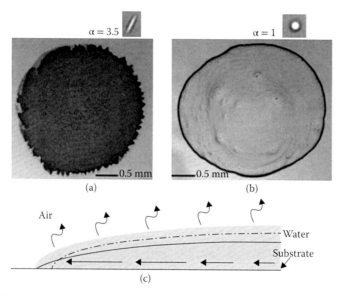

FIGURE 9.8

Image of film with ellipsoidal nanoparticles (a) and spherical nanoparticles (b), where the coffee ring effect can be clearly seen due to the outward flow in the droplet (c). (Reprinted with permission from Macmillan Publishers Ltd. *Nature*, Yunker et al. 2011, Copyright 2011.)

reduced and thus the capillary force is reduced. Inorganic nanocubes and nanooctahedra have also been shown to self-assemble into different structures through controlled evaporation by placing the substrate with a droplet of colloidal solution in a desiccator at ambient, under vacuum, or saturated environments (Quan and Fang 2010).

It has been observed that solvent molecules can become occluded to result in non-close-packed assemblies (Goodfellow et al. 2011). It has also been shown that host–guest chemistry can be used to create non-close-packed nanostructures (Nagaoka et al. 2012). For example, PbSe nanoparticle dispersions mixed with guest molecules squalane, squalene, or polymers such as polyisoprene, were allowed to evaporate on substrate, resulting in non-close-packed structures such as AB_{13}. Without the inclusion of guest molecules, a close-packed crystal is obtained.

To prevent the coffee ring effect, the dispersion can be contained within a cavity and allowed to evaporate. An example of this procedure is to place a Teflon ring on a substrate, add a polymer nanoparticle dispersion inside the ring, and allow the solution to evaporate. Often, the result is a close-packed colloidal crystal (Peng and Dinsmore 2007). Aside from typical polymer nanoparticles polystyrene (PS)/PMMA, mesoporous silica particles filled with conjugated polymers can also be assembled via dispersant evaporation. The silica can be etched away, leaving a polymeric nanostructured material (Kelly et al. 2008). Using PS nanoparticles self-assembled on a substrate constrained by a ring, the effect of polydispersity on the order of the

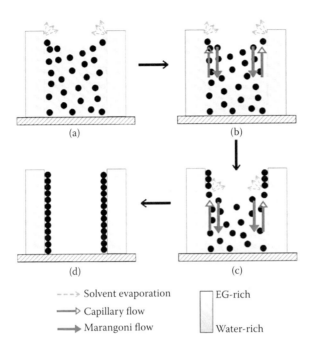

FIGURE 9.9
The formation of nanoparticle assemblies due to solvent evaporation, Marangoni flow, and capillary flow in EG/H$_2$O dispersant. (a) Solvent evaporation occurs at the contact line. (b) Marangoni flow is induced and flows opposite of capillary flow. (c) Assembly of nanoparticles occurs due to van der Waals interactions between nanoparticles, and between nanoparticles and the substrate to produce (d) a self-assembled nanoparticle film once all of the solvent has evaporated. (Reprinted with permission from Zheng et al. 2010, 16730–16736. Copyright 2010 American Chemical Society.)

assembled colloidal crystals was demonstrated. When a colloidal solution has wide size dispersity, it was shown that there are "non-uniformities in the filling fraction" (Cerdán et al. 2013).

Solvent evaporation/capillary flow–induced self-assembly of poly(9,9-di-*n*-octylfluorenyl-2,7-diyl) (PFO) and MEH-PPV and a blend of the two polymer nanoparticles has also been demonstrated in a confined well (Zheng et al. 2010). This was done in a 600-µm gap between two vertically aligned ITO substrates. The spacing between the two ITO substrates was filled with the nanoparticle dispersion (30 and 74 nm, respectively) in 40% (v/v) ethylene glycol (EG)/H$_2$O. The surface tension gradient creates a Marangoni flow opposite and parallel to the capillary forces. The Marangoni force can be tuned by varying amount of EG such that it reduces the capillary flow, and thus circumvents the nanoparticle concentration gradient in dried films (Figure 9.9). The nanoparticles self-assemble because of particle–particle interactions, particle–ITO interactions, and entropic effects.

Ordered structures have also been obtained using binary nanoparticle systems where the size of the small nanoparticle is less than or close to the size of the void created by the hexagonally packed large nanoparticles (Singh et al. 2011; Wang et al. 2009). In this example, the horizontal deposition has been shown to produce large binary colloidal crystals of PS where the small nanoparticles fill in the voids of the packed large nanoparticles. Similar binary and ternary nanoparticle colloidal crystals have also been obtained through layer-by-layer assembly. In this method, a single nanoparticle solution of PS or SiO_2 in water was confined on a substrate and allowed to evaporate. Then, a different-size nanoparticle dispersion was directly applied onto the aforementioned assembly and allowed to evaporate.

Assembly of inorganic nanoparticles through solvent evaporation has been demonstrated to produce a diverse number of ordered structures (Figure 9.10) with natural analogues, including but not limited to NaCl, CsCl, NiAs, CuAu, FeCr alloy, orth-AB, AlB_2, $MgZn_2$, $MgNi_2$, Cu_3Au, Fe_4C, $CaCu_5$, CaB_6, $NaZn_{13}$, CdSe, cub-AB_{13}, ico-AB_{13} (Bodnarchuk et al. 2010, 2011b; Chen et al. 2010, 2007; Evers et al. 2010; Kovalenko et al. 2010; Pichler et al. 2011; Redl et al. 2003; Shevchenko et al. 2006a, 2006b, 2007, 2008; Talapin et al. 2009, 2007; Urban et al. 2007; Ye et al. 2011), and A_6B_{19}, which interestingly has no natural analogue (Boneschanscher et al. 2013). The technique, most notably utilized by Murray, Talapin, and coworkers, is the solvent evaporation of single, binary, or ternary nanoparticle dispersions on an angled substrate at elevated temperatures and reduced pressure (Figure 9.11).

Murray and coworkers exemplified the effects of radius ratio, particle volume fractions, and the particles' relative charge on the self-assembly of nanoparticles, by tuning these properties to obtain a large variety of structures. It was demonstrated that the superlattices formed by a colloidal dispersion could be changed by adding small amounts of carboxylic acids, trioctylphosphine oxide (TOPO), or dodecylamine. This result is another example of complexity of the interplay of various interactions, such as the van der Waals, electrostatic, steric repulsion, dipolar, and entropic forces that dictate the structure of nanoparticle assemblies (Chen et al. 2007; Shevchenko et al. 2006a).

Talapin and coworkers used temperature as a tool to control the ordering of nanoparticles into diverse structures (Bodnarchuk et al. 2010). This method has been consistently proven to be a versatile technique to self-assemble inorganic nanoparticles into a variety of ordered nanostructures, and has also been extended to organic nanoparticles (Bayer et al. 2010). Using this method, it has been demonstrated that a binary dispersion of PS nanoparticles in $CHCl_3$ (radius ratio between 0.52 and 0.85) can self-assemble into ordered crystals. The charge on the nanoparticles is tuned by incorporating complementary functional groups during the synthesis of the nanoparticles to increase interparticle attraction.

Solvent composition plays a major role in determining whether nanoparticles self-assemble into glassy structures or ordered superlattices. The solvent plays an important role in mediating interparticle interactions (Fendler and Dekany 1996; Murray et al. 2000). Similar to small molecule crystallization, the key to obtaining an ordered crystal is a slow transition

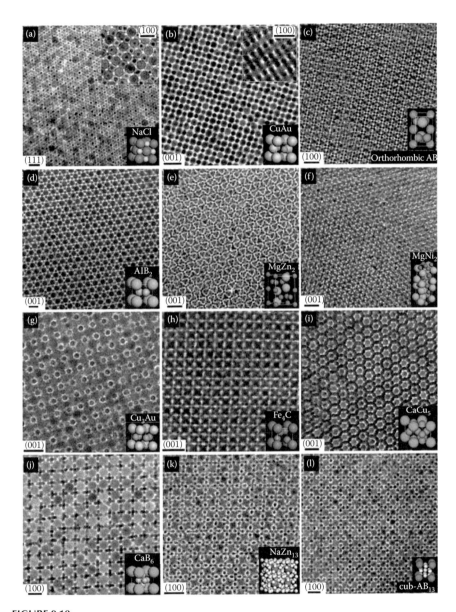

FIGURE 9.10

TEM images of binary nanoparticle assemblies and modeled unit cells, exemplifying the structural diversity of (a) NaCl structure from 13.4 nm γ-Fe_2O_3 and 5 nm Au; (b) CuAu structure from 7.6 nm PbSe and 5 nm Au; (c) *ortho*-AB structure from 6.2 nm PbSe and 3 nm Pd; (d) AlB_2 from 6.7 nm PbSe and 3 nm Pd; (e) $MgZn_2$ structure from 6.2 nm PbSe and 3 nm Pd; (f) $MgNi_2$ structure from 5.8 nm PbSe and 3 nm Pd; (g) Cu_3Au structure from 7.2 nm PbSe and 4.2 nm Ag; (h) Fe_4C structure from 6.2 nm PbSe and 3 nm Pd; (i) $CaCu_5$ structure from 0.2 nm PbSe and 5 nm Au; (j) CaB_6 structure from 5.8 nm PbSe and 3 nm Pd; (k) $NaZn_{13}$ structure from 7.2 nm PbSe and 4.2 nm Ag; and (l) *cub*-AB_{13} structure from 6.2 nm PbSe and 3 nm Pd nanoparticles. (Reprinted with permission from Macmillan Publishers Ltd. *Nature*, Shevchenko et al. 2006a, Copyright 2006.)

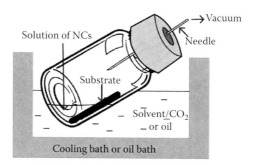

FIGURE 9.11
General apparatus setup for inorganic nanoparticle self-assembly. (Reprinted with permission from Bodnarchuk et al. 2010, 11967–11977. Copyright 2010 American Chemical Society.)

to the destabilization of the colloidal dispersion to establish equilibrium for particles to form a nanocrystal. A common strategy is to use a multicomponent dispersant with a low-boiling, good-dispersing solvent and a high-boiling, weak-dispersing solvent. As the good solvent evaporates, the weak-dispersing solvent causes particles to aggregate and thus nucleate nanocrystal growth. This technique was demonstrated with CdSe nanoparticles, where 95% hexane and 5% octane mixture produced a glassy crystal, whereas 95% octane and 5% octanol produced ordered superlattices (Murray 1997; Murray et al. 1995, 2000). Another interesting use of multicomponent dispersant was with PbS, CdSe, and CoPt$_3$ nanoparticles, where the colloidal solution in hexane was placed in a test tube with a vertically aligned substrate. i-PrOH was added carefully to prevent mixing. The slow diffusional intermixing of the solvents caused the growth of nanocrystals on the substrate (Podsiadlo et al. 2010). Multicomponent dispersant evaporation is a versatile technique and provides opportunity to easily tune interparticle interactions during self-assembly.

Another method that has been used to obtain colloidal crystals is the continuous motorized deposition technique developed by Nagayama and Dimitrov (Dimitrov and Nagayama 1996). In this method, nanoparticles were assembled into hexagonal close-packed arrays by the convective force generated by dispersant evaporation. Polystyrene nano/submicron particles (79 to 2106 nm) were suspended in water with trace amounts of SDS, NaCl, octanol, and milk casein protein to tune the wetting properties of the dispersion. A glass slide was mechanically lifted with a gearbox vertically out of the nanoparticle dispersion at the same rate at which the array is formed, so that the leading edge of the assembly is at a constant position with respect to the colloidal solution. The rate of convective assembly is described by the equation

$$v_c = \frac{\beta l j_e \varphi}{h(1-\varepsilon)(1-\varphi)}$$

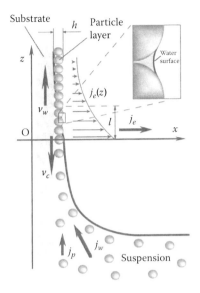

FIGURE 9.12

Schematic of continuous convective assembly where v_w is the substrate withdrawal rate, v_c is the array growth rate, j_w is the water influx, j_p is the respective particle influx, j_e is the water evaporation flux, and h is the thickness of the array. (Reprinted with permission from Dimitrov and Nagayama 1996, 1303–1311. Copyright 2010 American Chemical Society.)

where v_c is the rate of crystal growth, β a constant from 0 to 1 that depends on particle interactions, j_e the dispersant evaporation flux, φ the product of the number of nanoparticles per volume and the volume occupied by a single particle, h the thickness of the assembly or the size of the nanoparticle in a monolayer, and ε the porosity of the packing of the assembly. (Figure 9.12).

This technique has been used with PS and SiO_2 to make colloidal crystals and inverse structures (Cai et al. 2012). This principle of rate of convective assembly was also demonstrated in experiments where a small volume of PS/Au nanoparticles were dragged horizontally across the substrate by a deposition plate controlled by a motor to create hexagonally packed lattices (Prevo and Velev 2004), and in binary nanoparticle systems of PS nanoparticles (140–300 nm) in a vertically oriented confined convective assembly apparatus controlled by a motorized dipping machine (Figure 9.13). Depending on the radius ratio and particle volume fraction, ordered lattices of AB, AB_2, AB_3, AB_4, and AB_5 were obtained (Kim et al. 2005).

9.4.2 Self-Assembly at the Liquid–Air Interface

Self-assembly at the liquid–air interface has been widely used to obtain colloidal crystals and superlattices. This method circumvents the nonequilibrium droplet and coffee ring effect by moving nanoparticles to the liquid–air

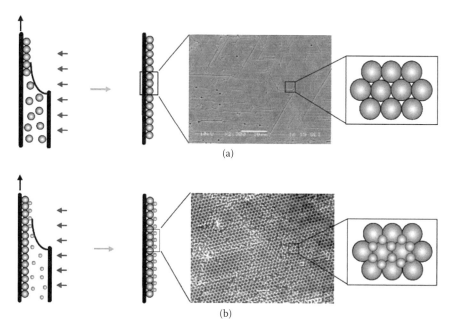

(a)

(b)

FIGURE 9.13

Schematic of confined assembly where a template is formed of larger particles (a) for the smaller particles to form binary assembly (b). (Kim, M.H. et al.,Im, S.H., and Park, O.O: Fabrication and structural analysis of binary colloidal crystals with two-dimensional superlattices. *Adv. Mater.* 2005. 17. 2501–2505. Copyright Wiley-VCH Verlag GmbH & Co. KGaA. Reproduced with permission.)

interface where equilibrium can be established (Bigioni et al. 2006). This section will discuss some of the methods that have been used to migrate nanoparticles from solution to the liquid–air interface to self-assemble.

Experiments with Au nanoparticles revealed that by adding dodecane-thiol to dispersions, the nanoparticles migrate to the liquid–air interface. Islands self-assemble at the interface, merge, and form long-range ordered lattices (Figure 9.14). There are two important elements to this technique: first, "rapid evaporation to segregate particles near the liquid–air interface," and second, an "attractive interaction between the particles and the liquid–air interface to localize them on the interface" (Bigioni et al. 2006; He et al. 2011; Jiang et al. 2010). This was further explored with various "alkanethiol/solvent combinations using decanethiol or dodecanethiol as the coating agent and hexane, octane, decane, or dodecane as the solvent to disperse the nanocrystals," where decahedra, icosahedra, and cubooctahedra crystals were formed (Sanchez-Iglesias et al. 2010).

Self-assembly at the liquid–air interface when the liquid is a "bad" dispersant for the nanoparticles has proven to be an interesting method for colloidal crystal formation. This technique was utilized by Murray and coworkers, where Fe_3O_4 and FePt nanoparticles dispersed in hexane were applied on top of an immiscible diethylene glycol (DEG) layer. As the hexane evaporated,

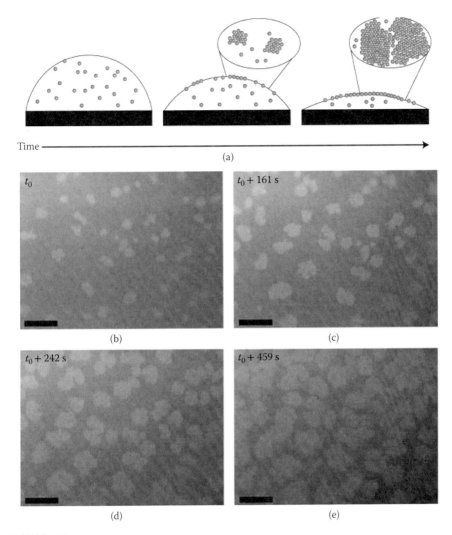

FIGURE 9.14

(See color insert.) Schematic of migration of nanoparticles to the liquid–air interface during drying (a) and optical images of nanoparticle island growth (b–e). (Reprinted with permission from Macmillan Publishers Ltd. *Nature Materials*, Bigioni et al. 2006, Copyright 2006.)

the nanoparticles self-assembled at the DEG–air interface, resulting in a large-scale AlB_2 superlattice that can be easily transferred from the interface to any substrate (Figure 9.15) (Dong et al. 2010). This technique has also been used to create additional binary and ternary superlattices (Dong et al. 2011), binary assemblies of Au nanospheres, rods, wires (Sanchez-Iglesias et al. 2010), CdSe nanoparticles (Murray et al. 1995), and anisotropic nanocrystals (Ye et al. 2010). Addition of a miscible solvent that is still a bad solvent for the nanoparticles also results in self-assembly at the liquid–air interface (Dong et al. 2011; Miszta et al. 2011; Wang et al. 2012).

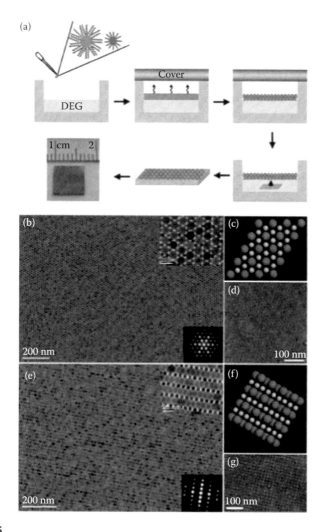

FIGURE 9.15

(See color insert.) Schematic of nanoparticle assembly at the DEG–air interface (a), and AlB2-type BNSL membranes self-assembled from 15-nm Fe_3O_4 and 6-nm FePt nanocrystals (b–g). (Reprinted with permission from Macmillan Publishers Ltd. *Nature*, Shevchenko et al. 2006b, Copyright 2010.)

Diverse binary assemblies at the air–dispersant interface have been obtained by Weiping Cai and coworkers using PS nanoparticles (Dai et al. 2012). Convective forces generated by the evaporation of the ethanol/water dispersant are used to assemble the nanoparticles at the liquid–air inter-face. In this method, nanoparticle dispersions of 1:1 ethanol/water contain-ing both the large and small nanoparticles are injected into a preformed water droplet on the substrate. The volatile alcohol carries the nanoparticles to the water–air interface where they self-assemble into ordered structures

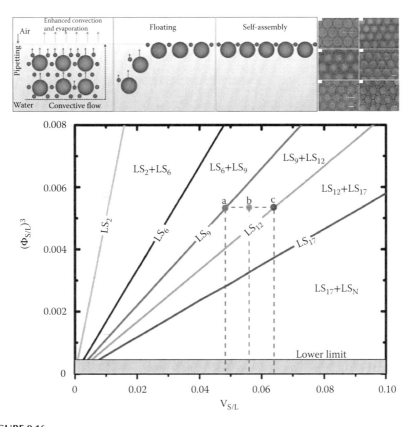

FIGURE 9.16

Scheme of ethanol-assisted transfer of nanoparticles to the liquid–air interface (top left), SEM image of binary assemblies (top right), and phase diagram of volume ratio vs. radius ratio cubed. (Adapted with permission from Dai et al. 2012, 6706–6716. Copyright 2012 American Chemical Society.)

(AB_2, AB_6, AB_9, AB_{12}, AB_{17}, and hybrid combinations of two structures). The rapid evaporation of ethanol increases the convective forces in the droplet carrying nanoparticles to the interface. The particles are confined to the interface because of hydrophobic effects and the surface tension of the dispersant (Figure 9.16 [top]). The structural diversity obtained through this method can be tuned with radius ratio, volume ratio, and concentration ratio of small to large nanoparticles (presented in a phase diagram in Figure 9.16 [bottom]). This is a useful and versatile technique; however, the maximum radius ratio in the phase diagram is 0.2, and has yet to be validated when the nanoparticles are similar in size.

The self-assembly of nanoparticles at the liquid–air interface is a versatile technique to obtain a wide array of morphologies. The most attractive feature of this method is the exit from the nonequilibrium—almost chaotic and uncontrollable—droplet on a substrate, to the interface where equilibrium can be established. A major concern is whether this concept can be extended to large-scale operations.

9.4.3 Self-Assembly at the Liquid–Liquid Interface

The liquid–liquid interface presents an interesting medium for nanoparticle self-assembly. Nanoparticles can quickly and easily assemble into the lowest free-energy configuration because of the equilibrium accomplished by the high mobility of nanoparticles at the liquid–liquid interface, as demonstrated with CdSe nanoparticles at the water–toluene interface (Kutuzov et al. 2007; Lin et al. 2003). Thin films of Au nanoparticles were self-assembled at the water–toluene interface using this method (Duan et al. 2004). Micelles of nanoparticle heterodimers (Fe_3O_4, Ag) were self-assembled at the interface between water and various organic solvents (Gu et al. 2005).

Furst and coworkers demonstrated the 2D self-assembly and size-sorting of positively charged PS nanoparticles near a water–oil interface (Park and Furst 2010). The authors reasoned that "as the interface confines colloids dispersed in the aqueous phase against a solid substrate, the disjoining pressure generated by the electrostatic repulsion with the charged particles causes the interface to deform locally, in the normal direction. This deformation generates long-range lateral capillary forces between the particles, which drive their rapid self-assembly." The resulting structures are "defect free" and ordered.

9.4.4 Other Types of Self-Assembly

Talapin and coworkers have used a microfluidic technique to assemble $CoFe_2O_4$, CdSe, PbS, Au, and Pd nanoparticles into single and binary nanoparticle superlattices (Bodnarchuk et al. 2011a). In this technique, the nanoparticle dispersion along with more dispersant (toluene) and a precipitant (miscible bad solvent, ethanol, *i*-propanol, *n*-butanol) are injected into the microfluidic capillary tubing with an immiscible bad solvent (perfluoro-tri-*n*-butylamine, perfluoro-di-*n*-butylmethylamine, or perfluoropentylamine) as the carrier fluid at a controlled injection rate. Nanoliter plugs are created in the tubing, and crystallization occurs either within the nanodroplet or at the liquid–liquid interface depending on the system, to create face-centered cubic, single-particle-size crystals, or AlB_2 binary crystals. The volumes of all three of the injected components can be varied to control the size of the plug, the NP concentration, and the amount of precipitant (Figure 9.17).

Template-assisted assembly has also used to obtain ordered nanoparticle assemblies. Nanoparticle aggregates of PS and SiO_2 nanoparticles were made by forcing them into templates with capillary flow (Lisunova et al. 2012). Also, inkjet printing colloidal poly(para-phenylene) or Au nanoparticle solutions into templates has been used to create ordered nanostructures (Fisslthaler et al. 2008). Furthermore, CdSe and Fe_2O_3 nanoparticles have been shown to

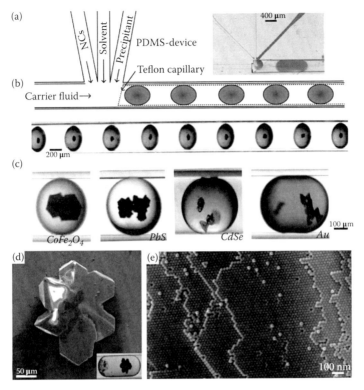

FIGURE 9.17
(See color insert.) Apparatus for microfluidic self-assembly (a) in nanoliter plugs (b). Crystallization within the plugs (c) results in ordered assemblies (d and e). (Reprinted with permission from Bodnarchuk et al. 2011a, 8956–8960. Copyright 2011 American Chemical Society.)

make up to 100-layer-thick ordered crystals using a patterned microfluidic cell (Akey et al. 2010; Liao et al. 2012).

Layer-by-layer spin coating is a simple technique that has been used to create binary colloidal crystals. This technique was demonstrated with 222- to 891-nm silica nanoparticles and works by spin coating nanoparticle dispersions in ethanol/ethylene glycol, first the larger-sized nanoparticles, then the smaller, and then repeat. This technique has been utilized to obtain AB_2 and AB_3 structures (Wang and Mohwald 2004). Another interesting nanoparticle self-assembly technique that also utilizes layer-by-layer spin coating is alternatively dipping the substrate into a polyelectrolyte and then into an oppositely charged colloidal dispersion. This creates a nanoparticle polyelectrolyte sandwich (Kotov et al. 1995; Srivastava and Kotov 2008), and has also been conducted with magnetic nanoparticles with the assistance of a magnetic field (Suda and Einaga 2009).

The use of biological molecules covalently or coordinately bound to nanoparticles has also drawn considerable interest for nanoparticle assembly

(Demers et al. 2001; Hill and Mirkin 2006; Jin et al. 2003; Jones et al. 2010; Kostiainen et al. 2013; Liu and Lu 2003; Mirkin et al. 1996; Noh et al. 2010; Park et al. 2000, 2001, 2008; Stoeva et al. 2005). For example, work by Mirkin and coworkers self-assembled gold nanoparticle-containing DNA oligonucleotide sticky ends. Complementary pieces of DNA bind to each other and create nanoparticle aggregates, which can be reversibly assembled and disassembled with heat by forming and breaking hydrogen bonds (Mirkin et al. 1996). One can imagine imparting functionality, which need not be biological, to rationally self-assemble nanoparticles.

9.5 Summary

In this chapter, we have summarized the various methods for the synthesis of organic nanoparticles and for the assembly of nanoparticles into ordered nanostructures. The assembly of nanoparticles into various structures provides a unique and novel approach to realize nanoscale morphologies. Thus far, the area of nanoparticle assemblies has been dominated by inorganic nanoparticles. To realize organic nanoparticle assemblies, several important challenges need to be tackled. The three main challenges are (1) fabrication of organic nanoparticles with low size dispersity, (2) fabrication of <20-nm organic nanoparticles in high concentrations for self-assembly, and (3) development of methods for self-assembly from aqueous dispersions. These challenges provide fertile ground for future research in the area of organic-based nanoparticle assemblies.

Acknowledgments

We thank the Polymer-Based Materials for Harvesting Solar Energy (PHaSe), an Energy Frontier Research Center funded by the U.S. Department of Energy, Office of Science, Basic Energy Sciences under Award no. DE-SC0001087 (financial support for TG) and the National Science Foundation Materials Research and Science Center on Polymers (DMR 0820506) (financial support for LR) for their support.

References

Aida, T., Meijer, E.W., and Stupp, S.I. Functional supramolecular polymers. *Science* 335(6070) (2012): 813–817.
Akey, A., Lu, C., Yang, L. et al. Formation of thick, large-area nanoparticle superlattices in lithographically defined geometries. *Nano Lett.* 10(4) (2010): 1517–1521.

Arias, A.C., MacKenzie, J.D., Stevenson, R. et al. Photovoltaic performance and morphology of polyfluorene blends: A combined microscopic and photovoltaic investigation. *Macromolecules* 34(17) (2001): 6005–6013.

Baier, M.C., Huber, J., and Mecking, S. Fluorescent conjugated polymer nanoparticles by polymerization in miniemulsion. *J. Am. Chem. Soc.* 131(40) (2009): 14267–14273.

Barry, E. and Dogic, Z. Entropy driven self-assembly of nonamphiphilic colloidal membranes. *Proc. Natl. Acad. Sci. USA* 107(23) (2010): 10348–10353.

Bartlett, P., Ottewill, R.H., and Pusey, P.N. Superlattice formation in mixtures of hard-sphere colloids. *Phys. Rev. Lett.* 68(25) (1992): 3801–3804.

Bates, F.S., Hillmyer, M.A., Lodge, T.P. et al. Multiblock polymers: Panacea or Pandora's box? *Science* 336(6080) (2012): 434–440.

Bayer, F.M., Hiltrop, K., and Huber, K. Hydrogen-bond-induced heteroassembly in binary colloidal systems. *Langmuir* 26(17) (2010): 13815–13822.

Bian, K., Choi, J.J., Kaushik, A. et al. Shape-anisotropy driven symmetry transformations in nanocrystal superlattice polymorphs. *ACS Nano* 5(4) (2011): 2815–2823.

Bigioni, T.P., Lin, X.M., Nguyen, T.T. et al. Kinetically driven self assembly of highly ordered nanoparticle monolayers. *Nat. Mater.* 5(4) (2006): 265–270.

Bodnarchuk, M.I., Kovalenko, M.V., Heiss, W. et al. Energetic and entropic contributions to self-assembly of binary nanocrystal superlattices: Temperature as the structure-directing factor. *J. Am. Chem. Soc.* 132(34) (2010): 11967–11977.

Bodnarchuk, M.I., Li, L., Fok, A. et al. Three-dimensional nanocrystal superlattices grown in nanoliter microfluidic plugs. *J. Am. Chem. Soc.* 133(23) (2011a): 8956–8960.

Bodnarchuk, M.I., Sheychenko, E.V., and Talapin, D.V. Structural defects in periodic and quasicrystalline binary nanocrystal superlattices. *J. Am. Chem. Soc.* 133(51) (2011b): 20837–20849.

Boneschanscher, M.P., Evers, W.H., Qi, W. et al. Electron tomography resolves a novel crystal structure in a binary nanocrystal superlattice. *Nano Lett.* 13(3) (2013): 1312–1316.

Cai, Z., Teng, J., Wan, Y., et al. An improved convective self-assembly method for the fabrication of binary colloidal crystals and inverse structures. *J. Colloid Interf. Sci.* 380 (2012): 42–50.

Cerdán, L., Costela, A., Enciso, E. et al. Random lasing in self-assembled dye-doped latex nanoparticles: Packing density effects. *Adv. Funct. Mater.* 23(31) (2013): 3916–3924.

Chen, J., Ye, X., and Murray, C.B. Systematic electron crystallographic studies of self-assembled binary nanocrystal superlattices. *ACS Nano* 4(4) (2010): 2374–2381.

Chen, J.T., Thomas, E.L., Ober, C.K. et al. Self-assembled smectic phases in rod-coil block copolymers. *Science* 273(5273) (1996): 343–346.

Chen, Z., Moore, J., Radtke, G. et al. Binary nanoparticle superlattices in the semiconductor-semiconductor system: CdTe and CdSe. *J. Am. Chem. Soc.* 129(50) (2007): 15702–15709.

Chen, Z. and O'Brien, S. Structure direction of Ii-Vi semiconductor quantum dot binary nanoparticle superlattices by tuning radius ratio. *ACS Nano* 2(6) (2008): 1219–1229.

Cheng, X.H., Das, M.K., Diele, S. et al. Influence of semiperfluorinated chains on the liquid crystalline properties of amphiphilic polyols: Novel materials with thermotropic lamellar, columnar, bicontinuous cubic, and micellar cubic mesophases. *Langmuir* 18(17) (2002): 6521–6529.

Chong, H., Nie, C.Y., Zhu, C.L. et al. Conjugated polymer nanoparticles for light-activated anticancer and antibacterial activity with imaging capability. *Langmuir* 28(4) (2012): 2091–2098.

Dai, Z., Li, Y., Duan, G. et al. Phase diagram, design of monolayer binary colloidal crystals, and their fabrication based on ethanol-assisted self-assembly at the air/water interface. *ACS Nano* 6(8) (2012): 6706–6716.

Damasceno, P.F., Engel, M., and Glotzer, S.C. Predictive self-assembly of polyhedra into complex structures. *Science* 337(6093) (2012): 453–457.

de Boer, B., Stalmach, U., van Hutten, P.F. et al. Supramolecular self-assembly and opto-electronic properties of semiconducting block copolymers. *Polymer* 42(21) (2001): 9097–9109.

Demers, L.M., Park, S.J., Taton, T.A. et al. Orthogonal assembly of nanoparticle building blocks on dip-pen nanolithographically generated templates of DNA. *Angew. Chem. Int. Edit.* 40(16) (2001): 3071–3073.

Dimitrov, A.S. and Nagayama, K. Continuous convective assembling of fine particles into two-dimensional arrays on solid surfaces. *Langmuir* 12(5) (1996): 1303–1311.

Dong, A., Ye, X., Chen, J. et al. A generalized ligand-exchange strategy enabling sequential surface functionalization of colloidal nanocrystals. *J. Am. Chem. Soc.* 133(4) (2011a): 998–1006.

Dong, A.G., Chen, J., Vora, P.M. et al. Binary nanocrystal superlattice membranes self-assembled at the liquid-air interface. *Nature* 466(7305) (2010): 474–477.

Dong, A.G., Ye, X.C., Chen, J. et al. Two-dimensional binary and ternary nanocrystal superlattices: The case of monolayers and bilayers. *Nano Lett.* 11(4) (2011b): 1804–1809.

Duan, H.W., Wang, D.Y., Kurth, D.G. et al. Directing self-assembly of nanoparticles at water/oil interfaces. *Angew. Chem. Int. Edit.* 43(42) (2004): 5639–5642.

Eldridge, M.D., Madden, P.A., and Frenkel, D. Entropy-driven formation of a super-lattice in a hard-sphere binary mixture. *Nature* 365(6441) (1993): 35–37.

Evers, W.H., De Nijs, B., Filion, L. et al. Entropy-driven formation of binary semiconductor-nanocrystal superlattices. *Nano Lett.* 10(10) (2010): 4235–4241.

Fendler, J.H. and Dekany, I. Nanoparticles in solids and solutions. Paper presented at the NATO Advanced Research Workshop on Nanoparticles in Solids and Solutions—An Integrated Approach to Their Preparation and Characterization, Szeged, Hungary, 1996.

Feng, L., Zhu, C., Yuan, H. et al. Conjugated polymer nanoparticles: Preparation, properties, functionalization and biological applications. *Chem. Soc. Rev.* 42(16) (2013): 6620–6633.

Fernando, L.P., Kandel, P.K., Yu, J.B. et al. Mechanism of cellular uptake of highly fluorescent conjugated polymer nanoparticles. *Biomacromolecules* 11(10) (2010): 2675–2682.

Fisslthaler, E., Bluemel, A., Landfester, K. et al. Printing functional nanostructures: A novel route towards nanostructuring of organic electronic devices via soft embossing, inkjet printing and colloidal self-assembly of semiconducting polymer nanospheres. *Soft Matter* 4(12) (2008): 2448–2453.

Fredrickson, G.H. and Bates, F.S. Dynamics of block copolymers: Theory and experiment. *Annu. Rev. Mater. Sci.* 26 (1996): 501–550.

Frenkel, D. Order through disorder: Entropy strikes back. *Phys. World* 6(2) (1993): 24–25.

Goodfellow, B.W., Patel, R.N., Panthani, M.G. et al. Melting and sintering of a body-centered cubic superlattice of PbSe nanocrystals followed by small angle X-ray scattering. *J. Phys. Chem. C* 115(14) (2011): 6397–6404.

Grzelczak, M., Vermant, J., Furst, E.M. et al. Directed self-assembly of nanoparticles. *ACS Nano* 4(7) (2010): 3591–3605.

Gu, H.W., Yang, Z.M., Gao, J.H. et al. Heterodimers of nanoparticles: Formation at a liquid-liquid interface and particle-specific surface modification by functional molecules. *J. Am. Chem. Soc.* 127(1) (2005): 34–35.

Gutacker, A., Adamczyk, S., Helfer, A. et al. All-conjugated polyelectrolyte block copolymers. *J. Mater. Chem.* 20(8) (2010): 1423–1430.

Halperin, A. Rod coil copolymers—Their aggregation behavior. *Macromolecules* 23(10) (1990): 2724–2731.

Hammond, P.T. Form and function in multilayer assembly: New applications at the nanoscale. *Adv. Mater.* 16(15) (2004): 1271–1293.

Hammond, P.T. Building biomedical materials layer-by-layer. *Mater. Today* 15(5) (2012): 196–206.

Hanrath, T., Choi, J.J., and Smilgies, D.-M. Structure/processing relationships of highly ordered lead salt nanocrystal superlattices. *ACS Nano* 3(10) (2009): 2975–2988.

Hawker, C.J. and Russell, T.P. Block copolymer lithography: Merging "bottom-up" with "top-down" processes. *MRS Bull.* 30(12) (2005): 952–966.

He, J., Lin, X.-M., Chan, H. et al. Diffusion and filtration properties of self-assembled gold nanocrystal membranes. *Nano Lett.* 11(6) (2011): 2430–2435.

Henderson, A.M.J., Saunders, J.M., Mrkic, J. et al. A new method for stabilising conducting polymer latices using short chain alcohol ethoxylate surfactants. *J. Mater. Chem.* 11(12) (2001): 3037–3042.

Hentze, H.P. and Antonietti, M. Template synthesis of porous organic polymers. *Curr. Opin. Solid State Mat. Sci.* 5(4) (2001): 343–353.

Hill, H.D. and Mirkin, C.A. The bio-barcode assay for the detection of protein and nucleic acid targets using Dtt-induced ligand exchange. *Nat. Protoc.* 1(1) (2006): 324–336.

Huber, J., Jung, C., and Mecking, S. Nanoparticles of low optical band gap conjugated polymers. *Macromolecules* 45(19) (2012): 7799–7805.

Huber, J. and Mecking, S. Processing of polyacetylene from aqueous nanoparticle dispersions. *Angew. Chem. Int. Edit.* 45(38) (2006): 6314–6317.

Ikkala, O. and ten Brinke, G. Hierarchical self-assembly in polymeric complexes: Towards functional materials. *Chem. Commun.* 19 (2004): 2131–2137.

Jang, J., Oh, J.H., and Stucky, G.D. Fabrication of ultrafine conducting polymer and graphite nanoparticles. *Angew. Chem. Int. Edit.* 41(21) (2002): 4016–4019.

Jiang, P., Bertone, J.F., Hwang, K.S. et al. Single-crystal colloidal multilayers of controlled thickness. *Chem. Mat.* 11(8) (1999): 2132–2140.

Jiang, Z., Lin, X.-M., Sprung, M. et al. Capturing the crystalline phase of two-dimensional nanocrystal superlattices in action. *Nano Lett.* 10(3) (2010): 799–803.

Jin, R.C., Wu, G.S., Li, Z. et al. What controls the melting properties of DNA-linked gold nanoparticle assemblies? *J. Am. Chem. Soc.* 125(6) (2003): 1643–1654.

Jones, B.H. and Lodge, T.P. Nanocasting nanoporous inorganic and organic materials from polymeric bicontinuous microemulsion templates. *Polym. J.* 44(2) (2012): 131–146.

Jones, M.R., Macfarlane, R.J., Lee, B. et al. DNA-nanoparticle superlattices formed from anisotropic building blocks. *Nat. Mater.* 9(11) (2010): 913–917.

Kelly, T.L., Yamada, Y., Che, S.P.Y. et al. Monodisperse poly(3,4-ethylenedioxythiophene)-silica microspheres: Synthesis and assembly into crystalline colloidal arrays. *Adv. Mater.* 20(13) (2008): 2616–2621.

Keum, C.G., Noh, Y.W., Baek, J.S. et al. Practical preparation procedures for docetaxel-loaded nanoparticles using polylactic acid-co-glycolic acid. *Int. J. Nanomed.* 6 (2011): 2225–2234.

Kietzke, T., Neher, D., Landfester, K. et al. Novel approaches to polymer blends based on polymer nanoparticles. *Nat. Mater.* 2(6) (2003): 408–412.

Kim, M.H., Im, S.H., and Park, O.O. Fabrication and structural analysis of binary colloidal crystals with two-dimensional superlattices. *Adv. Mater.* 17(20) (2005): 2501–2505.

Kimber, R.G.E., Walker, A.B., Schroder-Turk, G.E. et al. Bicontinuous minimal surface nanostructures for polymer blend solar cells. *Phys. Chem. Chem. Phys.* 12(4) (2010): 844–851.

Klok, H.A. and Lecommandoux, S. Supramolecular materials via block copolymer self-assembly. *Adv. Mater.* 13(16) (2001): 1217–1229.

Kostiainen, M.A., Hiekkataipale, P., Laiho, A. et al. Electrostatic assembly of binary nanoparticle superlattices using protein cages. *Nat. Nanotechnol.* 8(1) (2013): 52–56.

Kotov, N.A., Dekany, I., and Fendler, J.H. Layer-by-layer self-assembly of polyelectrolyte-semiconductor nanoparticle composite films. *J. Phys. Chem.* 99(35) (1995): 13065–13069.

Kovalenko, M.V., Bodnarchuk, M.I., and Talapin, D.V. Nanocrystal superlattices with thermally degradable hybrid inorganic-organic capping ligands. *J. Am. Chem. Soc.* 132(43) (2010): 15124–15126.

Kuehne, A. J. C., Gather, M. C., and Sprakel, J. Monodisperse conjugated polymer particles by Suzuki-Miyaura dispersion polymerization. *Nat. Comm.* 3 (2012), 1088.

Kurokawa, N., Yoshikawa, H., Hirota, N. et al. Size-dependent spectroscopic properties and thermochromic behavior in poly (substituted thiophene) nanoparticles. *Chem. Phys.Chem.* 5(10) (2004): 1609–1615.

Kutuzov, S., He, J., Tangirala, R. et al. on the kinetics of nanoparticle self-assembly at liquid/liquid interfaces. *Phys. Chem. Chem. Phys.* 9(48) (2007): 6351–6358.

Labastide, J.A., Baghgar, M., Dujovne, I. et al. Polymer nanoparticle super lattices for organic photovoltaic applications. *J. Phys. Chem. Lett.* 2(24) (2011): 3085–3091.

Landfester, K. The generation of nanoparticles in miniemulsions. *Adv. Mater.* 13(10) (2001): 765–768.

Landfester, K., Bechthold, N., Tiarks, F. et al. Miniemulsion polymerization with cationic and nonionic surfactants: A very efficient use of surfactants for heterophase polymerization. *Macromolecules* 32(8) (1999): 2679–2683.

Landfester, K., Montenegro, R., Scherf, U. et al. Semiconducting polymer nanospheres in aqueous dispersion prepared by a miniemulsion process. *Adv. Mater.* 14(9) (2002): 651–655.

Lecommandoux, S., Klok, H.A., Sayar, M. et al. Synthesis and self-organization of rod-dendron and dendron-rod-dendron molecules. *J. Polym. Sci. Pol. Chem.* 41(22) (2003): 3501–3518.

Lee, M., Cho, B.K., Jang, Y.G. et al. Spontaneous organization of supramolecular rod-bundles into a body-centered tetragonal assembly in coil-rod-coil molecules. *J. Am. Chem. Soc.* 122(31) (2000): 7449–7455.

Lee, M., Cho, B.K., and Zin, W.C. Supramolecular structures from rod-coil block copolymers. *Chem. Rev.* 101(12) (2001): 3869–3892.

Lee, S.H., Lee, Y.B., Park, D.H. et al. Tuning optical properties of poly(3-hexylthiophene) nanoparticles through hydrothermal processing. *Sci. Technol. Adv. Mater.* 12(2) (2011).

Li, F., Josephson, D.P., and Stein, A. Colloidal assembly: The road from particles to colloidal molecules and crystals. *Angew. Chem. Int. Edit.* 50(2) (2011): 360–388.

Liao, J., Li, X., Wang, Y. et al. Patterned close-packed nanoparticle arrays with controllable dimensions and precise locations. *Small* 8(7) (2012): 991–996.

Lin, S.H., Ho, C.C., and Su, W.F. Cylinder-to-gyroid phase transition in a rod-coil diblock copolymer. *Soft Matter* 8(18) (2012): 4890–4893.

Lin, Y., Skaff, H., Emrick, T. et al. Nanoparticle assembly and transport at liquid-liquid interfaces. *Science* 299(5604) (2003): 226–229.

Lisunova, M., Holland, N., Shchepelina, O. et al. Template-assisted assembly of the functionalized cubic and spherical microparticles. *Langmuir* 28(37) (2012): 13345–13353.

Liu, A.J. and Fredrickson, G.H. Phase separation kinetics of rod/coil mixtures. *Macromolecules* 29(24) (1996): 8000–8009.

Liu, C.L., Lin, C.H., Kuo, C.C. et al. Conjugated rod-coil block copolymers: synthesis, morphology, photophysical properties, and stimuli-responsive applications. *Prog. Polym. Sci.* 36(5) (2011): 603–637.

Liu, J.W. and Lu, Y. A colorimetric lead biosensor using dnazyme-directed assembly of gold nanoparticles. *J. Am. Chem. Soc.* 125(22) (2003): 6642–6643.

Matsen, M.W. and Bates, F.S. Unifying weak- and strong-segregation block copolymer theories. *Macromolecules* 29(4) (1996): 1091–1098.

Millstone, J.E., Kavulak, D.F.J., Woo, C.H. et al. Synthesis, properties, and electronic applications of size-controlled poly(3-hexylthiophene) nanoparticles. *Langmuir* 26(16) (2010): 13056–13061.

Mirkin, C.A., Letsinger, R.L., Mucic, R.C. et al. A DNA-based method for rationally assembling nanoparticles into macroscopic materials. *Nature* 382(6592) (1996): 607–609.

Miszta, K., de Graaf, J., Bertoni, G. et al. Hierarchical self-assembly of suspended branched colloidal nanocrystals into superlattice structures. *Nat. Mater.* 10(11) (2011): 872–876.

Murray, C.B. Synthesis and characterization of II-VI quantum dots and their assembly into 3D quantum dot superlattices. *Abstr. Pap. Am. Chem. Soc.* 213 (1997): 246–PHYS.

Murray, C.B., Kagan, C.R., and Bawendi, M.G. Self-organization of cdse nanocrystallites into three-dimensional quantum dot superlattices. *Science* 270(5240) (1995): 1335–1338.

Murray, C.B., Kagan, C.R., and Bawendi, M.G. Synthesis and characterization of monodisperse nanocrystals and close-packed nanocrystal assemblies. *Annu. Rev. Mater. Sci.* 30 (2000): 545–610.

Nagaoka, Y., Chen, O., Wang, Z. et al. Structural control of nanocrystal superlattices using organic guest molecules. *J. Am. Chem. Soc.* 134(6) (2012): 2868–2871.

Nagarjuna, G., Baghgar, M., Labastide, J.A. et al. Tuning aggregation of poly(3-hexyl-thiophene) within nanoparticles. *ACS Nano* 6(12) (2012): 10750–10758.

Noh, H., Choi, C., Hung, A.M. et al. Site-specific patterning of highly ordered nano-crystal super lattices through biomolecular surface confinement. *ACS Nano* 4(9) (2010): 5076–5080.

Olsen, B.D., Li, X.F., Wang, J. et al. Thin film structure of symmetric rod-coil block copolymers. *Macromolecules* 40(9) (2007): 3287–3295.

Olsen, B.D. and Segalman, R.A. Structure and thermodynamics of weakly segregated rod-coil block copolymers. *Macromolecules* 38(24) (2005): 10127–10137.

Olsen, B.D. and Segalman, R.A. Self-assembly of rod-coil block copolymers. *Mater. Sci. Eng. R-Rep.* 62(2) (2008): 37–66.

Olsen, B.D., Shah, M., Ganesan, V. et al. Universalization of the phase diagram for a model rod-coil diblock copolymer. *Macromolecules* 41(18) (2008): 6809–6817.

Palmer, L.C. and Stupp, S.I. Molecular self-assembly into one-dimensional nanostruc-tures. *Accounts Chem. Res.* 41(12) (2008): 1674–1684.

Park, B.J. and Furst, E.M. Fluid-interface templating of two-dimensional colloidal crystals. *Soft Matter* 6(3) (2010): 485–488.

Park, C., Yoon, J., and Thomas, E.L. Enabling nanotechnology with self assembled block copolymer patterns. *Polymer* 44(22) (2003): 6725–6760.

Park, E.J., Erdem, T., Ibrahimova, V. et al. White-emitting conjugated polymer nanoparticles with cross-linked shell for mechanical stability and controllable photometric properties in color-conversion led applications. *ACS Nano* 5(4) (2011): 2483–2492.

Park, S.J., Lazarides, A.A., Mirkin, C.A. et al. The electrical properties of gold nanoparticle assemblies linked by DNA. *Angew. Chem. Int. Edit.* 39(21) (2000): 3845–3848.

Park, S.J., Lazarides, A.A., Mirkin, C.A. et al. Directed assembly of periodic mate-rials from protein and oligonucleotide-modified nanoparticle building blocks. *Angew. Chem. Int. Edit.* 40(15) (2001): 2909–2912.

Park, S.Y., Lytton-Jean, A.K.R., Lee, B. et al. DNA-programmable nanoparticle crystal-lization. *Nature* 451(7178) (2008): 553–556.

Pauling, L. The principles determining the structure of complex ionic crystals. *J. Am. Chem. Soc.* 51 (1929): 1010–1026.

Pease, R.F. and Chou, S.Y. Lithography and other patterning techniques for future electronics. *Proc. IEEE* 96(2) (2008): 248–270.

Pecher, J. and Mecking, S. Nanoparticles of conjugated polymers. *Chem. Rev.* 110(10) (2010): 6260–6279.

Pelaz, B., Jaber, S., de Aberasturi, D.J. et al. The state of nanoparticle-based nanosci-ence and biotechnology: Progress, promises, and challenges. *ACS Nano* 6(10) (2012): 8468–8483.

Peng, X.T. and Dinsmore, A.D. Light propagation in strongly scattering, random colloidal films: The role of the packing geometry. *Phys. Rev. Lett.* 99(14) (2007): 143902.

Phillips, C.L. and Glotzer, S.C. Effect of nanoparticle polydispersity on the self-assembly of polymer tethered nanospheres. *J. Chem. Phys.* 137(10) (2012).

Pichler, S., Bodnarchuk, M.I., Kovalenko, M.V. et al. Evaluation of ordering in single-component and binary nanocrystal superlattices by analysis of their autocor-relation functions. *ACS Nano* 5(3) (2011): 1703–1712.

Podsiadlo, P., Krylova, G., Lee, B. et al. The role of order, nanocrystal size, and capping ligands in the collective mechanical response of three-dimensional nanocrystal solids. *J. Am. Chem. Soc.* 132(26) (2010): 8953–8960.

Potai, R. and Traiphol, R. Controlling chain organization and photophysical properties of conjugated polymer nanoparticles prepared by reprecipitation method: The effect of initial solvent. *J. Colloid Interf. Sci.* 403 (2013): 58–66.

Prevo, B.G. and Velev, O.D. Controlled, rapid deposition of structured coatings from micro- and nanoparticle suspensions. *Langmuir* 20(6) (2004): 2099–2107.

Puntes, V.F., Krishnan, K.M., and Alivisatos, A.P. Colloidal nanocrystal shape and size control: The case of cobalt. *Science* 291(5511) (2001): 2115–2117.

Quan, Z. and Fang, J. Superlattices with non-spherical building blocks. *Nano Today* 5(5) (2010): 390–411.

Ramanathan, M. and Darling, S. B. Mesoscale morphologies in polymer thin films. *Prog. Polym. Sci.* 36(6) (2011): 793–812.

Redl, F.X., Cho, K.S., Murray, C.B. et al. Three-dimensional binary superlattices of magnetic nanocrystals and semiconductor quantum dots. *Nature* 423(6943) (2003): 968–971.

Ryu, J.H. and Lee, M. Liquid crystalline assembly of rod-coil molecules. In *Liquid Crystalline Functional Assemblies and Their Supramolecular Structures*, edited by Kato, T., pp. 63–98. Berlin, Germany: Springer-Verlag Berlin, 2008.

Sanchez-Iglesias, A., Grzelczak, M., Perez-Juste, J. et al. Binary self-assembly of gold nanowires with nanospheres and nanorods. *Angew. Chem. Int. Edit.* 49(51) (2010): 9985–9989.

Scherf, U., Adamczyk, S., Gutacker, A. et al. All-conjugated, rod-rod block copolymers: Generation and self-assembly properties. *Macromol. Rapid Commun.* 30(13) (2009): 1059–1065.

Scherf, U., Gutacker, A., and Koenen, N. All-conjugated block copolymers. *Accounts Chem. Res.* 41(9) (2008): 1086–1097.

Segalman, R.A., McCulloch, B., Kirmayer, S. et al. Block copolymers for organic optoelectronics. *Macromolecules* 42(23) (2009): 9205–9216.

Sharma, V., Xia, D.Y., Wong, C.C. et al. Templated self-assembly of non-close-packed colloidal crystals: Toward diamond cubic and novel heterostructures. *J. Mater. Res.* 26, no. 2 (2011): 247–253.

Shen, H.W. and Eisenberg, A. Morphological phase diagram for a ternary system of block copolymer ps310-b-paa(52)/dioxane/H_2O. *J. Phys. Chem. B* 103(44) (1999): 9473–9487.

Shevchenko, E.V., Kortright, J.B., Talapin, D.V. et al. Quasi-ternary nanoparticle superlattices through nanoparticle design. *Adv. Mater.* 19(23) (2007): 4183–4188.

Shevchenko, E.V., Ringler, M., Schwemer, A. et al. Self-assembled binary superlattices of CdSe and Au nanocrystals and their fluorescence properties. *J. Am. Chem. Soc.* 130(11) (2008): 3274–3275.

Shevchenko, E.V., Talapin, D.V., Kotov, N.A. et al. Structural diversity in binary nanoparticle superlattices. *Nature* 439(7072) (2006a): 55–59.

Shevchenko, E.V., Talapin, D.V., Murray, C.B. et al. Structural characterization of self-assembled multifunctional binary nanoparticle superlattices. *J. Am. Chem. Soc.* 128(11) (2006b): 3620–3637.

Silva, A.T., Alien, N., Ye, C.M. et al. Conjugated polymer nanoparticles for effective sirna delivery to tobacco by-2 protoplasts. *BMC Plant Biol.* 10 (2010): 291.

Simon, P.F.W., Ulrich, R., Spiess, H.W. et al. Block copolymer-ceramic hybrid materials from organically modified ceramic precursors. *Chem. Mat.* 13(10) (2001): 3464–3486.

Singh, G., Pillai, S., Arpanaei, A. et al. Electrostatic and capillary force directed tunable 3d binary micro- and nanoparticle assemblies on surfaces. *Nanotechnology* 22(22) (2011).

Srivastava, S. and Kotov, N.A. Composite layer-by-layer (lbl) assembly with inorganic nanoparticles and nanowires. *Accounts Chem. Res.* 41(12) (2008): 1831–1841.

Stoeva, S.I., Huo, F.W., Lee, J.S. et al. Three-layer composite magnetic nanoparticle probes for DNA. *J. Am. Chem. Soc.* 127(44) (2005): 15362–15363.

Suda, M. and Einaga, Y. Sequential assembly of phototunable ferromagnetic ultrathin films with perpendicular magnetic anisotropy. *Angew. Chem. Int. Edit.* 48(10) (2009): 1754–1757.

Sun, S.H., Murray, C.B., Weller, D. et al. Monodisperse FePt nanoparticles and ferromagnetic FePt nanocrystal superlattices. *Science* 287(5460) (2000): 1989–1992.

Szymanski, C., Wu, C.F., Hooper, J. et al. Single molecule nanoparticles of the conjugated polymer MEH-PPV, preparation and characterization by near-field scanning optical microscopy. *J. Phys. Chem. B* 109(18) (2005): 8543–8546.

Talapin, D.V. Lego materials. *ACS Nano.* 2(6) (2008): 1097–1100.

Talapin, D.V., Shevchenko, E.V., Bodnarchuk, M.I. et al. Quasicrystalline order in self-assembled binary nanoparticle superlattices. *Nature* 461(7266) (2009): 964–967.

Talapin, D.V., Shevchenko, E.V., Murray, C.B. et al. Dipole–dipole interactions in nanoparticle superlattices. *Nano Lett.* 7(5) (2007): 1213–1219.

Tschierske, C. Non-conventional liquid crystals—The importance of micro-segregation for self-organisation. *J. Mater. Chem.* 8(7) (1998): 1485–1508.

Tuncel, D., and Demir, H.V. Conjugated polymer nanoparticles. *Nanoscale* 2(4) (2010): 484–494.

Urban, J.J., Talapin, D.V., Shevchenko, E.V. et al. Synergismin binary nanocrystal superlattices leads to enhanced p-type conductivity in self-assembled PbTe/Ag-2 Te thin films. *Nat. Mater.* 6(2) (2007): 115–121.

Velev, O.D. and Lenhoff, A.M. Colloidal crystals as templates for porous materials. *Curr. Opin. Colloid Interface Sci.* 5(1–2) (2000): 56–63.

Vincent, B. Electrically conducting polymer colloids and composites. *Polym. Advan. Technol.* 6(5) (1995): 356–361.

Walker, D.A., Browne, K.P., Kowalczyk, B. et al. Self-assembly of nanotriangle super lattices facilitated by repulsive electrostatic interactions. *Angew. Chem. Int. Edit.* 49(38) (2010): 6760–6763.

Wang, D.Y. and Mohwald, H. Rapid fabrication of binary colloidal crystals by stepwise spin-coating. *Adv. Mater.* 16(3) (2004): 244–247.

Wang, L., Wan, Y., Li, Y. et al. Binary colloidal crystals fabricated with a horizontal deposition method. *Langmuir* 25(12) (2009): 6753–6759.

Wang, T., Zhuang, J., Lynch, J. et al. Self-assembled colloidal superparticles from nanorods. *Science* 338(6105) (2012): 358–363.

Wiesenauer, B. R. and Gin, D. L. Nanoporous polymer materials based on self-organized, bicontinuous cubic lyotropic liquid crystal assemblies and their applications. *Polym. J.* 44(6) (2012): 461–468.

Wu, C., Bull, B., Szymanski, C. et al. Multicolor conjugated polymer dots for biological fluorescence imaging. *ACS Nano* 2(11) (2008): 2415–2423.

Wu, M., Frochot, C., Dellacherie, E. et al. Well-defined poly(butyl cyanoacrylate) nanoparticles via miniemulsion polymerization. *Macromol. Symp.* 281 (2009): 39–46.

Xia, D.Y., Ku, Z.Y., Lee, S.C. et al. Nanostructures and functional materials fabricated by interferometric lithography. *Adv. Mater.* 23(2) (2011): 147–179.

Xia, Y.N. and Whitesides, G.M. Soft lithography. *Annu. Rev. Mater. Sci.* 28 (1998): 153–184.

Xie, Y., Guo, S., Guo, C. et al. Controllable two-stage droplet evaporation method and its nanoparticle self-assembly mechanism. *Langmuir* 29(21) (2013): 6232–6241.

Xu, L., Ma, W., Wang, L. et al. Nanoparticle assemblies: Dimensional transformation of nanomaterials and scalability. *Chem. Soc. Rev.* 42(7) (2013): 3114–3126.

Yang, S.M., Kim, S.H., Lim, J.M. et al. Synthesis and assembly of structured colloidal particles. *J. Mater. Chem.* 18(19) (2008): 2177–2190.

Yang, X.M., Wan, L., Xiao, S.G. et al. Directed block copolymer assembly versus electron beam lithography for bit-patterned media with areal density of 1 terabit/inch(2) and beyond. *ACS Nano* 3(7) (2009): 1844–1858.

Ye, X., Collins, J.E., Kang, Y. et al. Morphologically controlled synthesis of colloidal upconversion nanophosphors and their shape-directed self-assembly. *Proc. Natl. Acad. Sci. USA* 107(52) (2010): 22430–22435.

Ye, X.C., Chen, J., and Murray, C.B. Polymorphism in self-assembled Ab(6) binary nanocrystal superlattices. *J. Am. Chem. Soc.* 133(8) (2011): 2613–2620.

Yodh, A.G., Lin, K.H., Crocker, J.C. et al. Entropically driven self-assembly and interaction in suspension. *Philos. Trans. R. Soc. Lond. Ser. A: Math. Phys. Eng. Sci.* 359(1782) (2001): 921–937.

Yunker, P.J., Still, T., Lohr, M.A. et al. Suppression of the coffee-ring effect by shape-dependent capillary interactions. *Nature* 476(7360) (2011): 308–311.

Zanella, M., Bertoni, G., Franchini, I.R. et al. Assembly of shape-controlled nanocrystals by depletion attraction. *Chem. Commun.* 47(1) (2011): 203–205.

Zeng, H., Li, J., Liu, J.P. et al. Exchange-coupled nanocomposite magnets by nanoparticle self-assembly. *Nature* 420(6914) (2002): 395–398.

Zhang, G. and Wang, D.Y. Colloidal lithography—The art of nanochemical patterning. *Chem. Asian J.* 4(2) (2009): 236–245.

Zhang, H., Edwards, E.W., Wang, D.Y. et al. Directing the self-assembly of nanocrystals beyond colloidal crystallization. *Phys. Chem. Chem. Phys.* 8(28) (2006): 3288–3299.

Zhang, J., Li, Y., Zhang, X. et al. Colloidal self-assembly meets nanofabrication: From two-dimensional colloidal crystals to nanostructure arrays. *Adv. Mater.* 22(38) (2010): 4249–4269.

Zhao, B. and Brittain, W.J. Polymer brushes: Surface-immobilized macromolecules. *Prog. Polym. Sci.* 25(5) (2000): 677–710.

Zheng, C., Xu, X., He, F. et al. Preparation of high-quality organic semiconductor nanoparticle films by solvent-evaporation-induced self-assembly. *Langmuir* 26(22) (2010): 16730–16736.

Index

Printed and bound by CPI Group (UK) Ltd, Croydon, CR0 4YY

22/10/2024

01777647-0003